A Panorama in Number Theory
or
The View from Baker's Garden

A Panorama in Number Theory
or
The View from Baker's Garden

edited by

Gisbert Wüstholz
ETH, Zürich

CAMBRIDGE
UNIVERSITY PRESS

CAMBRIDGE
UNIVERSITY PRESS

University Printing House, Cambridge CB2 8BS, United Kingdom

Cambridge University Press is part of the University of Cambridge.

It furthers the University's mission by disseminating knowledge in the pursuit of education, learning and research at the highest international levels of excellence.

www.cambridge.org
Information on this title: www.cambridge.org/9780521807999

First published 2002

A catalogue record for this publication is available from the British Library

ISBN 978-0-521-80799-9 Hardback

Contents

Contributors *page* vii
Introduction xi
1 One Century of Logarithmic Forms *G. Wüstholz* 1
2 Report on p-adic Logarithmic Forms *Kunrui Yu* 11
3 Recent Progress on Linear Forms in Elliptic Logarithms *Sinnou David & Noriko Hirata-Kohno* 26
4 Solving Diophantine Equations by Baker's Theory *Kálmán Győry* 38
5 Baker's Method and Modular Curves *Yuri F. Bilu* 73
6 Application of the André–Oort Conjecture to some Questions in Transcendence *Paula B. Cohen & Gisbert Wüstholz* 89
7 Regular Dessins, Endomorphisms of Jacobians, and Transcendence *Jürgen Wolfart* 107
8 Maass Cusp Forms with Integer Coefficients *Peter Sarnak* 121
9 Modular Forms, Elliptic Curves and the ABC-Conjecture *Dorian Goldfeld* 128
10 On the Algebraic Independence of Numbers *Yu.V. Nesterenko* 148
11 Ideal Lattices *Eva Bayer-Fluckiger* 168
12 Integral Points and Mordell–Weil Lattices *Tetsuji Shioda* 185
13 Forty Years of Effective Results in Diophantine Theory *Enrico Bombieri* 194
14 Points on Subvarieties of Tori *Jan-Hendrik Evertse* 214
15 A New Application of Diophantine Approximations *G. Faltings* 231
16 Search Bounds for Diophantine Equations *D.W. Masser* 247
17 Regular Systems, Ubiquity and Diophantine Approximation *V.V. Beresnevich, V.I. Bernik & M.M. Dodson* 260
18 Diophantine Approximation, Lattices and Flows on Homogeneous Spaces *Gregory Margulis* 280

19 On Linear Ternary Equations with Prime Variables – Baker's Constant and Vinogradov's Bound *Ming-Chit Liu & Tianze Wang* 311
20 Powers in Arithmetic Progression *T.N. Shorey* 325
21 On the Greatest Common Divisor of Two Univariate Polynomials, I *A. Schinzel* 337
22 Heilbronn's Exponential Sum and Transcendence Theory *D.R. Heath-Brown* 353

Contributors

Eva Bayer-Fluckiger, Departement de Mathématiques Ecole Polytechnique Federale de Lausanne 1015 Lausanne Switzerland
eva.bayer@epfl.ch

V.V. Beresnevich, Institute of Mathematics of the Belarus Academy of Sciences, 220072, Surganova 11, Minsk, Belarus
beresnevich@im.bas-net.by

V.I. Bernik, Institute of Mathematics of the Belarus Academy of Sciences, 220072, Surganova 11, Minsk, Belarus
bernik@im.bas-net.by

Yuri F. Bilu, A2X Université Bordeaux I, 351, cours de la Liberation, 33405 Talence, France
yuri@math.u-bordeaux.fr

Enrico Bombieri, School of Mathematics, Institute for Advanced Study, Einstein Drive, Princeton NJ 08540, USA
eb@math.ias.edu

Paula B. Cohen, MR AGAT au CNRS, UFR de Mathématiques, Bâtiment M2, Université des Sciences et Technologies de Lille, 59655 Villeneuve d'Ascq cedex, France
Paula.Cohen@univ-lille1.fr

Sinnou David, Université P. et M. Curie (Paris VI), Institut Mathématique de Jussieu, Problèmes Diophantiens, Case 247, 4, Place Jussieu, 75252 Paris CEDEX 05, France
david@math.jussieu.fr

M.M. Dodson, Department of Mathematics, University of York, York YO10 5DD, UK
mmd1@york.ac.uk

Jan–Hendrik Evertse, Universiteit Leiden, Mathematisch Instituut, Postbus 9512, 2300 RA Leiden, The Netherlands
evertse@math.leidenuniv.nl

G. Faltings, Max-Planck-Institut für Mathematik, Vivatgasse 7, 53111 Bonn, Germany
faltings@mpim-bonn.mpg.de

Dorian Goldfeld, Columbia University, Department of Mathematics, New York, NY 10027, USA
goldfeld@columbia.edu

Kálmán Győry, Institute of Mathematics and Informatics, University of Debrecen, H-4010 Debrecen, P.O. Box 12, Hungary
gyory@math.klte.hu

D.R. Heath-Brown Mathematical Institute, Oxford, UK
rhb@maths.ox.ac.uk

Noriko Hirata-Kohno, Department of Mathematics, College of Science and Technology, Nihon University, Suruga-Dai, Kanda, Chiyoda, Tokyo 101-8308, Japan
hirata@math.cst.nihon-u.ac.jp

Ming-Chit Liu, Department of Mathematics, The University of Hong Kong, Pokfulam, Hong Kong; *and* PO Box 625, Alhambra, California CA 91802, USA
matmcliu@yahoo.com

G. Margulis, Yale University, Department of Mathematics, PO Box 208283, New Haven CT 06520-8283, USA
margulis@math.yale.edu

D.W. Masser, Mathematisches Institut, Universität Basel, Rheinsprung 21, 4051 Basel, Switzerland
masser@math.unibas.ch

Yu. Nesterenko, Faculty of Mechanics and Mathematics, Main Building, MSU, Vorobjovy Gory, Moscow, 119899, Russia
nest@trans.math.msu.su

Peter Sarnak, Princeton University, Department of Mathematics, Fine Hall, Princeton NJ 08544-1000, USA
sarnak@math.princeton.edu

Andrzej Schinzel, Institute of Mathematics, Polish Academy of Sciences, ul. Śniadeckich 8, PO Box 137, 00-950 Warszawa, Poland
schinzel@impan.gov.pl

T. Shioda, Department of Mathematics, Rikkyo University, Nishi-Ikebukuro, Toshima-ku, Tokyo 171, Japan
`shioda@rkmath.rikkyo.ac.jp`

T.N. Shorey, School of Mathematics, Tata Institute of Fundamental Research, Homi Bhabha Road, Mumbai 400 005, India
`shorey@math.tifr.res.in`

Tianze Wang, Department of Mathematics, Henan University, Kaifeng 475001, China
`wangtz@henu.edu.cn`

Jürgen Wolfart, Mathematisches Seminar der Johann Wolfgang Goethe-Universität, Robert-Mayer-Str. 6–10, D–60054 Frankfurt a.M., Germany
`wolfart@math.uni-frankfurt.de`

G. Wüstholz, ETH Zürich, ETH Zentrum, D-MATH, HG G 66.3 Raemistrasse 101, CH-8092 Zürich, Switzerland.
`wustholz@math.ethz.ch`

Kunrun Yui, Department of Mathematics, Hong Kong University of Science and Technology, Clear Water Bay, Kowloon, Hong Kong
`makryu@ust.hk`

Introduction

The millennium, together with Alan Baker's 60th birthday offered a singular occasion to organize a meeting in number theory and to bring together a leading group of international researchers in the field; it was generously supported by ETH Zurich together with the Forschungsinstitut für Mathematik. This encouraged us to work out a programme that aimed to cover a large spectrum of number theory and related geometry with particular emphasis on Diophantine aspects. Almost all selected speakers we able to accept the invitation; they came to Zurich from many parts of the world, gave lectures and contributed to the success of the meeting. The London Mathematical Society was represented by its President, Professor Martin Taylor, and it sent greetings to Alan Baker on the occasion of his 60th birthday.

This volume is dedicated to Alan Baker and it offers a panorama in number theory. It is as exciting as the scene we enjoyed, during the conference, from the cafeteria on top of ETH overlooking the town of Zurich, the lake and the Swiss mountains as well as the spectacular view that delighted us on our conference excursion to Lake Lucerne in central Switzerland. The mathematical spectrum laid before us in the lectures ranged from sophisticated problems in elementary number theory through to diophantine approximations, modular forms and varieties, metrical diophantine analysis, algebraic independence, arithmetic algebraic geometry and, ultimately, to the theory of logarithmic forms, one of the great achievements in mathematics in the last century. The articles here document the present state of the art and suggest possible new directions for research; they can be expected to inspire much further activity. Almost all who were invited to contribute to the volume were able to prepare an article; with very few exceptions, the promised papers were eventually submitted and the result turns out itself to be like a very colourful panoramic picture of mathematics taken on a beautiful clear day in the autumn of the year 1999.

It is not easy to group together the different contributions in a systematic way. Indeed we appreciate that any attempt at categorisation can certainly be disputed and may invoke criticism. Nonetheless, for the reader who is not an expert in this area, we think that it would be helpful to have some guidelines. Accordingly we shall now discuss briefly the various subjects covered in the book.

Since one of the main motivations for the conference – as already said at the beginning – was the 60th birthday of Alan Baker, the theory of **logarithmic forms** was a very important and significant part of the proceedings. The article *One Century of Logarithmic Forms* by Gisbert Wüstholz is an overview of the evolution of the subject beginning with the famous seventh problem of Hilbert. It describes the history of its solution, the subsequent development of the theory of logarithmic forms and then goes on to explain how the latter is now regarded as an integral part of a general framework relating to group varieties. This overview should be seen as a homage to Alan Baker and his work. Several contributions are directly connected with it: we mention Yu Kunrui's paper *Report on p-adic Logarithmic Forms* which surveys the now very extensive p-adic aspects of the theory, and the article *Recent Progress on Linear Forms in Elliptic Logarithms* by Sinnou David and Noriko Hirata-Kohno which gives a detailed exposition of elliptic aspects, especially with regard to important quantitative results. The theory of logarithmic forms has found numerous applications in very different areas. One of the earliest and most direct of these has been to diophantine equations of classical type coming in part from problems in algebraic number theory. This side of the subject is explained in Kalman Győry's paper *Solving Diophantine Equations by Baker's Theory*.

One domain, seemingly very far from the area of logarithmic forms but in fact surprisingly strongly related to it, is **modular forms and varieties**. A good illustration is provided by the article *Baker's Method and Modular Curves* by Yuri F. Bilu, which shows how Baker's theory can be applied in the context of Siegel's theorem to give effective estimates for the heights of integral points on a large class of modular curves. Another example is the paper *Application of the André–Oort Conjecture to Some Questions in Transcendence* by Paula Cohen and Gisbert Wüstholz. Here it is not so much the classical logarithmic theory that is involved but the considerably wider framework mentioned earlier on group varieties. And Jürgen Wolfart's contribution *Regular Dessins, Endomorphisms of Jacobians, and Transcendence* connects modular geometry with modern logarithmic theory, in particular with abelian varieties. Quite differently, Peter Sarnak's article *Maass Cusp Forms with Integer Coefficients* spans the bow from cuspidal eigenforms of Laplacians for congruence subgroups of $SL(2, \mathbb{Z})$ and automorphic cuspidal representations to classical logarithmic theory and transcendence.

In the articles of Wüstholz and of Yu Kunrui mentioned above, the very intimate relation between the theory of linear forms in logarithms and the so-called *abc*-conjecture is explained. The latter is now recognised as one of the most central problems in mathematics. Dorian Goldfeld succeeds in giving an extraordinarily broad picture of the topic in his fine paper *Modular Forms, Elliptic Curves and the abc-Conjecture*. And in Yu.V. Nesterenko's survey article *On Algebraic Independence of Numbers* we have an excellent reference for research and achievements during the last millennium on algebraic independence questions; the emphasis is on recent significant progress concerning modular forms and their connection with hypergeometric functions and it can be confidently predicted that the article will be very influential for further investigations.

Another important subject is the theory of **lattices** and the contribution of Eva Bayer-Fluckiger on *Ideal Lattices* can be seen as a splendid introduction. Applications are described, for instance, to Knot theory and Arakelov theory. Another series of applications can be found in the paper of Tetsuji Shioda entitled *Integral Points and Mordell–Weil Lattices*. Here the author demonstrates a close connection with Lie theory by explaining, amongst other things, how integral points on elliptic curves can be regarded as roots of root lattices associated with Lie groups of exceptional type like E_8.

Diophantine approximations and equations are the general subject of a further series of papers. We mention the very interesting overview of Enrico Bombieri *Forty Years of Effective Results in Diophantine Theory*. Bombieri has for many years sought to square the circle in the sense of making the Thue–Siegel–Dyson–Schneider–Roth theory effective, and he has met with much success. A substantial part of his article is devoted to describing the state of the art here and, in particular, how it relates to Baker's theory. Complementing Bombieri's point of view, the paper *Points on Subvarieties of Tori* by Jan-Hendrik Evertse explains how the non-effective theory has made progress in the last four decades especially with regard to diophantine geometry. Gerd Faltings' article *A New Application of Diophantine Approximation* indicates the path along which some very modern diophantine theory might develop in the near future; it contains very exciting new geometrical ideas and tools and it can be expected to serve as a valuable source for further research. Finally there is the contribution of David Masser entitled *Search Bounds for Diophantine Equations*; here the author takes a very fundamental point of view which leads to an important new topic, namely the existence of *a priori* (or what Masser calls *search*) bounds for the solutions of equations. It seems likely to attract the interest not only of number theorists but also of theoretical and possibly even practical computer scientists.

A totally different point of view on diophantine approximation is taken by the so-called **metrical theory**. This deals with approximations on very general classes of manifolds and spaces typically concerning points on a geometric space outside some fixed sets of measure zero. In an early paper Alan Baker considered such questions and there has been considerable progress in the field since then. Apart from its significance to number theory, applications have been given in the context of Hausdorff dimensions and so-called small denominator questions related to stability problems for dynamical systems. In *Regular Systems, Ubiquity and Diophantine Approximation*, V.V. Beresnevich, V.I. Bernik and M.M. Dodson present a careful and valuable report on the development of the theory. The article by Gregory Margulis entitled *Diophantine Approximation, Lattices and Flows on Homogeneous Spaces* connects with this and these two contributions taken together can be expected to be very influential for future research. In Margulis' paper the direction is the study of homogeneous spaces rather than arbitrary manifolds and it furnishes the framework for investigating orbits in the space of lattices. This point of view has made it possible to successfully apply techniques from differential geometry and Lie theory and it shows again the remarkable range of tools and techniques currently used in number theory.

Broadly speaking, one can place under the heading **analytic number theory** the article of Ming-Chit Liu and Tianze Wang *On Linear Ternary Equations with Prime Variables – Baker's Constant and Vinogradov's Bound*, the paper by T.N. Shorey *Powers in Arithmetic Progression*, the contribution 'On the Greatest Common Divisor of Two Univariate Polynomials' by Andrzej Schinzel and the short note of D.R. Heath-Brown entitled *Heilbronn's Exponential Sum and Transcendence Theory*. Apart from the classical sphere of ideas which one traditionally associates with analytic number theory, these papers have the extra quality of bringing in methods from outside the field. The Heath-Brown contribution gives a particularly good example in which transcendence techniques similar to those introduced by Stepanov in the context of the Weil conjectures become central for the study of exponential sums.

Returning to the beautiful Swiss landscape, in the same way that it invites one to climb this or that grand mountain or to explore some of the host of picturesque features, so we hope that the panorama exhibited in this volume invites visits to and explorations of a more abstract but nonetheless beautiful and colourful region of mathematics. Many of the exciting sites owe their existence to Alan Baker.

I express my gratitude to the Schulleitung of ETH and especially to Alain Sznitman, the director of the Forschungsinstitut when the conference took place, for making possible the whole project which has led to this volume. The assistance of Renate Leukert has been invaluable in the editing of the work. Finally, I am sincerely grateful to my secretary Hedi Oehler and the secretaries of the Forschungsinstitut at the time, in particular Ruth Ebel, without whose help the organization of the conference would certainly have been impossible.

1

One Century of Logarithmic Forms

G. Wüstholz

1 Introduction

At the turn of any century it is very natural on the one hand for us to look back and see what were great achievements in mathematics and on the other to look forward and speculate about which directions mathematics might take. One hundred years ago Hilbert was in a similar situation and he raised on that occasion a famous list of 23 problems that he believed would be very significant for the future development of the subject. Hilbert's article on future problems in mathematics published in the *Comptes Rendus du Deuxième Congrès International des Mathématiciens* stimulated tremendous results and an enormous blossoming of the mathematical sciences overall. A significant part of Hilbert's discussion was devoted to number theory and Diophantine geometry and we have seen some wonderful achievements in these fields since then. In this survey, we shall recall how transcendence and arithmetical geometry have grown into beautiful and far-reaching theories which now enhance many different aspects of mathematics. Very surprisingly three of Hilbert's problems, which at first seemed very distant from each other, have now come together and have provided the catalyst for a vast interplay between the subjects in question. We shall concentrate on one of them, namely the seventh, and describe the principal developments in transcendence theory which it has initiated. This will lead us to the theory of linear forms in logarithms and to the generalization of the latter in the context of commutative group varieties. The theory has evolved to be the most crucial instrument towards a solution of the tenth problem of Hilbert on the effective solution of diophantine equations as well as many other well-known questions. The intimate relationships in this field become especially evident through a simple conjecture, the *abc*-conjecture, which seems to hold the key to much of the future direction of number theory. We shall discuss this at the end of this article.

2 Hilbert's seventh problem

Hilbert remarked in connection with the seventh problem that he believed that the proof of the transcendence of α^β for algebraic $\alpha \neq 0, 1$ and algebraic irrational β would be extremely difficult and that certainly the solution of this and analogous problems would lead to valuable new methods. Surprisingly, the problem was eventually solved independently, by different methods, by Gelfond and Schneider in 1934. Gelfond and Kuzmin had solved some particular cases of the conjecture a few years earlier and the solutions of Gelfond and Schneider used similar methods together with techniques introduced by Siegel in his well-known investigations on Bessel functions. The Gelfond–Schneider theorem shows that for any non-zero algebraic numbers α_1 and α_2 with $\log \alpha_1$ and $\log \alpha_2$ linearly independent over the rationals we have

$$\beta_1 \log \alpha_1 + \beta_2 \log \alpha_2 \neq 0.$$

In 1935 Gelfond considered the problem of establishing a lower bound for the absolute value of the linear form $L = \beta_1 T_1 + \beta_2 T_2$ evaluated at $(\log \alpha_1, \log \alpha_2)$ and succeeded in proving that its value Λ is bounded below by

$$\log |\Lambda| \gg -h(L)^\kappa$$

where $h(L)$ denotes the logarithmic height of the linear form and $\kappa > 5$. It was realized by Gelfond around 1940 that an extension of the theorem to linear forms in more than two variables would enable one to solve some of the most challenging problems in number theory and in the theory of diophantine equations. We mention here the *Liouville problem* of establishing effective lower bounds for the approximation of an algebraic number by rationals sharper than the bound obtained by Liouville himself. Other examples were the *Thue equation* and effective bounds for the size of solutions in *Siegel's great theorem on integral points* on algebraic curves. To great surprise one of the oldest and most exciting problems in number theory, Euler's famous *numeri idonii* problem, has also turned out to be intimately related to the theory of logarithmic forms.

The Liouville problem was mentioned by Davenport to Baker as a research topic in the early 60s. Baker's early papers made the first breakthrough in this area. The approach was through hypergeometric functions and Padé approximation theory and was related to some work of Thue and Siegel. After several significant results in diverse branches of transcendence theory, Baker was led to the famous class number problem of Gauss. A careful study of the work of Heilbronn, Gelfond, Linnik and others in this field convinced him that the most promising approach to this and many other fundamental questions in number theory was through linear forms in logarithms. Despite the fact that no significant progress had been made in this subject for many years, Baker succeeded in 1966 in establishing a definitive result; namely if $\alpha_1, \ldots, \alpha_n$ are non-zero algebraic numbers such that $\log \alpha_1, \ldots, \log \alpha_n$ are linearly independent over the rationals, then $1, \log \alpha_1, \ldots, \log \alpha_n$ are linearly independent over the field of all algebraic numbers. The result and the method of proof was one of the most significant advances in number theory made in 20th century. Baker's theorem includes both the Hermite–Lindemann and the Gelfond–Schneider theorems as special cases. Baker's original paper also contained a quantitative result on lines similar to Gelfond's two-variable estimate mentioned earlier; this was sufficient to deal with the class number problem, the Thue problem, the Liouville problem, the elliptic curve problem and clearly had great potential for future research. It soon became clear that further progress on many critical problems would depend on sharp estimates for logarithmic forms, and between 1966 and 1975 Baker wrote a series of important papers on this subject. It was here that many of the instruments now familiar to specialists in the field were introduced, among them Kummer theory, the so-called Kummer descent and delta functions. In this context one should mention that both Stark and Feldman made substantial contributions.

3 Elliptic theory

Having published his solution to Hilbert's seventh problem, Schneider started to study elliptic and later also abelian functions. He had been motivated by a paper of Siegel on periods of elliptic functions. There Siegel had proved that for an elliptic curve with algebraic invariants g_2, g_3 not all periods can be algebraic. In particular he obtained the transcendence of non-zero periods when the elliptic curve has complex multiplication. In a fundamental paper Schneider established the transcendence of elliptic integrals of the first and second kind taken between algebraic points. As a special instance one obtains the transcendence of the value of the linear form $L = \alpha T_1 + \beta T_2$ at $(\omega, \eta(\omega))$ where ω is a non-zero period and $\eta(\omega)$ the corresponding quasi-period. Plainly the result gives Siegel's theorem without the additional hypothesis on complex multiplication. Schneider also applied the result to the modular j-function and found that it takes algebraic values at algebraic arguments if and only if the argument is imaginary quadratic. As was realized only recently, this opened a close connection with Hilbert's twelfth problem which emerged out of the famous *Jugendtraum* of Kronecker. For some forty years there was no obvious progress; then in 1970 Baker realized that the method which he had developed for dealing with linear forms in logarithms could be adapted to give the transcendence of non-zero values of linear forms in two elliptic periods and their associated quasi-periods. With this fundamental contribution he opened a new field of research. Masser and Coates succeeded a few years later in including the period $2\pi i$ and determining the dimension of the vector space generated by the numbers $1, 2\pi i, \omega_1, \omega_2, \eta(\omega_1), \eta(\omega_2)$. The main difficulty in going further and extending the results to an arbitrary number of periods lay in the use of determinants in Baker's method. It became clear that to achieve a breakthrough such aspects had to be modified significantly.

The situation was very similar in the case of abelian varieties which was first studied by Schneider in 1939 and which was subsequently investigated by Masser, Coates and Lang between 1975 and 1980. Again difficulties relating to the use of determinants presented a severe obstacle for progress. It was realized about that time that transcendence theory has much to do with algebraic groups and with the exponential map of a Lie group in particular. Lang was the first to consider a reformulation of the Gelfond–Schneider theorem and other classical results in the language of group varieties. With advice from Serre this was taken up by Waldschmidt and, amongst other things, he interpreted Schneider's result on elliptic integrals in the new language. This prepared the ground for the first successful attack, by Laurent, on Schneider's third problem concerning elliptic integrals of the third kind. However the difficulties relating to determinants referred to earlier blocked the passage to a complete solution. Likewise, for abelian integrals, Schneider had raised the question of extending his elliptic theorems to abelian varieties. As Arnold pointed out in his monograph on Newton, Hooke and Huyghens, the question is closely related to an unsolved problem of Leibniz emerging from celestical mechanics.

4 Group varieties

The situation changed completely when it was realized almost simultaneously by Brownawell, Chudnovsky, Masser, Nesterenko and Wüstholz that one way to deal with the difficulties referred to in the previous section was to use commutative algebra. Masser and Wüstholz started to apply the theory to commutative group varieties and the breakthrough was obtained by Wüstholz in 1981 when he succeeded in establishing the correct multiplicity estimates for group varieties. As a consequence he was able to formulate and prove the Analytic Subgroup Theorem. It says that if G is a commutative and connected algebraic group defined over $\overline{\mathbb{Q}}$ then an analytic subgroup defined over $\overline{\mathbb{Q}}$ contains a non-trivial algebraic point if and only if it contains a non-trivial algebraic subgroup over $\overline{\mathbb{Q}}$. The theorem generalizes that of Baker in a natural way and hence includes, as special cases, the classical theorems of Hermite, Lindemann and Gelfond–Schneider. It also implies directly the results of Schneider, Baker, Masser, Coates, Lang and Laurent mentioned above on elliptic and abelian functions and integrals. It further includes complete solutions to problems raised by Baker relating to elliptic logarithmics, to a problem mentioned by Waldschmidt on analytic homomorphisms as well as the outstanding third and fourth problems listed by Schneider at the end of his well-known book. The Analytic Subgroup Theorem was therefore one of the most significant results in modern transcendence theory; all the above consequences can be seen as questions concerning generalised logarithms defined in terms of suitable commutative algebraic groups. There are many other applications of the Theorem, in particular the work of Wolfart and Wüstholz on a question of Lang about the complex uniformisation of algebraic curves. There are also nice areas of application to Siegel modular functions, as studied by Cohen, Shiga and Wolfart, and to hypergeometric theory, as investigated by Wolfart, Beukers, Cohen and Wüstholz.

The multiplicitiy estimates on group varieties referred to earlier which were crucial to the proof of the Analytic Subgroup Theorem also lead to an improvement in the basic linear form estimate of Baker. This was noted independently by Wüstholz and by Philippon & Waldschmidt. Subsequently, in 1993, Baker & Wüstholz used the multiplicitiy estimates to establish a very sharp bound for logarithmic forms; it remains the best to date and it has served as an indispensible reference for practical applications to diophantine equations.

5 The quantitative theory

In the previous section we discussed the most natural version of the qualitative theory of logarithmic forms in the context of algebraic groups. Many of the most important applications, however, involve a quantitative form of the theory and this we shall discuss now. We begin with a report on the latest results concerning linear forms in ordinary logarithms. As mentioned earlier their derivation depends critically on the theory of multiplicity estimates on group varieties. Let now $\alpha_1, \ldots, \alpha_n$ be algebraic numbers, not 0 or 1, and let $\log \alpha_1, \ldots, \log \alpha_n$ be fixed determinations of the logarithms. Let K be the field generated by $\alpha_1, \ldots, \alpha_n$ over the rationals and let d be the degree of K. For each α in K and any given determination of $\log \alpha$ we define the modified height $h'(\alpha)$ by

$$h'(\alpha) = \frac{1}{d} \max(h(\alpha), |\log \alpha|, 1),$$

where $h(\alpha)$ is the logarithm of the standard Weil height of α. We consider the linear form

$$L = b_1 z_1 + \cdots + b_n z_n,$$

where $b_1, \ldots, b_n$ are integers, not all 0, and put

$$h'(L) = \frac{1}{d} \max(h(L), 1),$$

where $h(L)$ is the logarithmic Weil height of L, that is $d \log \max(|b_j|/b)$ with b given by the highest common factor of $b_1, \ldots, b_n$. In their 1993 paper Baker & Wüstholz showed that if $\Lambda = L(\log \alpha_1 \ldots, \log \alpha_n) \neq 0$ then

$$\log |\Lambda| > -C(n, d) h'(\alpha_1) \ldots h'(\alpha_n) h'(L),$$

where

$$C(n, d) = 18(n+1)!\, n^{n+1} (32d)^{n+2} \log(2nd).$$

As we already indicated, the result gives the best lower bound for Λ known to date, and it is essential for computational diophantine theory. In this context we mention the important work of Győry and others who have applied the theory to large classes of diophantine equations; these include the so-called norm form, index form and discriminant form equations in particular. Recently a striking application was given by Bilu, Hanrot & Voutier (1999) in the realm of primitive divisors of Lucas and Lehmer numbers. Here the precision of the Baker–Wüstholz bound was critical in making the computations feasible.

We now discuss briefly the extent to which the classical quantitative theory of logarithmic forms has been carried over to deal with the general situation of commutative group varieties. The latest and most precise work in this field is due to Hirata-Kohno (1991) and the precision of her results is close to that established in the classical case. To indicate the form of Hirata-Kohno's main result, let K be a number field and G be a commutative group variety of dimension n defined over K. The basic theory refers to a non-vanishing linear form $L(z_1, \ldots, z_n) = \beta_1 z_1 + \cdots + \beta_n z_n$ as above with coefficients in K. The linear form is evaluated at a point $z_1 = u_1, \ldots, z_n = u_n$, where $u = (u_1, \ldots, u_n)$ is an element in the Lie algebra of G such that $\alpha = \exp_G(u)$ belongs to $G(K)$. Then Hirata-Kohno's work shows that there exists a positive constant C independent of u and L which can be determined effectively and which has the property that if $\Lambda = L(u) \neq 0$ then

$$\log |\Lambda| > -C\big(h'(\alpha)\big)^n \big(h'(L) + h'(\alpha)\big)\big(\max(1, \log(h'(L)h'(\alpha))\big)^{n+1}.$$

The result can be applied, in particular, to yield an alternative approach to Siegel's theorem on integral points on curves. It can also be used to give an effective bound for the height $h(P)$ of an integral point P depending on the height of a set of generators for the Mordell–Weil group; consequently one sees that in the cases when a suitable set of generators can be determined explicitly, all the integral points P on the curve can be fully computed. The method was successfully applied in the elliptic case by Stroeker & Tzanakis (1994) to give the complete set of solutions in integers x, y of the equation

$$y^2 = (x + 337)(x^2 + 337^2),$$

and by Gebel, Pethö & Zimmer (1994) to deal with the instance

$$y^2 = x^3 - 1642032x + 628747920.$$

Here an essential ingredient is some work of David (1992) which furnishes, in the elliptic case, an explicit estimate for the constant occurring in Hirata-Kohno's general result.

It emerged unexpectedly from the early work of Baker on logarithmic forms that if there is a rational linear dependence relation satisfied by logarithms of algebraic numbers then there exists such a relation with coefficients bounded in terms of the heights of the numbers. Motivated by Faltings' famous work on the Mordell conjecture, Masser & Wüstholz realized that Baker's observation, appropriately generalized to abelian varieties, could be applied to yield an effective Isogeny Theorem. The result significantly improves upon an essential aspect of Faltings' 1983 paper and it has initiated a substantial body of new theory on arithmetical properties of abelian varieties. For instance, Masser & Wüstholz obtained in this way an effective version of the well-known Tate conjecture, which was crucial to Faltings' work; they established discriminant estimates for endomorphism algebras; and they derived a solution to a problem of Serre on representations of Galois groups. This is currently a very active area of research.

6 The *abc*-conjecture

When Richard Mason became a graduate student in Cambridge in the early 80s, Baker suggested to him as a research topic the problem of generalizing the theory of logarithmic forms to function fields. This led Mason to an assertion about equations in polynomials of the form

$$a + b = c$$

which we now recognize as being the analogue of the *abc*-conjecture in the function field setting. The conjecture itself relates to relatively prime integers

a, b, c satisfying the above equation and it asserts that, for any $\epsilon > 0$,

$$\max(|a|, |b|, |c|) \ll N^{1+\epsilon},$$

where N, the conductor or radical of abc, denotes the product of all distinct prime factors of abc, and the constant implied by $\ll$ depends only on ϵ. The origin of the assertion lies in a conjecture of Szpiro on discriminants and conductors of elliptic curves; this was adapted by Oesterlé to give a conjecture as above but in a weaker form and it was Masser who formulated the precise statement as we recognize it today. The conjecture enables one in principle to resolve in integers x, y, z, l, m, n and given r, s, t, the exponential diophantine equation

$$rx^l + sy^m + tz^n = 0,$$

where l, m, n are positive and subject to $(1/l) + (1/m) + (1/n) < 1$; this includes the celebrated cases of Fermat and Catalan. Other consequences of the conjecture are the famous theorems of Roth and Faltings, and, if one assumes a generalised version for number fields, then it resolves, in principle, the problem of the non-existence of the Siegel zero for Dirichlet L-functions. The only non-trivial estimate for $\max(|a|, |b|, |c|)$ to date is due to Stewart & Yu Kunrui (1991). Their work is based on the Baker–Wüstholz archimedean bound for logarithmic forms quoted in Section 5 and on a natural non-archimedean analogue established by Yu Kunrui (1998).

In a very interesting recent paper, Baker (1998) described an intimate connnection between the *abc*-conjecture and the theory of logarithmic forms. He began by suggesting two refinements to the *abc*-conjecture, first

$$\max(|a|, |b|, |c|) \ll N(\log N)^{\omega}/\omega!,$$

and, secondly, for some absolute constant κ,

$$\max(|a|, |b|, |c|) \ll \epsilon^{-\kappa\omega(ab)} N^{1+\epsilon},$$

where the constants implied by $\ll$ are absolute, and where $\omega(n)$ signifies the number of distinct prime factors of the integer n and $\omega = \omega(abc)$. Baker went on to relate the second refinement with an estimate for the logarithmic form

$$\Lambda = u_1 \log v_1 + \cdots + u_n \log v_n,$$

in positive integers $v_1, \ldots, v_n$ and integers $u_1, \ldots, u_n$, not all 0. Indeed he showed that the second refinement is equivalent to the lower bound

$$\Xi \gg (N(v))^{-1}(\epsilon^{\kappa\omega(v)} a^{-\epsilon})^{1/(1+\epsilon)},$$

for the expression

$$\Xi = \min(1, |\Lambda|) \prod \min(1, p|\Lambda|_p),$$

where the product is taken over all primes p; here $v = v_1 \cdots v_n$ and $N(v)$ denotes the radical of v.

A slightly weaker version of the inequality is given by

$$\log \Xi \gg -\log u \log v,$$

with $u = \max |u_j|$, and the latter can be compared with the Baker–Wüstholz theorem which gives

$$\log |\Lambda| \gg -\log u \log v_1 \cdots \log v_n$$

with an implied constant depending only on n. Thus we see that a result in the direction of the *abc*-conjecture sufficient for all the major applications would follow if one could replace $|\Lambda|$ by Ξ and also the product $\log v_1 \times \cdots \times \log v_n$ by the sum $\log v_1 + \cdots + \log v_n$. Bearing in mind what has been achieved in connection with non-archimedean valuations, this would seem to present the most feasible line of attack for the future.

Bibliography

Baker, A. (1975), *Transcendental Number Theory*, Cambridge University Press, 1st ed., 1975; (3rd ed., Math. Library Series, 1990).

Baker, A. (1998), Logarithmic forms and the *abc*-conjecture. In *Number Theory: Diophantine, Computational and Algebraic Aspects*, de Gruyter, 37–44.

Baker, A. & A. Schinzel (1971), On the least integers represented by the genera of binary quadratic forms, *Acta Arith.*, **18**, 137–144.

Baker, A. & G. Wüstholz (1993), Logarithmic forms and group varieties, *J. Reine Angew. Math.*, **442**, 19–62.

Baker, A. & G. Wüstholz (1999), Number theory, transcendence and diophantine geometry in the next millennium. In *Mathematics: Frontiers and Perspectives*, American Mathematical Society, 1–12.

Bilu, Y., G. Hanrot & P.M. Voutier (1999), Existence of primitive divisors of Lucas and Lehmer numbers (with an appendix by M. Mignotte). Preprint.

Beukers, F. & J. Wolfart (1988), Algebraic values of hypergeometric functions. In *New Advances in Transcendence Theory*, A. Baker (ed.), Cambridge University Press, 68–81.

Cohen, Paula B. (1996), Humbert surfaces and transcendence properties of automorphic functions, *Rocky Mountain J. Math.*, **26**, 987–1001.

David, S. (1992), Minorations de formes linéaires de logarithmes elliptiques, *Publ. Math. Univ. Pierre et Marie Curie, Problèmes Diophantiens*, **106**, (1991/92), exposé no. 3.

Dvornicich, R. & U. Zannier (1994), Fields containing values of algebraic functions, *Ann. Scuola Norm. Sup. Pisa Cl. Sci.*, **21**, 421–443.

Faltings, G. (1983), Endlichkeitssätze für abelsche Varietäten über Zahlkörpern, *Invent. Math.*, **73**, 349–366.

Gebel, J., A. Pethö & H.G. Zimmer (1994), Computing integer points on elliptic curves, *Acta Arith.*, **68**, 171–192.

Hirata-Kohno, N. (1991), Formes linéaires de logarithmes de points algébriques sur les groupes algébriques, *Invent. Math.*, **104**, 401–433.

Masser, D.W. & G. Wüstholz (1993), Isogeny estimates for abelian varieties and finiteness theorems, *Annals Math.*, **137**, 459–472.

Mignotte, M. & Y. Roy (1997), Minorations pour l'équation de Catalan, *C. R. Acad. Sci. Paris*, **324**, 377–380.

Ribenboim, P. (1994), *Catalan's Conjecture*, Academic Press.

Shiga, H. & J. Wolfart (1995), Criteria for complex multiplication and transcendence properties of automorphic functions, *J. Reine Angew. Math.*, **463**, 1–25.

Stewart, C.L. & Kunrui Yu (1991) On the *abc* conjecture, *Math. Ann.*, **291**, 225–230.

Stroeker, R.J. & N. Tzanakis (1994), Solving elliptic diophantine equations by estimating linear forms in elliptic logarithms, *Acta Arith.*, **67**, 177–196.

Wolfart, J. (1988), Werte hypergeometrischer Funktionen, *Invent. Math.*, **92**, 187–216.

Wüstholz, G. (1989), Algebraische Punkte auf analytischen Untergruppen algebraischer Gruppen, *Annals Math.*, **129** (1989), 501–517.

Yu, Kunrui (1998), p-adic logarithmic forms and group varieties I, *J. Reine Angew. Math.*, **502**, 29–92.

2

Report on p-adic Logarithmic Forms

Kunrui Yu

1 Historical introduction

The p-adic theory of logarithmic forms has a long history, following closely the results in the complex domain; and it has been applied to Leopoldt's conjecture on p-adic regulators (for abelian extensions of $\mathbf{Q}$, see Ax 1965 and Brumer 1967), to polynomial and exponential Diophantine equations, to the problem of the greatest prime divisors of polynomials or binary forms, to linear recurrence sequences (see Shorey & Tijdeman 1986), to knot theory (see Riley 1990) and to the abc-conjecture (see Stewart & Tijdeman 1986, Stewart & Yu 1991, 2001), etc. The present report will emphasize the evolution of the theory of p-adic logarithmic forms and its application to the abc-conjecture.

Mahler (1932) proved the p-adic analogue of the Hermite–Lindemann theorem. In 1935, he obtained a p-adic analogue of the Gel'fond–Schneider Theorem. During the course of this work, he founded the p-adic theory of analytic functions.

Gel'fond (1940) proved a quantitative result on linear forms in *two* p-adic logarithms in analogy with his classic work on Hilbert's seventh problem relating to *two* complex logarithms. Schinzel (1967) refined Gel'fond's results, giving completely explicit bounds.

At the end of his 1952 book, Gel'fond wrote 'Nontrivial lower bounds for linear forms, with integral coefficients, of an arbitrary number of logarithms of algebraic numbers, obtained effectively by methods of the theory of transcendental numbers, will be of extraordinarily great significance in the solution of very difficult problems of modern number theory.' For many years after Gel'fond and Schneider succeeded independently in giving a complete answer to Hilbert's seventh problem, the above problem proposed by Gel'fond seemed resistant to attack but it was eventually solved by Baker in 1966. Between then and 1968 Baker published his first series of papers on linear forms in n (an *arbitrary* positive integer) logarithms of algebraic numbers, and thus made a fundamental and far-reaching breakthrough. Baker's method has subsequently been employed to the investigation on linear forms in n p-adic logarithms of algebraic numbers. To begin with, Brumer (1967) obtained a p-adic analogue of Baker's (1966). Sprindžuk (1967), (1968) obtained p-adic analogues of Baker's (1966, 1967a,b, 1968) results. Independently, Coates (1969) obtained a quantitative p-adic analogue following Baker (1968). Kaufman (1971) obtained a p-adic analogue of Feldman (1968). Further Baker & Coates (1975) obtained a p-adic analogue of Baker's Sharpening II (Baker 1973) for the case $n = 2$.

Heights of algebraic numbers. For an algebraic number α, let

$$P(x) = a_0 x^\delta + a_1 x^{\delta-1} + \cdots + a_\delta = a_0(x - \alpha^{(1)}) \cdots (x - \alpha^{(\delta)})$$

be its minimal polynomial over $\mathbf{Z}$. We call $A(\alpha) = \max(|a_0|, \ldots, |a_\delta|)$ the classical height of α, and

$$h_0(\alpha) = \delta^{-1} \log \left(a_0 \max(1, |\alpha^{(1)}|) \cdots \max(1, |\alpha^{(\delta)}|)\right)$$

the absolute logarithmic Weil height of α. We have (see Baker & Wüstholz (1993), p. 22)

$$h_0(\alpha) \le (2\delta)^{-1} \log \left(a_0^2 + \cdots + a_\delta^2\right) \le \log \left(\sqrt{2} A(\alpha)\right).$$

Now we state the result in Baker (1977) in the fundamental rational case. Let $\alpha_1, \ldots, \alpha_n$ be non-zero algebraic numbers; $K = \mathbf{Q}(\alpha_1, \ldots, \alpha_n)$; $d = [K : \mathbf{Q}]$; $A_j \ge \max(A(\alpha_j), 4)$ with $A_n = \max_{1 \le j \le n} A_j$; $\Omega = \log A_1 \cdots \log A_n$ and $\Omega' = \Omega / \log A_n$. For $L(z_1, \ldots, z_n) = b_1 z_1 + \cdots + b_n z_n$ with $b_1, \ldots, b_n$ in $\mathbf{Z}$, not all zero, let $B \ge \max(|b_1|, \ldots, |b_n|, 4)$ and $\Lambda = L(\log \alpha_1, \ldots, \log \alpha_n)$.

Theorem 1 *If* $\Lambda \ne 0$ *and* $\log \alpha_1, \ldots, \log \alpha_n$ *have their principal values, then*

$$\log |\Lambda| > -(16nd)^{200n}\, \Omega \log \Omega' \log B.$$

$\wp$-adic valuation. Let K be a number field with $[K : \mathbf{Q}] = d$, let $\wp$ be a prime ideal of the ring $\mathcal{O}_K$ of algebraic integers in K, let p be the unique prime number contained in $\wp$, and let $e_\wp$ and $f_\wp$ be the ramification index and the residue class degree of $\wp$, respectively. We define $\mathrm{ord}_\wp 0 = \infty$ and $\mathrm{ord}_\wp \alpha$ for $\alpha \in K$, $\alpha \ne 0$, to be the maximal exponent to which $\wp$ divides the fractional ideal generated by α in K. Set $\mathrm{ord}_p \alpha = e_\wp^{-1} \mathrm{ord}_\wp \alpha$, $|\alpha|_p = p^{-\mathrm{ord}_p \alpha}$, so that $|p|_p = p^{-1}$. The completion of K with respect to $|\ |_p$ is written as $K_\wp$ (the completion of $\mathrm{ord}_\wp$ is denoted again by $\mathrm{ord}_\wp$). Let $\bar{\mathbf{Q}}_p$ be an algebraic closure of $\mathbf{Q}_p$ and let $\mathbf{C}_p$ be the completion of $\bar{\mathbf{Q}}_p$ with respect to the valuation of $\bar{\mathbf{Q}}_p$, which is the unique extension of the valuation $|\ |_p$ of $\mathbf{Q}_p$. According to Hasse (1980), pp. 298–302, we can embed $K_\wp$ into $\mathbf{C}_p$: there exists a $\mathbf{Q}$-isomorphism σ from K into $\bar{\mathbf{Q}}_p$ such that $K_\wp$ is value-isomorphic to $\mathbf{Q}_p(\sigma(K))$, whence we can identify $K_\wp$ with $\mathbf{Q}_p(\sigma(K))$.

We note that if we formulate estimates for p-adic logarithmic forms as a lower bound for $|\Lambda|_p$ with

$$\Lambda = b_1 \log_p \alpha_1 + \cdots + b_n \log_p \alpha_n,$$

where $\log_p \alpha_j$ signifies the p-adic logarithm of α_j defined in $K_\wp$ by

$$\log_p \alpha_j = \sum_{k=1}^{\infty} (-1)^{k-1} \frac{(\alpha_j - 1)^k}{k},$$

then it demands, *a priori*, $|\alpha_j - 1|_p < 1$. Hence it is *too restrictive*. So it is more convenient to study the lower bound for $|\alpha_1^{b_1} \cdots \alpha_n^{b_n} - 1|_p$ or the upper bound for $\mathrm{ord}_\wp(\alpha_1^{b_1} \cdots \alpha_n^{b_n} - 1)$ (see Section 3), as it is well-known that an equivalent formulation of Theorem 1 is a lower bound for $|\alpha_1^{b_1} \cdots \alpha_n^{b_n} - 1|$.

Intending to prove p-adic analogues of Theorem 1 and of Baker's Sharpening II (Baker 1973), van der Poorten (1977) developed a strategy based on introducing a root of unity and, as a consequence, on modifying the auxiliary polynomials, and for which he acknowledged that he was in part indebted to a conversation with Mahler. Unfortunately, he overlooked several fundamental facts in algebraic number theory related to the structures of group of $\wp$-adic units and group of roots of unity in $\wp$-adic fields, and related to cyclotomic extensions of $\wp$-adic fields, thereby missing technical difficulties:

(i) For instance, the use of a primitive $G_\wp$th root of unity $\zeta_{G_\wp}$ where $G_\wp = p^{f_\wp g_\wp}(p^{f_\wp} - 1)$ with $g_\wp = \left[\frac{1}{2} + e_\wp/(p-1)\right]$, is impossible in general, as one can see from the following example. Take $K = \mathbf{Q}(\zeta_{p^a})$, where ζ_m denotes a primitive mth root of unity and $a \geq 3$. Then $p\mathcal{O}_K = \wp^{e_\wp}$ with $e_\wp = \phi(p^a)$ and $f_\wp = 1$, where ϕ is the Euler's ϕ-function. Thus $g_\wp = p^{a-1}$ and $G_\wp = p^{p^{a-1}}(p-1)$. Clearly $\zeta_{G_\wp}$ is out of $K_\wp$, since it is well-known that $[K_\wp : \mathbf{Q}_p] = e_\wp f_\wp = \phi(p^a)$ and $[\mathbf{Q}_p(\zeta_{p^{p^{a-1}}}) : \mathbf{Q}_p] = \phi(p^{p^{a-1}})$, and the latter is greater than the former.

(ii) Moreover the use of congruences mod $G_\wp$ requires that $qx \equiv b \pmod{G_\wp}$, where q is a prime with $q \neq p$, $x = r_1\lambda_1 + \cdots + r_n\lambda_n$, $b = r_0 - (r_1\lambda_1' + \cdots + r_n\lambda_n')$ with all r_i and λ_j' being fixed integers, has a unique solution $x \bmod G_\wp$, i.e.,

$$(q, G_\wp) = 1.$$

However, removing the Kummer condition (i.e., Kummer descent), which is based on Lemma 3 of Baker & Stark (1971), requires that $\zeta_q \in K$, which together with $q \neq p$ implies that $q|(p^{f_\wp}-1)$ by Hasse (1980), p. 220, whence $q|G_\wp$. Evidently, one can not have both $(q, G_\wp)=1$ and $q \mid G_\wp$.

2 A successful strategy

As a result of a careful analysis of the problems mentioned at the end of Section 1, Yu (1989, 1990, 1994) succeeded in developing a modified strategy, thereby establishing p-adic analogues of Theorem 1 and Baker (1973). The strategy consists of the following points:

(i) the correct choice of a root of unity and the adjustment of $\alpha_1, \ldots, \alpha_n$ for optimal p-adic convergence;
(ii) the right choice of moduli for congruences used in the auxiliary polynomials and in the Kummer descent;
(iii) the introduction of the (p, q)-Capelli–Kummer descent.

Now we explain (i)–(iii) as a whole. Subject to some cost – see Yu (1990), pp. 97–98 – we may assume that $\alpha_1, \ldots, \alpha_n$ are $\wp$-adic units. It is well-known that the multiplicative group of the residue class field of $K_\wp$ is a cyclic group of order

$$G = \Phi(\wp) = p^{f_\wp} - 1,$$

and that it is generated by the residue class represented by ζ, where ζ is a primitive Gth root of unity in $K_\wp$. So it is a natural choice to use this root of unity. Thus we can find $r_1', \ldots, r_n' \in \mathbf{Z}$ with $0 \leq r_j' < G$ such that

$$\operatorname{ord}_\wp\left(\alpha_j \zeta^{r_j'} - 1\right) \geq 1 \quad (1 \leq j \leq n).$$

That is, $\alpha_j \zeta^{r_j'}$ $(1 \leq j \leq n)$ are principal $\wp$-adic units in $K_\wp$. We need a further p-adic device.

Lemma 1 (See Yu 1994) *Let $\kappa \geq 0$ be the rational integer determined by*

$$p^{\kappa-1}(p-1) \leq 2e_\wp < p^\kappa(p-1).$$

If $\beta \in K_\wp$ is a principal $\wp$-adic unit, then

$$\operatorname{ord}_p(\beta^{p^\kappa} - 1) \geq \vartheta + 1/(p-1) \text{ with } \vartheta = p^\kappa/(2e_\wp).$$

Setting $r_j = p^\kappa r'_j$, by Lemma 1 we have

$$\left|\alpha_j^{p^\kappa}\zeta^{r_j} - 1\right|_p \leq p^{-\vartheta-1/(p-1)}.$$

This makes the p-adic series

$$(\alpha_j^{p^\kappa}\zeta^{r_j})^z := \exp\left(z\log(\alpha_j^{p^\kappa}\zeta^{r_j})\right)$$

convergent in a larger region

$$|z|_p < p^\vartheta$$

in $\mathbf{C}_p$ which contains strictly the unit disk $|z|_p < 1$, so that it is convenient to handle. (We call ϑ the supernormality of the function $(\alpha_j^{p^\kappa}\zeta^{r_j})^z$; for the significance of supernormality, see Yu (1989), Sections 1.2 and 1.3, and Yu (1998), Lemma 2.1.) Thus, in Yu (1989), we managed to obtain an upper bound for $\mathrm{ord}_\wp \Xi$, where $\Xi = \alpha_1^{b_1}\cdots\alpha_n^{b_n} - 1 \neq 0$ and $\alpha_1, \ldots, \alpha_n \in K$ are $\wp$-adic units satisfying

$$\left[K\left(\alpha_1^{1/q}, \ldots, \alpha_n^{1/q}\right) : K\right] = q^n \tag{1}$$

for a prime q with

$$\left(q, p(p^{f_\wp} - 1)\right) = 1, \tag{2}$$

which is needed for guaranteeing that $qx \equiv b \pmod{G}$ with $G = p^{f_\wp} - 1$ has a unique solution mod G. The upper bound for $\mathrm{ord}_\wp \Xi$ in Yu (1989) contains a factor q^{2n}. Hence we need to obtain a good upper bound for the least prime q satisfying (2). As Hugh L. Montgomery pointed out to the author, one has $q \leq c_1 f_\wp \log p$, since $\vartheta(q-1) \leq \log(p(p^{f_\wp} - 1))$, where $\vartheta(x) = \Sigma \log p'$ with p' ranging over all primes $\leq x$. Further, one can not have essentially better upper bound, because by Heath-Brown (1992) there exists a prime p with $p \equiv 1 \pmod{\Pi q'}$ with q' running over all primes $< q$, such that $p < c_2 \exp(5.5\vartheta(q-1))$. The factor $(f_\wp \log p)^{2n}$ is, of course, not desirable in any upper bound for $\mathrm{ord}_\wp \Xi$. Further, Kummer descent requires ζ_q be in K, that is, a field extension of large degree in general, which one wishes to avoid. Thus, in order to overcome the essential difficulty in Kummer descent, we are forced, in Yu (1990, 1994), to appeal to the following

Corollary to the Vahlen–Capelli Theorem *Let q be a prime, k a positive integer, and E a field. When $q = 2$ and $k \geq 2$ we suppose further that $\zeta_4 \in E$. If $a \in E$ and $a \notin E^q$, then the polynomial $x^{q^k} - a$ is irreducible in $E[x]$.* See Capelli (1901) and Rédei (1967).

Now we choose the 'Kummer prime' q as

$$q = \begin{cases} 3 & \text{if } p = 2 \\ 2 & \text{if } p > 2. \end{cases} \tag{3}$$

Then Kummer descent requires that $\zeta_3 \in K$ when $p = 2$; and $\zeta_2 \in K$ (when $p > 2$) holds trivially. In order to be able to apply the above Corollary with the prime q given by (3), we may simply assume, in addition, that $\zeta_4 \in K$ when $p > 2$ (see Yu 1990). But this causes worse dependence on p, e.g., the dependence is a factor p^2 in the p-adic estimates when $\alpha_1, \ldots, \alpha_n \in \mathbf{Q}$, and not p, as one would expect. To apply the above Corollary efficiently, we suppose that K is a number field containing $\alpha_1, \ldots, \alpha_n$, which satisfies

$$\begin{cases} \zeta_3 \in K & \text{if } p = 2 \\ \text{either } \; p^{f_\wp} \equiv 3 \pmod 4 \text{ or } \zeta_4 \in K & \text{if } p > 2, \end{cases} \tag{4}$$

see Yu (1994). Let

$$\begin{cases} u = \max\{k \in \mathbf{Z}_{\geq 0} \mid \zeta_{q^k} \in K\}, \quad \alpha_0 = \zeta_{q^u}, \\ \mu = \operatorname{ord}_q G, \; \text{ with } G = p^{f_\wp} - 1. \end{cases} \tag{5}$$

By (3)–(5), and Hasse (1980), p. 220, we have $1 \leq u \leq \mu$, and obviously

$$(G/q^\mu \, , \, q) = 1.$$

Now we are ready to proceed to prove the main results for $\wp$-adic units $\alpha_1, \ldots, \alpha_n \in K$ which satisfy, as we may call it, the (p, q)-Capelli–Kummer condition

$$\left[K\left(\alpha_0^{1/q}, \alpha_1^{1/q}, \ldots, \alpha_n^{1/q}\right) : K\right] = q^{n+1}. \tag{6}$$

Here we indicate two points.

(a) We use a congruence mod G_1 with $G_1 = G/q^\mu$ in the construction of auxiliary polynomials. As the congruence

$$qx \equiv b \pmod{G_1}$$

has a unique solution mod G_1, the inductive steps go through smoothly. We also use the elementary fact that every congruence class mod G_1 can be partitioned into $q^{\mu-u}$ congruence classes mod G/q^u and into $q^{\mu+1}$ congruence classes mod qG.

(b) In the study of fractional points s/q ($s \in \mathbf{Z}$, $(s, q) = 1$), we need the fact that

$$[K(\zeta_{q^{\mu+1}})(\alpha_1^{1/q}, \ldots, \alpha_n^{1/q}) : K(\zeta_{q^{\mu+1}})] = q^n,$$

which is a consequence of (6), since the Corollary to the Vahlen–Capelli Theorem implies that the polynomial $x^{q^{\mu-u+1}} - \alpha_0$ is irreducible over $K(\alpha_1^{1/q}, \ldots, \alpha_n^{1/q})$.

3 New developments

Between 1982 and 1989, Wüstholz developed the theory of multiplicity estimates on group varieties (see Wüstholz 1989) and, in a series of papers (Wüstholz 1987, 1988) gave a new approach to Baker's theory. This was used by Baker & Wüstholz in their fundamental (1993) memoir, which represents a new stage of the theory of logarithmic forms. Recall that in that memoir, $K = \mathbf{Q}(\alpha_1, \ldots, \alpha_n)$, $d = [K : \mathbf{Q}]$, $h'(\alpha_j) = \max\big(h_0(\alpha_j), |\log\alpha_j|/d, 1/d\big)$, $A_j = \max\big(A(\alpha_j), e\big)$, $L(z_1, \ldots, z_n) = b_1 z_1 + \cdots + b_n z_n$, $\Lambda = L(\log\alpha_1, \ldots, \log\alpha_n)$,

$$h'(L) = \max\left(\log\left(\frac{\max(|b_1|, \ldots, |b_n|)}{\gcd(b_1, \ldots, b_n)}\right), \frac{1}{d}\right).$$

Theorem 2 (Baker & Wüstholz 1993) *If* $\Lambda \neq 0$ *then*

$$\log|\Lambda| > -C(n,d)h'(\alpha_1)\cdots h'(\alpha_n)h'(L),$$

where

$$C(n,d) = 18(n+1)!n^{n+1}(32d)^{n+2}\log(2nd).$$

Baker & Wüstholz proved that if $\Lambda \neq 0$ and $\log\alpha_1, \ldots, \log\alpha_n$ have their principal values, then Theorem 2 implies that

$$\log|\Lambda| > -(16nd)^{2(n+2)}\log A_1 \cdots \log A_n \log B.$$

In 1998, we succeeded in bringing the p-adic theory more in line with the Archimedean theory as in Baker & Wüstholz (1993). Let K be a number field containing $\alpha_1, \ldots, \alpha_n$, which satisfies (4). Set q by (3), u and α_0 by (5). Define

$$h'(\alpha_j) = \max\big(h_0(\alpha_j), f_{\wp}(\log p)/d\big), \quad B = \max(|b_1|, \ldots, |b_n|, 3).$$

Let $\omega_2(n) = \omega_3(n) = 1$ for $n = 1, 2$ and for $n > 2$

$$\omega_2(n) = 4^{s-n} \cdot \frac{(n+s+1)!}{(2s+1)!},$$

$$\omega_3(n) = 6^{t-n} \cdot \frac{(n+2t+1)!}{(3t+1)!},$$

where

$$s = \left[1/4 + \sqrt{n + (17/16)}\right]$$

and t is the unique rational integer such that

$$g(t) := 9t^3 - 8t^2 - (8n+5)t - 2n(n+1) \le 0 \text{ and } g(t+1) > 0.$$

Hence $t = [x_n]$, where x_n is the unique real zero of $g(x)$, which can be determined explicitly by the Cardano's formula.

Theorem 3 (Yu 1998) *Suppose that* $\operatorname{ord}_{\wp}\alpha_j = 0\ (1 \le j \le n)$. *If* $\Xi = \alpha_1^{b_1} \cdots \alpha_n^{b_n} - 1 \ne 0$, *then*

$$\operatorname{ord}_{\wp} \Xi < C_1(n, d, \wp) h'(\alpha_1) \cdots h'(\alpha_n) \log B,$$

where

$$\begin{aligned} C_1(n, d, \wp) &= ca^n \cdot \frac{n^n (n+1)^{n+2}}{n!} \omega_q(n) \frac{p^{f_\wp} - 1}{q^u} \\ &\quad \times \left(\frac{d}{f_\wp \log p}\right)^{n+2} \max\left(f_\wp \log p, \log\left(e^4 (n+1) d\right)\right), \end{aligned}$$

with

$$\begin{aligned} &a = 16,\ c = 1544, && \text{if } p > 2, \\ &a = 32,\ c = 81, && \text{if } p = 2; \end{aligned}$$

furthermore we can take $a = 8(p-1)/(p-2)$ *when* $p \ge 5$ *with* $e_\wp = 1$.

Finally if (6) *is satisfied then* $C_1(n, d, \wp)$ *can be replaced by* $C_1(n, d, \wp)/\omega_q(n)$.

Let $K = \mathbf{Q}(\alpha_1, \ldots, \alpha_n)$, $d = [K : \mathbf{Q}]$, $h_j = \max\left(h_0(\alpha_j), \log p\right)\ (1 \le j \le n)$. It is indicated in Yu (1998) that if $\Xi = \alpha_1^{b_1} \cdots \alpha_n^{b_n} - 1 \ne 0$, then Theorem 3 implies that

$$\operatorname{ord}_{\wp} \Xi < C_1'(n, d, \wp) h_1 \cdots h_n \log B,$$

where

$$C_1'(n, d, \wp) = 12 \left(6(n+1) d / \sqrt{\log p}\right)^{2(n+1)} \left(p^{f_\wp} - 1\right) \log(e^5 n d).$$

Note that in the above consequence of Theorem 3, we do not assume that $\operatorname{ord}_{\wp}\alpha_j = 0\ (1 \le j \le n)$ and we have removed assumption (4) on K.

Remark concerning Yu (1998).

(i) The proof of Theorem 3 follows Baker & Wüstholz (1993). Wüstholz's multiplicity estimates on group varieties is critical for removing a factor of the type $\log(\log A_1 \cdots \log A_n)$ from the results in Yu (1989, 1990, 1994). Proposition 6.1 of Yu (1998) is a modified version of the multiplicity estimates proved in Baker & Wüstholz (1993).

(ii) The strategy explained in Section 2 is used in Yu (1998).

(iii) As suggested by Alan Baker, we have used Schnirelman integral (Schnirelman 1938, see also Adams 1966) to replace Hermite interpolation, thus obtaining better (than in Yu 1989, 1990, 1994) lemmas for extrapolation and interpolation.

(iv) Instead of considering the valuation of $\mathcal{K} = K(\alpha_0^{1/q}, \alpha'^{1/q}_1, \ldots, \alpha'^{1/q}_r)$ at a single prime ideal of $\mathcal{O}_\mathcal{K}$ lying above $\wp$, we investigate the valuations at *all* prime ideals of $\mathcal{O}_\mathcal{K}$ lying above $\wp$.

These lead to better numerical values of a and c in Theorem 3.

Stimulated by Matveev's (1996) Oberwolfach lecture but via a different approach, and in substance quite independently, we obtained a refinement of Theorem 3.

Theorem 4 (Yu 1999) *In Theorem* 3, *we can replace* $C_1(n, d, \wp)$ *by*

$$C_2(n, d, \wp) = C_1(n, d, \wp)(c_\wp/n)^{n-1} \text{ with } c_\wp = (4e)e_\wp f_\wp \log p.$$

Here we get a refinement of Theorem 3 when $c_\wp < n$. For a detailed statement of the refinement, see Yu (1999), Theorem 1.

Remark concerning Yu (1999). The crucial new idea in our 1999 paper is to consider an equivalence relation defined by

$$\prod_{i=1}^{r} \left(\alpha'^{p^\kappa}_i \zeta^{a'_i}\right)^{\lambda_i} \equiv \prod_{i=1}^{r} \left(\alpha'^{p^\kappa}_i \zeta^{a'_i}\right)^{\lambda'_i} \pmod{\wp^{m_0+m}}$$

(see Yu 1999, (10.15)) on a certain set of integral points $(\lambda_1, \ldots, \lambda_r)$, and then to apply the pigeon-hole principle to that set, thereby constructing improved auxiliary rational functions.

So far we have reviewed the development of the theory of p-adic logarithmic forms, following Baker's method. We now report briefly on the development following Schneider's method. Dong Pingping (1995) obtained, for the first time, a sharp lower bound for *simultaneous* linear forms in p-adic logarithms. Philippon & Waldschmidt (1989) also deal with simultaneous linear forms, but in complex logarithms and with Baker's method, while Dong Pingping used an extension of Schneider's method to several variables, following Waldschmidt (1991, 1993). As far as the estimate for a single linear form in p-adic logarithms is concerned – see Corollaire 1.4 in Dong Pingping (1995), p. 39 – it has now been superseded by Yu (1998). For linear forms in two p-adic logarithms, Bugeaud & Laurent (1996) proved a p-adic analogue of Laurent, Mignotte &

Nesterenko (1995). The latest and the most precise result (obtained by Schneider's method) on linear forms in two p-adic logarithms is given in Bugeaud & Laurent (1996).

4 The application to the *abc*-conjecture

Masser (1985) proposed a refinement of a conjecture formulated by Oesterlé, conjecturing that *for any given* $\varepsilon > 0$ *there exists a positive number* $c_3(\varepsilon)$ *depending only on* ε*, such that for all positive integers* x*,* y *and* z *with*

$$x + y = z \text{ and } (x, y, z) = 1, \tag{7}$$

we have

$$z < c_3(\varepsilon) N^{1+\varepsilon},$$

where N *is the product of all the distinct prime divisors of* xyz*.* The conjecture is now known as the *abc*-conjecture and it has profound consequences (see Baker 1998, Elkies 1991, Lang 1990, Langevin 1993, Philippon 1999, Stewart & Tijdeman 1986, Vojta 1987).

Stewart & Tijdeman (1986) proved that there exists an effectively computable constant c_4 such that for all positive integers x, y, z with (7),

$$z < \exp\left(c_4 N^{15}\right). \tag{8}$$

The proof depends on the theory of linear forms in logarithms due to Baker; more specifically, it utilizes the developments in the p-adic domain due to van der Poorten (1977). Stewart & Yu (1991) strengthened (8). They proved, by combining the best p-adic and Archimedean estimates then available, due to Yu (1990) and Waldschmidt (1980) respectively, that there exists an effectively computable constant c_5 such that for all positive integers x, y, z, with $z > 2$, satisfying (7),

$$z < \exp\left(N^{2/3 + c_5/\log\log N}\right). \tag{9}$$

Recently, Stewart & Yu (2001) obtained two further improvements on (9).

Theorem 5 (Stewart & Yu 2001) *There exists an effectively computable positive number* c_6 *such that for all positive integers* x*,* y *and* z *with* $x + y = z$ *and* $(x, y, z) = 1$*,*

$$z < \exp\left(c_6 N^{1/3} (\log N)^3\right). \tag{10}$$

The key new ingredient in our proof of Theorem 5 is Theorem 4 (for a more precise formulation, see Yu (1999) Theorem 1) which, as indicated above, has a better dependence on the number of terms in the linear form than previous p-adic estimates. We employ Yu (1999) Theorem 1 in order to control the p-adic order of x, y and z at the small primes p dividing x, y and z. Next we combine the contributions from these small primes in order to reduce the number of terms in our linear forms. We conclude with a further application of estimates for logarithmic forms in a fashion similar to Stewart & Yu (1991). Here we appeal to Theorem 3 for the p-adic estimates and Theorem 2 for the Archimedean estimates.

An examination of our proof reveals that the hindrance to a further refinement of Theorem 5 is *the dependence on the prime ideal* $\wp$ in the p-adic estimates. (Currently, the dependence is a factor $\Phi(\wp) = p^{f_\wp} - 1$ in the p-adic estimates; this happens even in the simplest case when $n = 1$; see Yu (1990) Lemma 1.4.) This fact is highlighted by our next result which shows that if the greatest prime divisor of one of x, y and z is small relative to N then (10) can be improved. In particular, let p_x, p_y and p_z denote the greatest prime divisors of x, y and z respectively with the convention that the greatest prime divisor of 1 is 1. Put $p' = \min\{p_x, p_y, p_z\}$. Denote the ith iterate of the logarithmic function by $\log_i$ so that $\log_1 t = \log t$ and $\log_i t = \log(\log_{i-1} t)$ for $i = 2, 3, \ldots$.

Theorem 6 (Stewart & Yu 2001) *There exists an effectively computable positive number c_7 such that for all positive integers x, y and z with $x + y = z$, $(x, y, z) = 1$ and $z > 2$,*

$$z < \exp\left(p' N^{c_7 \log_3 N^* / \log_2 N}\right), \tag{11}$$

where $N^* = \max(N, 16)$.

Thus, for each $\varepsilon > 0$ there exists a number $c_8(\varepsilon)$, which is effectively computable in terms of ε, such that for all positive integers x, y and z with $x+y = z$ and $(x, y, z) = 1$,

$$z < \exp\left(c_8(\varepsilon) p' N^{\varepsilon}\right).$$

Observe that $p' \le (p_x p_y p_z)^{1/3} \le N^{1/3}$, and so we immediately obtain

$$z < \exp\left(c_8(\varepsilon) N^{1/3+\varepsilon}\right),$$

a slightly weaker version of Theorem 5. On the other hand, if p' is appreciably smaller than $N^{1/3}$, (11) gives a sharper upper bound than (10).

Acknowledgment The author would like to express his cordial gratitude to Gisbert Wüstholz for inviting him to speak at the Conference and to the Forschungsinstitut für Mathematik of ETH Zürich for its support.

References

Adams, W.W. (1966), Transcendental numbers in the p-adic domain, *Amer. J. Math.* **88**, 279–308.

Ax, J. (1965), On the units of an algebraic number field, *Illinois J. Math.* **9**, 584–589.

Baker, A. (1966), Linear forms in the logarithms of algebraic numbers I, *Mathematika* **13**, 204–216.

Baker, A. (1967a), Linear forms in the logarithms of algebraic numbers II, *Mathematika* **14**, 102–107.

Baker, A. (1967b), Linear forms in the logarithms of algebraic numbers III, *Mathematika* **14**, 220–228.

Baker, A. (1968), Linear forms in the logarithms of algebraic numbers IV, *Mathematika* **15**, 204–216.

Baker, A. (1973), A sharpening of the bounds for linear forms in logarithms II, *Acta Arith.* **24**, 33–36.

Baker, A. (1977), The theory of linear forms in logarithms, in *Transcendence Theory: Advances and Applications*, A. Baker and D.W. Masser (eds.), Academic Press, 1–27.

Baker, A. (1998), Logarithmic forms and the abc-conjecture, in *Number Theory (Eger, 1996)*, de Gruyter, 37–44.

Baker, A. and J. Coates (1975), Fractional parts of powers of rationals, *Math. Proc. Camb. Phil. Soc.* **77**, 269–279.

Baker, A. and H.M. Stark (1971), On a fundamental inequality in number theory, *Ann. Math.* **94**, 190–199.

Baker, A. and G. Wüstholz (1993), Logarithmic forms and group varieties, *J. Reine Angew. Math.* **442**, 19–62.

Brumer, A. (1967), On the units of algebraic number fields, *Mathematika* **14**, 121–124.

Bugeaud, Y. and M. Laurent (1996), Minoration effective de la distance p-adique entre puissances de nombres algébriques, *J. Number Theory* **61** (2), 311–342.

Capelli, A. (1901), Sulla riduttibilità della funzione $x^n - A$ in un campo qualunque di razionalità (Auszug aus einem an Herrn F. Klein gerichteten Briefe), *Math. Ann.* **54**, 602–603.

Coates, J. (1969), An effective p-adic analogue of a theorem of Thue I: The greatest prime factor of a binary form, *Acta Arith.* **15**, 279–305.

Coates, J. (1970), An effective p-adic analogue of a theorem of Thue II: The greatest prime factor of a binary form, *Acta Arith.* **16**, 399–412.

Dong Pingping (1995), Minorations de combinaisons linéaires de logarithmes p-adiques de nombres algébriques, *Dissertationes Math.* **343**, 97 pp.

Elkies, N.D. (1991), ABC implies Mordell, *Int. Math. Res. Notices* no. 7, 99–109.

Feldman, N.I. (1968), An improvement of the estimate of a linear form in the logarithms of algebraic numbers, *Mat. Sbornik* **77**, 423–436; English translation in *Math. USSR Sbornik* **6** (1968), 393–406.

Gel'fond, A.O. (1940), Sur la divisibilité de la difference des puissances de deux nombres entières par une puissance d'un idéal premier, *Mat. Sbornik* **7**, 7–26.

Gel'fond, A.O. (1952), *Transcendental and Algebraic Numbers*, Moscow; English translation, Dover (1960).

Hasse, H. (1980), *Number Theory*, Springer-Verlag.

Heath-Brown, D.R. (1992), Zero-free regions for Dirichlet L-functions, and the least prime in an arithmetic progression, *Proc. London Math. Soc.* **64** (3), 265–338.

Kaufman, R.M. (1971), Bounds for linear forms in the logarithms of algebraic numbers with p-adic metric, *Vestnik Moskov. Univ. Ser. I,* **26**, 3–10.

Lang, S. (1990), Old and new conjectured Diophantine inequalities, *Bull. (New Ser.) Amer. Math. Soc.* **23** (1), 37–75.

Langevin, M. (1993), Cas d'égalité pour le Théoréme de Mason et application de la conjecture (*abc*), *Comptes Rendus Acad. Sci. Paris* **317**, 441–444.

Laurent, M., M. Mignotte and Y. Nesterenko (1995), Formes linéaires en deux logarithmes et déterminants d'interpolation, *J. Number Theory,* **55** (2), 285–321.

Mahler, K. (1932), Ein Beweis der Transzendenz der P-adischen Exponentialfunktion, *J. Reine Angew. Math.* **169**, 61–66.

Mahler, K. (1935), Über transzendente P-adische Zahlen, *Compositio. Math.* **2**, 259–275.

Masser, D.W. (1985), Open problems, In *Proc. Symp. Analytic Number Theory, London, Imperial College 1985*, W.W.L. Chen (ed.).

Matveev, E.M. (1996), Elimination of the multiple $n!$ from estimates for linear forms in logarithms, Tagungsbericht 11/1996, Diophantine Approximations, 17.03–23.03.1996, Oberwolfach.

Matveev, E.M. (1998), An explicit lower bound for a homogeneous rational linear form in logarithms of algebraic numbers, *Izvestiya Math.* **62** (4), 723–772.

Philippon, P. (1999), Quelques remarques sur des questions d'approximation diophantienne, *Bull. Austral. Math. Soc.* **59**, 323–334.

Philippon, P. and M. Waldschmidt (1989), Formes linéaires de logarithmes simultanées sur les groupes algébriques commutatifs, in *Séminaire de Théorie des Nombres, Paris 1986–87*, Birkhäuser, 313–347.

van der Poorten, A.J. (1977), Linear forms in logarithms in the p-adic case, in *Transcendence Theory: Advances and Applications*, A. Baker and D.W. Masser (eds.), Academic Press, 29–57.

Rédei, L. (1967), *Algebra* Vol. 1, Akadémiai Kiadó, Budapest.

Riley, R. (1990), Growth of order of homology of cyclic branched covers of knots, *Bull. London Math. Soc.* **22**, 287–297.

Schinzel, A. (1967), On two theorems of Gel'fond and some of their applications, *Acta Arith.* **13**, 177–236.

Schnirelman, L.G. (1938), On functions in normed algebraically closed fields, *Izv. Akad. Nauk SSSR, Ser. Mat.* 5/6, **23**, 487–496.

Shorey, T.N. and R. Tijdeman (1986), *Exponential Diophantine Equations*, Cambridge University Press.

Sprindžuk, V.G. (1967), Concerning Baker's theorem on linear forms in logarithms, *Dokl. Akad. Nauk BSSR,* **11**, 767–769.

Sprindžuk, V.G. (1968), Estimates of linear forms with p-adic logarithms of algebraic numbers, *Vesci Akad. Nauk BSSR, Ser. Fiz-Mat.*, (4), 5–14.

Stewart, C.L. and R. Tijdeman (1986), On the Oesterlé–Masser conjecture, *Monatsh. Math.* **102**, 251–257.

Stewart, C.L. and Kunrui Yu (1991), On the *abc* conjecture, *Math. Ann.* **291**, 225–230

Stewart, C.L. and Kunrui Yu (2001), On the *abc* conjecture, II, *Duke Math. J.* **108**, 169–181.

Vojta, P. (1987), *Diophantine Approximations and Value Distribution Theory*, Lect. Notes Math. **1239**, Springer-Verlag.

Waldschmidt, M. (1980), A lower bound for linear forms in logarithms, *Acta Arith.* **37**, 257–283.

Waldschmidt, M. (1991), Nouvelles méthodes pour minorer des combinaisons linéaires de logarithmes de nombres algébriques, in *Sém. Théorie Nombres Bordeaux* **3**, 129–185.

Waldschmidt, M. (1993), Minorations de combinaisons linéaires de logarithmes de nombres algébriques, *Canadian J. Math.* **45** (1), 176–224.

Wüstholz, G. (1987), A new approach to Baker's theorem on linear forms in logarithms I, II, in *Diophantine Problems and Transcendence Theory*, Lect. Notes Math. **1290**, 189–211.

Wüstholz, G. (1988), A new approach to Baker's theorem on linear forms in logarithms III, in *New Advances in Transcendence Theory*, A. Baker (ed.), Cambridge University Press, 399–410.

Wüstholz, G. (1989), Multiplicity estimates on group varieties, *Ann. Math.* **129**, 471–500.

Yu, Kunrui (1989), Linear forms in p-adic logarithms, *Acta Arith.* **53**, 107–186.

Yu, Kunrui (1990), Linear forms in p-adic logarithms II, *Compositio Math.* **74**, 15–113.

Yu, Kunrui (1994), Linear forms in p-adic logarithms III, *Compositio Math.* **91**, 241–276.

Yu, Kunrui (1998), P-adic logarithmic forms and group varieties I, *J. Reine Angew. Math.* **502**, 29–92.

Yu, Kunrui (1999), p-adic logarithmic forms and group varieties II, *Acta Arith.* **89**, 337–378.

3

Recent Progress on Linear Forms in Elliptic Logarithms

Sinnou David and Noriko Hirata-Kohno

1 Introduction

In this article, we describe recent progress on the theory of linear forms in logarithms associated with elliptic curves defined over a number field. In this set-up, and without any extra hypothesis (e.g. complex multiplication), we obtain the first best possible dependence in the height of the linear form. We shall also briefly describe the main ideas (which date back to G.V. Chudnovsky in the late 1970s) leading to this refinement. The complete proof of this result will be published in our forthcoming article (David & Hirata-Kohno 2002).

Let us first start with a short account of the history of the theory of linear forms in elliptic logarithms.

Let K be an algebraic number field of degree D over the rational number field $\mathbb{Q}$. We denote by $\overline{\mathbb{Q}}$ the algebraic closure of $\mathbb{Q}$ in $\mathbb{C}$. Let k be a rational integer ≥ 1. Let $\mathcal{E}_1, \ldots, \mathcal{E}_k$ be k elliptic curves defined over K. We assume that these curves are defined by Weierstraß' equations, normalized as follows†:

$$y^2 = 4x^3 - g_{2,i}x - g_{3,i} \ : \qquad g_{2,i}, g_{3,i} \in K, \quad 1 \leq i \leq k.$$

We denote by $\wp_i$, for $1 \leq i \leq k$, (resp. σ_i, for $1 \leq i \leq k$), the Weierstraß elliptic functions (resp. the Weierstraß sigma functions), associated with the underlying period lattice $\Lambda_i = \omega_{1,i}\mathbb{Z} + \omega_{2,i}\mathbb{Z}, \ 1 \leq i \leq k$.

For each $1 \leq i \leq k$, let $u_i \in \mathbb{C}$ satisfy

$$\gamma_i \ := (\sigma_i^3(u_i), \sigma_i^3(u_i)\wp_i(u_i), \sigma_i^3(u_i)\wp_i'(u_i)) \in \mathcal{E}_i(\overline{\mathbb{Q}}).$$

When u_i is a pole of $\wp_i$, we consider $\gamma_i = (0, 0, 1)$.

Such complex numbers $u_1, \ldots, u_k$ are called elliptic logarithms (of rational points).

† Such a normalization is not strictly necessary and any model would do; we however fix the latter for convenience and easier comparisons between earlier works.

Thus, clearly, any point in the period lattice is an elliptic logarithm.

Let $N \geq 1$ be an integer and $P = (x_0, \ldots, x_N) \in \mathbb{P}^N(\overline{\mathbb{Q}})$. We introduce the absolute logarithmic projective height on $\mathbb{P}^N$. Let L be a number field containing all coordinates of the point P. Put

$$h(P) = \frac{1}{[L:\mathbb{Q}]} \sum_v n_v \log(\max\{|x_0|_v, \ldots, |x_N|_v\}),$$

where v runs over the set of absolute values of L which are normalised such that for all $x \in L, x \neq 0$, we have $\sum_v n_v \log |x|_v = 0$ and $\sum_{v|\infty} n_v = d$. Here, we denote by $n_v = [K_v : \mathbb{Q}_v]$ the local degree at each v. Because of the extension formula, it is well known that $h(P)$ is independent of the choice of the field L, and the product formula ensures on the other hand that the definition does not depend on the choice of projective coordinates of P.

The study of linear forms in elliptic logarithms derives from an analogy with the theory of linear forms in usual logarithms, simply by viewing the Weierstraß elliptic $\wp$-function with algebraic invariants as an exponential map of an elliptic curve (i.e. a commutative algebraic group) defined over a number field.

A basic question is to ask whether non-zero elliptic logarithms of rational points are transcendental. An answer was first given by Siegel (1932). For $k = 1$, we write $u = u_1$, $\Lambda = \Lambda_1$, and $\wp = \wp_1$, in our above notation. Siegel showed that there exists at least one element of Λ which is transcendental over $\mathbb{Q}$. If $\wp$ has complex multiplication (CM), it is well known that the ratio of two non-zero elements of Λ belongs to the corresponding quadratic imaginary field $\mathfrak{K}$. Thus, in the case of CM, Siegel's result implies that any non-zero element in Λ is transcendental. Schneider (1937) proved more generally that any elliptic logarithm u is either zero or transcendental without any hypothesis of complex multiplication. Now consider the case $k = 2$ with $\mathcal{E}_1 = \mathcal{E}_2$, $\wp := \wp_1 = \wp_2$. Schneider also showed that the quotient of two non-zero elliptic logarithms u_1, u_2 is transcendental if and only if the two functions $\wp(u_1 z)$ and $\wp(u_2 z)$ are algebraically independent. If $\wp$ has no complex multiplication, this happens precisely when u_1/u_2 is not rational. If $\wp$ has complex multiplication, this happens only when u_1/u_2 does not belong to the corresponding quadratic imaginary field.

Baker (1970a) proved, using the method he developed for the study of linear forms in usual logarithms (see Baker 1975), that, when $k = 2$, $u_1 \in \Lambda_1$ and $u_2 \in \Lambda_2$, the linear form $\beta_1 u_1 + \beta_2 u_2$ with algebraic coefficients β_1, β_2 is either zero or transcendental (see also related results together with quasi-periods and $2\pi i$ by S. Lang, J. Coates and by D.W. Masser, mentioned in Masser 1977).

Masser (1975) succeeded in proving a generalization to arbitrary k elliptic logarithms $u_1, \ldots, u_k$ when $\mathcal{E}_1 = \cdots = \mathcal{E}_k$, provided that $\wp := \wp_1 = \cdots = \wp_k$ has CM over $\mathfrak{K}$: if $u_1, \ldots, u_k$ are linearly independent over $\mathfrak{K}$, then $1, u_1, \ldots, u_k$ are linearly independent over $\overline{\mathbb{Q}}$ (see Chapter 7 with Appendix 3 of Masser 1975a). This was extended in 1980, to the non-CM case by D. Bertrand and Masser: suppose that $\wp$ has no CM and that $u_1, \ldots, u_k$ are linearly independent over $\mathbb{Q}$. Then $1, u_1, \ldots, u_k$ are linearly independent over $\overline{\mathbb{Q}}$ (see Bertrand & Masser 1980a).

Generalizations in the abelian case were treated by Schneider (1941) for abelian integrals, more generally by Lang and by Masser (see Lang 1964, Masser 1975b, Lang 1975a, Masser 1976a, b). Masser proved the linear independence of 'abelian' logarithms over $\overline{\mathbb{Q}}$ under a hypothesis of complex multiplication (with a quantitative version of exponential magnitude: see below). The non-CM case was presented in 1980 by Bertrand & Masser (see Bertrand & Masser 1980b); they however needed real multiplication.

Let us consider the linear independence problem of elliptic logarithms without the simplifying hypothesis $\mathcal{E}_1 = \cdots = \mathcal{E}_k$, nor assuming complex multiplication. More generally, consider the corresponding problem on a connected commutative algebraic group defined over a number field. The linear independence over $\overline{\mathbb{Q}}$ of 1 and 'generalized abelian' logarithms was proven by G. Wüstholz (1989), where we can deduce all qualitative results mentioned above as corollaries.

We now give an account of the history of quantitative estimates†.

In 1951, N.I. Fel'dman obtained a Diophantine approximation measure of an elliptic logarithm by an algebraic number. Precisely, it concerns the case $k = 1$, $u := u_1 \neq 0$ in our above notation. Let B be a real number ≥ 3. He proved that there exists an effective constant $c > 0$ which is independent of B such that for any $\beta \in \overline{\mathbb{Q}}$ with‡ $h(\beta) \leq \log B$ we have

$$\log|u - \beta| \geq -\log B \cdot \exp\{c(\log\log B)^{1/2}\};$$

he refined the estimate for a non-zero period $u \in \Lambda := \Lambda_1$ to obtain

$$\log|u - \beta| \geq -c \cdot \log B \cdot (\log\log B)^4.$$

The case of a quotient of two non-zero elliptic logarithms was also treated by him (see Feldman 1951, 1958, 1968) (in fact, he used a classical height, but it can be translated to the logarithmic height; see the relation between various heights in Waldschmidt 1979).

† As a convention, we shall always specialize such estimates in the elliptic case, and merely mention, if needed, that they are valid in a more general set-up.

‡ where $h(\beta)$ stands for $h(1, \beta)$.

Let $\mathcal{L}(\mathbf{z}) = \beta_0 z_0 + \cdots + \beta_k z_k$ be a non-zero linear form on $\mathbb{C}^{k+1}$ with coefficients in K. We write $\mathbf{v} = (1, u_1, \ldots, u_k)$. Let B be a real number satisfying $B \geq e$.

Baker proved a positive lower bound of $|\mathcal{L}(\mathbf{v})|$ in 1970 (see Baker (1970b) for $k = 2$, $\mathcal{E}_1 = \mathcal{E}_2, u_1, u_2 \in \Lambda := \Lambda_1 = \Lambda_2$ and $\beta_0 \neq 0$. Masser (1975a) showed the following estimate for arbitrary k, $\mathcal{E}_1 = \cdots = \mathcal{E}_k$ and $\beta_0 = 0$ under a hypothesis of complex multiplication over $\mathfrak{K}$; assume that $u_1, \ldots, u_k$ are linearly independent over $\mathfrak{K}$. For any $\epsilon > 0$, there exists an effective constant $c > 0$ which depends on ϵ and other data, but independent of B, such that for any $\beta_1, \ldots, \beta_k \in K$ satisfying $h(\beta_i) \leq \log B$, $1 \leq i \leq k$, we have $\log |\mathcal{L}(\mathbf{v})| \geq -c \cdot B^{\epsilon}$ (see also the abelian cases in Lang 1975a; Masser 1975b, 1976a,b; the estimates in Masser 1975a, 1976b are of the same magnitude). Also assuming complex multiplication, Coates & Lang (1976) refined this estimate, actually in the abelian case, to get $\log |\mathcal{L}(\mathbf{v})| \geq -c \cdot (\log B)^{8k+6+\epsilon}$. M. Anderson (1977) refined this estimate and proved, in the not necessarily homogeneous case but still assuming complex multiplication on elliptic curves, that $\log |\mathcal{L}(\mathbf{v})| \geq -c \cdot \log B \cdot (\log\log B)^{k+1+\epsilon}$, where $h(\beta_i) \leq \log B$, $0 \leq i \leq k$, and $\log B \geq e$. Some related results were treated by Brownawell & Masser (1980), by Reyssat (1980) and by Kunrui Yu (1985).

Philippon & Waldschmidt (1988) gave the first such estimate without any hypothesis of complex mulplication. Let us put $\mathcal{W} = \ker(\mathcal{L})$. Suppose that for any connected algebraic subgroup† $\mathbb{G}'$ of $\mathbb{G} := \mathbb{G}_a \times \mathcal{E}_1 \times \cdots \times \mathcal{E}_k$ with $T_{\mathbb{G}'}(\mathbb{C}) \subset \mathcal{W}$ we have $\mathbf{v} \notin T_{\mathbb{G}'}(\mathbb{C})$ (here we denote by $T_{\mathbb{G}'}(\mathbb{C})$ the tangent space of $\mathbb{G}$ at the origin). Let B be a real number satisfying $\log B \geq \max\{1,\ h(\beta_i)\ ;\ 0 \leq i \leq k\}$. Then Philippon & Waldschmidt obtained a lower bound of the form

$$|\mathcal{L}(\mathbf{v})| \geq \exp\left(-c \cdot (\log B)^{k+1}\right).$$

They did not assume $\mathcal{L}(\mathbf{v}) \neq 0$ as was often done; thus we can deduce also qualitative linear independence or transcendence results from this quantitative one (such a lower bound clearly implies that $\mathcal{L}(\mathbf{v}) \neq 0$). In fact, they proved a result in the general case where $\mathbb{G}$ is any connected commutative algebraic group. This estimate was refined in Hirata-Kohno (1990, 1991) with $\log B \geq e$ to get

$$\log |\mathcal{L}(\mathbf{v})| \geq -c \cdot \log B \cdot (\log\log B)^{k+1}$$

in the case of connected commutative algebraic group also, relying upon an idea originally due to Feld'man (1951) and used as well in Reyssat (1980) but by introducing a 'redundant variable'.

† As usual, $\mathbb{G}_a$ stands for the additive group.

David (1995) then gave a completely explicit version in the elliptic case of this result, with c made explicit as a function of all given data. Here, the dependence of $|u_i|$ with $1 \leq i \leq k$ is better than the previous results when these quantities are small. Ably (1998) showed in the elliptic case an estimate of the form

$$\log |\mathcal{L}(\mathbf{v})| \geq -c \cdot \log B$$

under a hypothesis of complex multiplication. For this purpose, he generalized Fel'dman's polynomials to quadratic fields and studied their properties. He was thus the first to obtain the optimal estimate in the elliptic case, albeit with the extra hypothesis of complex multiplication. A little later, in 1999, a special case related with periods and quasi-periods of an elliptic function was treated by Bruiltet (2001), where one part corresponds in fact to a statement announced by Chudnovsky (1984). We would also like to mention current work by E. Gaudron, which aims to provide an estimate of the same optimal shape, i.e. $-c \cdot \log B$ for any commutative algebraic group, by studying the arithmetic properties of infinitesimal neighbourhoods of the origin on suitable integral models.

Our contribution basically originates from an idea of G.V. Chudnovsky, which says that local parameters have better arithmetic properties than the complex uniformization, though they do not have a good analytic behaviour. We therefore build on his idea of 'variable change' (see Chapter 8 on algebraic independence measure of Chudnovsky 1984) to the case of elliptic logarithms, which are not necessarily in the period lattice, and we work with the parameters coming from the so-called formal group (see, for example, chapter IV of Silverman 1986).

2 New result

We put $\tau_i = \omega_{2,i}/\omega_{1,i}$, $1 \leq i \leq k$. It is no restriction to assume that τ_i belongs to the upper half plane $\mathfrak{H}$, or even to the usual fundamental domain $\mathfrak{F}$ of $\mathfrak{H}$ by the action of $SL_2(\mathbb{Z})$; for this, we choose a suitable basis of Λ_i and this does not change the invariants $g_{2,i}, g_{3,i}$, $1 \leq i \leq k$.

We denote by $h = \max\{1, h(1, g_{2,i}, g_{3,i})\ ;\ 1 \leq i \leq k\}$ the height of our elliptic curves.

We also denote by $\hat{h}(\gamma_i)$ the Néron–Tate height of γ_i defined as in Silverman (1986); namely, $\hat{h}(\gamma_i) = \lim_{n\to\infty} \frac{h(n\gamma_i)}{n^2}$.

Finally we put $\mathbb{G} = \mathbb{G}_a \times \mathcal{E}_1 \times \cdots \times \mathcal{E}_k$ which is a connected commutative algebraic group. Write $T_{\mathbb{G}}(\mathbb{C})$ for the tangent space of $\mathbb{G}$ at the origin, which we shall identify with $\mathbb{C}^{k+1}$. We denote by $T_{\mathbb{G}'}(\mathbb{C})$ the tangent space at the origin of an algebraic subgroup $\mathbb{G}'$ of $\mathbb{G}$.

Now we present our result.

Theorem 1 *There exists an effective function $C > 0$ of k, with the following property. Let $\mathcal{L}(\mathbf{z}) = \beta_0 z_0 + \cdots + \beta_k z_k$ be a non-zero linear form on $\mathbb{C}^{k+1}$ with coefficients in K; we put $\mathcal{W} = \ker(\mathcal{L})$; let moreover $u_1, \ldots, u_k$ be complex numbers such that $\gamma_i = (1, \wp_i(u_i), \wp_i'(u_i)) \in \mathcal{E}_i(K) \subset \mathbb{P}^2(K)$ if $u_i \notin \Lambda_i$, and $\gamma_i = (0, 0, 1)$ if $u_i \in \Lambda_i$ for $1 \leq i \leq k$. We write $\mathbf{v} = (1, u_1, \ldots, u_k)$. Let B, E, $V_1, \ldots, V_k$ be real numbers satisfying the following conditions:*

$$\log B \geq \max\{1, h(\beta_i)\ ;\ 0 \leq i \leq k\}$$

$$V_1 \geq \cdots \geq V_k$$

$$\log V_i \geq \max\left\{e, \hat{h}(\gamma_i)\,,\ \frac{|u_i|^2}{D|\omega_{1,i}|^2 \Im m\ \tau_i}\right\}, \quad 1 \leq i \leq k$$

$$e \leq E \leq \min\left\{\frac{|\omega_{1,i}|\,(\Im m\ \tau_i \cdot D \log V_i)^{\frac{1}{2}}}{|u_i|}\ ;\ 1 \leq i \leq k\right\}.$$

Suppose that for any connected algebraic subgroup $\mathbb{G}'$ of $\mathbb{G}$ with $T_{\mathbb{G}'}(\mathbb{C}) \subset \mathcal{W}$, we have $\mathbf{v} \notin T_{\mathbb{G}'}(\mathbb{C})$.

Then we have $\mathcal{L}(\mathbf{v}) \neq 0$ and

$$\begin{aligned}\log|\mathcal{L}(\mathbf{v})| \quad \geq \quad & -C \cdot D^{2k+2} \times (\log E)^{-2k-1} (\log B + \log(DE) + h + \log\log V_1) \\ & \times (\log(DE) + h + \log\log V_1)^{k+1} \prod_{i=1}^{k} (h + \log V_i)\,.\end{aligned}$$

Thus we obtain here a lower bound of the form

$$\log|\mathcal{L}(\mathbf{v})| \geq -c \cdot \log B$$

without any hypothesis of complex multiplication. However, the dependence in V_i, $1 \leq i \leq k$) of our estimate is weaker than that of Philippon & Waldschmidt (1988), by a factor $\log\log V_1$.

3 Key estimate for the refinement

We now present two propositions, one analytic and the other arithmetic. They allow us to remove the $(\log\log B)^{k+1}$ factor in David (1995), Hirata-Kohno (1991): they concern indeed an estimate of the height at the origin of derivatives of a polynomial in elliptic functions.

Let us consider a variable change that is based upon the formal group of the elliptic curves. Namely, for each $1 \le i \le k$, let $x = \wp_i(z_i)$, $y = \wp_i'(z_i)$ satisfy the equation

$$y^2 = 4x^3 - g_{2,i}x - g_{3,i} : \quad g_{2,i}, g_{3,i} \in K.$$

Now we introduce a local parameter $t_i = -\frac{2\wp_i(z_i)}{\wp_i'(z_i)}$ and write $w_i(t_i) = -\frac{2}{\wp_i'(z_i)}$. One easily sees that $w_i(t_i)$ is a formal power series in t_i with coefficients in the ring $\mathbb{Z}[g_{2,i}/4, g_{3,i}/4]$, and we see that the series has a positive radius of convergence around the origin, i.e. $w_i(t_i)$ can be identified with its Taylor series. Furthermore, for $1 \le i \le k$, put $z_i = z_i(t_i) = \int \Omega_i(t_i)$, where $\Omega_i(t_i)$ is a differential form in the local parameter t_i, see Silverman (1986), Chapter IV, Section 5; thus

$$z_i = z_i(t_i) = \int \Omega_i(t_i) = \int \frac{\frac{d}{dt_i}\left(\frac{t_i}{w_i(t_i)}\right)}{-\frac{2}{w_i(t_i)}} dt_i,$$

defined around $t_i = 0$. Hence we can describe a behaviour of an 'elliptic logarithmic function' $z_i(t_i)$.

Proposition 1 *Let L_0, L, T be rational integers ≥ 1. Let $\beta_1, \dots, \beta_k \in \mathbb{C}$. Let $P(X_0, \mathbf{X}_1, \dots, \mathbf{X}_k) \in \mathbb{C}[X_0, \mathbf{X}_1, \dots, \mathbf{X}_k]$ be a polynomial of degree $\le L_0$ in X_0 and of homogeneous degree $\le L$ in $\mathbf{X}_i = (X_{0,i}, X_{1,i}, X_{2,i})$, $1 \le i \le k$. Put*

$$\begin{aligned} F(\mathbf{z}) &= F(z_1, \dots, z_k) \\ &= P\left(\beta_1 z_1 + \dots + \beta_k z_k, \left(\frac{1}{\wp_1'(z_1)}, \frac{\wp_1(z_1)}{\wp_1'(z_1)}, 1\right),\right. \\ &\qquad \left.\dots, \left(\frac{1}{\wp_k'(z_k)}, \frac{\wp_k(z_k)}{\wp_k'(z_k)}, 1\right)\right) \end{aligned}$$

which is a meromorphic function, analytic at $\mathbf{z} = (z_1, \dots, z_k) = \mathbf{0}$. Put also

$$\begin{aligned} G(\mathbf{t}) &= P\left(\beta_1 z_1(t_1) + \dots + \beta_k z_k(t_k), (w_1(t_1), t_1, -2),\right. \\ &\qquad \left.\dots, (w_k(t_k), t_k, -2)\right), \end{aligned}$$

a meromorphic function, analytic at $\mathbf{t} = (t_1, \dots, t_k) = \mathbf{0}$. For $\tau_1, \dots, \tau_k \in \mathbb{Z}$, $\tau_1, \dots, \tau_k \ge 0$, put $\tau = (\tau_1, \dots, \tau_k)$, $|\tau| = \tau_1 + \dots + \tau_k$. We define

$$\Delta_{\mathbf{z}}^{\tau} F(\mathbf{z}) = \frac{1}{\tau_1! \dots \tau_k!} \left(\frac{\partial}{\partial z_1}\right)^{\tau_1} \circ \dots \circ \left(\frac{1}{\partial z_k}\right)^{\tau_k} F(\mathbf{z}),$$

and similarly

$$\Delta_{\mathbf{t}}^{\tau} G(\mathbf{t}) = \frac{1}{\tau_1! \dots \tau_k!} \left(\frac{\partial}{\partial t_1}\right)^{\tau_1} \circ \dots \circ \left(\frac{\partial}{\partial t_k}\right)^{\tau_k} G(\mathbf{t}).$$

Then we have the following two properties.

(i) *If* $\Delta_{\mathbf{z}}^{\tau} F(\mathbf{0}) = 0$ *for* $|\tau| < T$ *then we have* $\Delta_{\mathbf{t}}^{\tau} G(\mathbf{0}) = 0$ *for* $|\tau| < T$.

(ii) *If* $\Delta_{\mathbf{z}}^{\tau} F(\mathbf{0}) = 0$ *for* $|\tau| < T$ *then we have* $\Delta_{\mathbf{z}}^{\tau} F(\mathbf{0}) = \Delta_{\mathbf{t}}^{\tau} G(\mathbf{0})$ *for* $|\tau| = T$.

Proposition 1 can be proved by direct induction on $|\tau|$ and using the formula of Faà-de-Bruno on derivatives of composed functions in several variables†.

When we consider the 'divided' derivatives as in Proposition 1, instead of the usual ones, the values of divided derivatives at the origin correspond to nothing but the coefficients of the Taylor expansion at 0 of the function. This fact avoids the necessity of introducing the usual $t!$ factor in the upper bound for the height of the tth derivatives. However, it does require an estimate of the height of the coefficients of the underlying Taylor series.

As we shall see below, the arithmetic behaviour of an 'elliptic logarithmic function' is a suitable one so that we can control well the coefficients of its Taylor expansion at 0. The whole point, since such an elliptic logarithm function is not entire, is to use the classical Weierstraß elliptic functions multiplied by a suitable power of sigma functions for the whole archimedean part of the argument, and switch back to elliptic logarithm functions at the end for arithmetic computations. Proposition 1 allows such back and forth movements.

Now let us state an arithmetic property of the elliptic logarithmic functions.

Proposition 2 *Let* L_0, L, T *be rational integers* ≥ 1, *with* $L_0 \leq T$, $L \leq T$. *Let* $\beta_1, \dots, \beta_k \in K$. *Let* $P(X_0, \mathbf{X}_1, \dots, \mathbf{X}_k) \in K[X_0, \mathbf{X}_1, \dots, \mathbf{X}_k]$ *be a polynomial of degree* $\leq L_0$ *in* X_0 *and of homogeneous degree* $\leq L$ *in* $\mathbf{X}_i = (X_{0,i}, X_{1,i}, X_{2,i})$, $1 \leq i \leq k$. *Put* $F(\mathbf{z}) = F(z_1, \dots, z_k)$ *and* $G(\mathbf{t}) = G(t_1, \dots, t_k)$ *as in Proposition 1. Suppose* $\Delta_{\mathbf{z}}^{\tau} F(\mathbf{0}) = 0$ *for* $|\tau| < T$. *Fix any* τ, *such that* $|\tau| = T$, *and put* $\gamma = \gamma_\tau = \Delta_{\mathbf{t}}^{\tau} G(\mathbf{0})$. *Then there exists an effective constant* $c > 0$ *depending only on* k *such that*

$$h(\gamma) \leq h(P) + c\,(Th + T \log L_0 + L_0 \log B)\,.$$

Proof Write $P = \sum_\lambda a_\lambda X_0^{\lambda_0} \mathbf{X}_1^{\lambda_1} \cdots \mathbf{X}_k^{\lambda_k}$ where λ stands for $(\lambda_0, \lambda_1, \dots, \lambda_k)$ with $\lambda_i = (\lambda_{0,i}, \lambda_{1,i}, \lambda_{2,i}) \in \mathbb{Z}^3$, $(1 \leq i \leq k)$, $\lambda_0 \in \mathbb{Z}$, $0 \leq \lambda_0 \leq L_0$,

† Of course, Proposition 1 holds for any function $F(z)$ analytic at the neighbourhood of the origin in $\mathbb{C}^k$ since $z_i(t_i) = O(t_i)$ (for (i)) and more precisely $z_i(t_i) = t_i + O(t_i^2)$ (for (ii)).

$\lambda_{0,i}, \lambda_{1,i}, \lambda_{2,i} \geq 0, 0 \leq \lambda_{0,i}+\lambda_{1,i}+\lambda_{2,i} \leq L$, $(1 \leq i \leq k)$. As we saw in David (1995), Hirata-Kohno (1990), it is no restriction to suppose that $\beta_0 = -1$. Then

$$\begin{aligned}\gamma &= \frac{1}{\tau_1! \dots \tau_k!}\left(\frac{\partial}{\partial t_1}\right)^{\tau_1} \circ \cdots \circ \left(\frac{\partial}{\partial t_k}\right)^{\tau_k} \\ &\quad \times \sum_\lambda a_\lambda (\beta_1 z_1(t_1) + \cdots + \beta_k z_k(t_k))^{\lambda_0} \\ &\quad \times \left(\left(\frac{-w_1(t_1)}{2}\right)^{\lambda_{0,1}} \left(\frac{-t_1}{2}\right)^{\lambda_{1,1}} \cdots \left(\frac{-w_k(t_k)}{2}\right)^{\lambda_{0,k}} \left(\frac{-t_k}{2}\right)^{\lambda_{1,k}}\right),\end{aligned}$$

evaluated at $(t_1, \dots, t_k) = (0, 0, \dots, 0)$, where $z_i(t_i)$ $(1 \leq i \leq k)$ is the elliptic logarithmic function defined above.

Put $w_i(t_i) = \sum_{s_i \geq 3}^{\infty} A_{i,s_i} t_i^{s_i}$, $z_i(t_i) = \sum_{s_i \geq 1}^{\infty} B_{i,s_i} t_i^{s_i}$; let us also fix a k-tuple $\tau = (\tau_1, \dots, \tau_k)$ of nonnegative integers such that $|\tau| = T$. Calculations by induction show that A_{i,s_i} is a homogeneous polynomial† of degree $\leq s_i - 3$ in $g_{2,i}/4$, $g_{3,i}/4$ with coefficients in $\mathbb{Z}$ of absolute value $\leq 3^3 \cdot 8^{s_i-3}$, and also that $B_{i,s_i} = \dfrac{C_{i,s_i}}{-2s_i}$ where C_{i,s_i} is a homogeneous polynomialof degree $\leq s_i - 1$ in $g_{2,i}/4$, $g_{3,i}/4$ with coefficients in $\mathbb{Z}$ of absolute value $\leq 10^{4s_i}$.

Hence γ is the coefficient of $t_1^{\tau_1} \dots t_k^{\tau_k}$ of the sum

$$\begin{aligned}&\sum_\lambda a_\lambda \sum_{(j_1,\dots,j_k)\in J} \frac{\lambda_0!}{j_1! \dots j_k!} \beta_1^{j_1} \cdots \beta_k^{j_k} \left(\sum_{s_1 \geq 1} B_{1,s_1} t_1^{s_1}\right)^{j_1} \cdots \left(\sum_{s_k \geq 1} B_{k,s_k} t_k^{s_k}\right)^{j_k} \\ &\quad \times \prod_{i=1}^{k} \left(\sum_{s_i \geq 3} A_{i,s_i} t_i^{s_i}\right)^{\lambda_{0,i}} (t_i)^{\lambda_{1,i}} \left(\frac{-1}{2}\right)^{\lambda_{0,i}+\lambda_{1,i}},\end{aligned}$$

with $J = J_{\lambda_0} = \{(j_1, \dots, j_k) \in \mathbb{Z} : j_1, \dots, j_k \geq 0,\ j_1 + \cdots + j_k = \lambda_0\}$.

By the estimates for the A_{i,s_i}, B_{i,s_i} and the formula above, one easily deduces that the archimedean part of the height of γ is bounded above by $c_1 T h$, for some constant $c_1 > 0$.

Now consider the set

$$E_{l_0,\tau_i} = \left\{s_1 \cdots s_{l_0} \ :\ s_1 + \cdots + s_{l_0} = \tau_i,\ s_1, \dots, s_{l_0} \in \mathbb{Z},\ s_1, \dots, s_{l_0} \geq 0\right\}.$$

By Prime Number Theory, we know that the least common multiple of all elements in $\{x : x \in E_{l_0,\tau_i},\ l_0 \leq L_0,\ \tau_i \leq T\}$ is less than

$$\exp(c_2 T \log(L_0 + 1)),$$

† with the usual weighted graduation

with an absolute constant $c_2 > 0$. Then, we use this fact to estimate the denominators of elliptic logarithmic functions, and to obtain, together with the estimate of B_{i,s_i}, that there exists a constant $c_3 > 0$ such that the non-archimedean contribution to the height of γ is bounded above by:

$$c_3(Th + T\log(L_0 + 1))$$

for $j_i \leq \lambda_0 \leq L_0$, $\tau_i \leq T$, $1 \leq i \leq k$. Putting together both estimates yields Proposition 2. □

Acknowledgements We are grateful to the referee, whose remarks helped us to improve the first draft of this article.

References

Ably, M. (2000), Formes linéaires de logarithmes de points algébriques sur une courbe elliptique de type CM, *Ann. de l'Institut Fourier* **50**, 1–33.

Anderson, M. (1977), Inhomogeneous linear forms in algebraic points of an elliptic function, in *Transcendence Theory: Advances and Applications*, A. Baker (ed.), Academic Press, 121–144.

Baker, A. (1970a), On the periods of the Weierstraß $\wp$-function, in *Symposia Math.* **IV**, INDAM Rome 1968, Academic Press, 155–174.

Baker, A. (1970b) An estimate for the $\wp$-function at an algebraic point, *Amer. J. Math.* **92**, 619–622.

Baker, A. (1975) *Transcendental Number Theory*, Cambridge University Press.

Bertrand, D. & D.W. Masser (1980a), Linear forms in elliptic integrals, *Inv. Math.* **58**, 283–288.

Bertrand, D. & D.W. Masser (1980b), Formes linéaires d'intégrales abéliennes, *C. R. Acad. Sc. Paris Série A* **290**, 725–727.

Brownawell, W.D. & D.W. Masser (198), Multiplicity estimates for analytic functions I, *J. Reine Angew. Math.* **314**, 200–216.

Bruiltet, S. (2001) D'une mesure d'approximation simultanée à une mesure d'irrationalité : $\Gamma(1/4)$ et $\Gamma(1/3)$, *Acta Arithmetica*.

Chudnovsky, G.V. (1984) *Contributions to the Theory of Transcendental Numbers*, Amer. Math. Soc. Math. Surveys Monographs **19**.

Coates, J. & S. Lang (1976), Diophantine approximation on Abelian varieties with complex multiplication, *Inv. Math.* **34**, 129–133.

David, S. (1995), Minorations de formes linéaires de logarithmes elliptiques, *Mémoires, Nouvelle série 62, Supplément au Bulletin de la Soc. Math. de France,* **123**, (3).

David, S. & N. Hirata-Kohno (2002), Formes linéaires de logarithmes elliptiques, *preprint*.

Fel'dman, N.I. (1951), Approximation of certain transcendental numbers II: the approximation of certain numbers associated with the Weierstraß $\wp$-function, *Izv. Akad. Nauk. SSSR. Ser. Mat.* **15**, 153–176; English translation in *Amer. Math. Trans. Ser. 2,* **59**, (1966), 246–270).

Fel'dman, N.I. (1958), Simultaneous approximation of the periods of an elliptic function by algebraic numbers, *Izv. Akad. Nauk. SSSR, Ser. Mat.* **22**, 563–576; English translation in *Amer. Math. Trans. Ser. 2,* **59**, (1966), 271–284).

Fel'dman, N.I. (1968), An elliptic analogue of an inequality of A.O. Gel'fond, *Trudy. Moskov* **18**, 65–76; English translation in *Trans. Moscow Math. Soc.* **18**, (1968), 71–84).

Hirata-Kohno, N. (1990), Formes linéaires d'intégrales elliptiques, in *Sém. de Théorie des Nombres, Paris, 1988/89*, C. Goldstein (ed). Birkhäuser, 1–23.

Hirata-Kohno, N. (1991), Formes linéaires de logarithmes de points algébriques sur les groupes algébriques, *Inv. Math.* **104**, 401–433.

Lang, S. (1964), Diophantine approximation on toruses, *Amer. J. Math.* **86**, 521–533.

Lang, S. (1975), Diophantine approximation on Abelian varieties with complex multiplication, *Adv. in Math.* **17**, 281–336.

Masser, D.W. (1975a), *Elliptic Functions and Transcendence*, Lecture Notes in Math. **437**, Springer.

Masser, D.W. (1975b), Linear forms in algebraic points of Abelian functions I, *Math. Proc. Camb. Phil. Soc.* **77**, 499–513.

Masser, D.W. (1976a), Linear forms in algebraic points of Abelian functions II, *Math. Proc. Camb. Phil. Soc.* **79**, 55–70.

Masser, D.W. (1976b), Linear forms in algebraic points of Abelian functions III, *Proc. London Math. Soc.* **33**, 549–564.

Masser, D.W. (1977), Some vector spaces associated with two elliptic functions, in *Transcendence Theory: Advances and Applications*, A. Baker (ed.), Academic Press, 101–120.

Philippon, P. & M. Waldschmidt (1988), Formes linéaires de logarithmes sur les groupes algébriques commutatifs, *Illinois J. Math.* **32**, 281–314.

Reyssat, E. (1980), Approximation algébrique de nombres liés aux fonctions elliptiques et exponentielle, *Bull. Soc. Math. France* **108**, 47–79.

Schneider, Th. (1937), Arithmetische Untersuchungen elliptischer Integrale, *Math. Annalen* **113**, 1–13.

Schneider, Th. (1941), Zur Theorie der Abelschen Funktionen und Integrale, *J. Reine Angew. Math.* **183**, 110–128.

Schneider, Th. (1957) *Einführung in die transzendenten Zahlen*, Springer.

Siegel, C.L. (1932), Über die Perioden elliptischer Funktionen, *J. Reine Angew. Math.* **167**, 62–69.

Silverman, J.H. (1986), *The arithmetic of elliptic curves*, Springer.

Waldschmidt, M. (1979), *Nombres transcendants et groupes algébriques*, Astérisque **69/70**.

Wüstholz, G. (1989), Algebraische Punkte auf analytischen Untergruppen algebraischer Gruppen, *Ann. Math.* **129**, 501–517.

Yu, Kunrui (1985), Linear forms in elliptic logarithms, *J. Number Theory* **20**, 1–69.

4

Solving Diophantine Equations by Baker's Theory

Kálmán Győry

Abstract

The purpose of this paper is to give a survey of some important applications of Baker's theory of linear forms in logarithms to diophantine equations. We shall mainly be concerned with Thue equations, elliptic equations, unit equations, discriminant form and index form equations, more general decomposable form equations and some related diophantine problems. A special emphasis will be laid on Baker's landmark results obtained through the theory of linear forms in logarithms as well as on some remarkable contributions of number theorists from Debrecen, including A. Pethő, Z.Z. Papp, B. Brindza, I. Gaál, Á. Pintér, L. Hajdu, T. Herendi, A. Bérczes and myself.

Introduction

In his celebrated series of papers (Baker 1966, 1967a, 1967b, 1968a) Baker made in the 1960s a major breakthrough in transcendental number theory by giving non-trivial explicit lower bounds for linear forms in logarithms of the form

$$\Lambda = b_1 \log \alpha_1 + \cdots + b_m \log \alpha_m \neq 0,$$

where $b_1, \ldots, b_m$ are rational integers, $\alpha_1, \ldots, \alpha_m$ are algebraic numbers different from 0 and 1, and $\log \alpha_1, \ldots, \log \alpha_m$ denote fixed determinations of the logarithms. His general effective estimates led to significant applications in number theory, and opened a new epoch in the theory of diophantine equations. Later several improvements, generalizations and analogues of Baker's lower bounds were established by Baker and others; for comprehensive accounts and extensive bibliographies the reader can consult Baker (1975), Baker & Masser (1977), Baker (1988) and the papers of Wüstholz, Yu and David in this volume.

To diophantine equations the first applications of Baker's estimates (Baker 1966, 1967a, 1967b, 1968a) were given by Baker himself (see Baker 1968b, 1968c, 1969) and Baker & Davenport (1969). In the last thirty years very extensive diophantine investigations were made by using Baker's theory on linear forms in logarithms. Effective finiteness theorems have been established for various general classes of equations. These provide explicit upper bounds on the solutions and make it possible, at least in principle, to determine all the solutions. Furthermore, Baker's theory has been combined with reduction algorithms and computational techniques to furnish practical methods for the numerical resolution of certain types of equation. A great many numerical results have been obtained, giving all solutions of equations.

For applications to diophantine problems, the best known general estimates concerning linear forms in logarithms are due to Baker & Wüstholz (1993), Waldschmidt (1993) and, in the p-adic case, K. Yu (1994, 1999). For linear forms in elliptic logarithms an explicit lower bound was given by David (1995). Baker & Wüstholz proved the inequality

$$|\Lambda| > \exp\{-c(m,d)\log A_1 \cdots \log A_m \log B\}, \tag{1}$$

where d denotes the degree of the number field generated by $\alpha_1, \ldots, \alpha_m$ over $\mathbb{Q}$, A_i ($\geq e$) is an upper bound for the ordinary height $H(\alpha_i)$ of α_i, B ($\geq e$) is an upper bound for $|b_i|$, $i = 1, \ldots, m$, and $c(m,d) = (16md)^{2(m+2)}$. In this sharp result the precision of $c(m,d)$ is particularly important for numerical resolution of equations. In Waldschmidt's estimate the corresponding constant $c(m,d)$ is larger in terms of m, $\log B$ is, however, replaced by $\log(2mB/\log A_m)$ which yields slightly better bounds in certain parameters on the solutions of some equations. In Matveev (1998)† such a lower bound is given which is weaker in $A_1, \ldots, A_m$, but contains no factor of the form m^m. For further perspectives of improvement and their connection with the *abc*-conjecture, see Baker (1998). Any real progress in this direction would have implications for the solutions of diophantine problems.

In this article we will present some remarkable applications of Baker's theory to Thue equations, Thue–Mahler equations, elliptic, hyperelliptic and superelliptic equations, unit and S-unit equations, discriminant form and index form equations, and decomposable form equations of more general type. The first part is devoted to general effective finiteness theorems. In the second part we are concerned with numerical resolution of concrete equations.

Many other types of equation can be studied by Baker's theory. Some important topics will not be discussed here and many references will be left out

† Added in proof. For an improvement, see also Matveev's paper in *Izvestiya Math.* **64** (2000), 1217–1269.

owing to lack of space. For instance, we shall not deal with numerical results concerning parametric families of equations. For further topics and references we shall refer to books and other survey papers.

1 General effective finiteness theorems

In the first decades of the 20th century Thue, Mordell, Siegel, Mahler and others obtained finiteness results on the integral solutions of polynomial diophantine equations in two unknowns. Siegel (1929) classified all irreducible algebraic curves over $\mathbb{Q}$ on which there are infinitely many integral points. He showed that these curves must be of genus 0 and have at most two infinite valuations. Various generalizations and analogues were later established. However, all these results have an ineffective character; their proofs which involve the Thue–Siegel method or its generalizations do not provide any algorithm for finding the solutions.

Important classes of polynomial equations in two unknowns are the Thue equations, Mordell equations, elliptic and superelliptic equations and equations of genus 1. Using his fundamental inequalities concerning linear forms in logarithms, Baker derived in the 1960s explicit upper bounds for the solutions of all these equations. These provided the first general algorithms for the solutions of such equations and, in case of these equations, furnished an affirmative answer to Hilbert's famous 10th problem.

Baker's quantitative results were later improved and generalized by himself and others. Furthermore, several other important types of equation were also investigated by means of Baker's theory and many remarkable effective and quantitative finiteness theorems were established; see e.g. the books Baker (1975) , Baker & Masser (1977), Baker (1988), Győry (1980b), Shorey & Tijdeman (1986), Smart (1998), Fel'dman & Nesterenko (1998) and the references given there. In the last three decades an *effective theory* of diophantine equations has developed.

The *general strategy* of deriving explicit upper bounds for the solutions by Baker's theory can be summarized as follows.

1. Reduce the equation if necessary to such equation(s) to which Baker's theory is applicable.
2. Reduce the (new) equation(s) to inequalities of the form

$$0 < |\alpha_1^{b_1} \cdots \alpha_r^{b_r} - \alpha_{r+1}| < c_1 \exp\{-c_2 B\} \tag{2}$$

   where $\alpha_1, \ldots, \alpha_{r+1}$ are non-zero algebraic numbers, $b_1, \ldots, b_r$ are unknown rational integers, $B = \max_i |b_i|$ and c_1, c_2 as well as c_3, c_4 below denote positive constants which are independent of $b_1, \ldots, b_r$ and can be given explicitly. If B is large, (2) implies that

$$|\Lambda| \leq c_3 \exp\{-c_2 B\} \tag{3}$$

where

$$\Lambda = b_1 \log \alpha_1 + \cdots + b_r \log \alpha_r - \log \alpha_{r+1}.$$

For simplicity, we assume here that $\alpha_1, \ldots, \alpha_{r+1}$ are real and positive.

3. The crucial step is the application of Baker's theory which gives

$$\exp\{-c_4 \log B\} \leq |\Lambda|.$$

Together with (3) this yields an explicit upper bound B_0 for B.

4. Deduce an explicit upper bound for the unknowns in the initial equation.

In the following sections we outline how Baker's theory can be applied to derive bounds for the solutions of diophantine equations mentioned in the Introduction. For reasons of exposition, the results and arguments will be formulated and illustrated in their simplest form.

1.1 Thue equations and Thue–Mahler equations

Let $F(X, Y)$ be an irreducible binary form of degree $n \geq 3$ with integer coefficients, and m a non-zero integer. In 1909 Thue showed that the equation

$$F(x, y) = m \quad \text{in } x, y \in \mathbb{Z}, \tag{4}$$

called now *Thue equation*, has only finitely many solutions. Thue's proof was ineffective.

The first general effective version of Thue's theorem was established in Baker (1968b). By means of his fundamental estimate for linear forms in logarithms he proved that all solutions x, y of (4) satisfy

$$\max(|x|, |y|) < \exp\left\{n^{\nu^2} H^{\nu n^3} + (\log |m|)^{2n+2}\right\},$$

where $\nu = 128n(n+1)$ and H denotes the height (the maximum absolute value of the coefficients) of F. As a consequence Baker gave the first effective improvement of Liouville's theorem on approximation of algebraic numbers.

The *main steps* in deriving a bound for X $= \max(|x|, |y|)$ are as follows. For simplicity, assume that the polynomial $F(\text{X}, 1)$ is monic. Put $K = \mathbb{Q}(\theta)$, where $F(\theta, 1) = 0$. Denote by $\vartheta^{(1)} = \vartheta, \vartheta^{(2)}, \ldots, \vartheta^{(n)}$ the field conjugates of any ϑ of K. Then (4) can be written in the form

$$m = \beta^{(1)} \cdots \beta^{(n)}, \quad \text{where} \quad \beta^{(i)} = x - \theta^{(i)} y, \quad i = 1, \ldots, n.$$

In K there are an algebraic integer γ with bounded height, an unknown unit μ, and a system of fundamental units $\varepsilon_1, \ldots, \varepsilon_r$ with small heights such that

$$\beta = \gamma \cdot \mu \text{ and } \mu = \zeta \varepsilon_1^{b_1} \cdots \varepsilon_r^{b_r}, \tag{5}$$

where ζ is a root of unity and $b_1, \ldots, b_r$ are unknown rational integers. If X is large then $B = \max_j |b_j|$ is also large. Further, at least one conjugate of β, say $\beta^{(1)}$ is small in absolute value and

$$|\beta^{(1)}| \leq c_1 \exp\{-c_2 B\}. \tag{6}$$

Here c_1, c_2 and c_3 below are positive constants which can be given explicitly in terms of F and m. There is another conjugate of β, say $\beta^{(2)}$, whose absolute value is not small. For latter purpose we write the identity

$$(\theta^{(2)} - \theta^{(i)})\beta^{(1)} + (\theta^{(1)} - \theta^{(2)})\beta^{(i)} + (\theta^{(i)} - \theta^{(1)})\beta^{(2)} = 0 \tag{7}$$

in the form

$$\lambda_1 \mu^{(1)} + \lambda_2 \mu^{(2)} + \lambda_i \mu^{(i)} = 0 \text{ for } i = 3, \ldots, n \tag{8}$$

with appropriate non-zero algebraic integers $\lambda_1, \lambda_2, \lambda_i$. We note that (8) is a special *unit equation*; these equations will be discussed in Section 1.3. Dividing (7) by $\beta^{(2)}$ and using (6), it follows that

$$0 < \left|\alpha_{i1}^{b_1} \cdots \alpha_{ir}^{b_r} - \alpha_{i,r+1}\right| < c_3 \exp\{-c_2 B\}, \tag{9}$$

where $\alpha_{ij} = \varepsilon_j^{(i)}/\varepsilon_j^{(2)}$, $j = 1, \ldots, r$, and $\alpha_{i,r+1}$ is an appropriate algebraic number, $i = 3, \ldots, n$. On applying now Baker's theory one can obtain an upper bound B_0 for B, and then an upper bound X_0 for X.

Baker's bound on the solutions of Thue equations was later improved by Baker, Feldman, Sprindžuk, Stark, Győry, Papp, Brindza, Evertse and Bugeaud. The main new ingredients in these improvements were, amongst other things, certain sharpenings of Baker's first estimates for linear forms in logarithms, the use of fundamental/independent units $\varepsilon_1, \ldots, \varepsilon_r$ having a good upper bound for $\log H(\varepsilon_1) \cdots \log H(\varepsilon_r)$, the involvement in the bounds of the regulator R_K and discriminant D_K of K as well as the discriminant $D(F)$ of F, and in certain generalizations, the employment of some sharp estimates for S-regulators. The best known bounds on the solutions of (4) are due to

Bugeaud & Győry (1996b) and Bugeaud (1998) who proved that all solutions x, y with $\mathrm{X} = \max(|x|, |y|)$ satisfy

$$\mathrm{X} \leq C_1(H \cdot |m|)^{c_1 R_K (\log^* R_K)}, \tag{10}$$

where $\log^* R_K = \max(\log R_K, 1)$, and $c_1 = c_1(n)$, $C_1 = C_1(n, R_K)$ denote positive constants which are given explicitly in terms of the parameters occurring in the parantheses. In fact (10) is proved in Bugeaud (1998) without the factor $\log^* R_K$, but in Bugeaud (1998), $c_1(n)$ is larger than in Bugeaud & Győry (1996b). Brindza, Evertse & Győry (1991) gave the bound

$$\mathrm{X} \leq C_2 H^{3/n} |D(F) \cdot m|^{c_2} \tag{11}$$

for the solutions, where c_2, C_2 are effectively computable and depend only on K. If H is large relative to $|D(F)|$, the estimate (11) is better than (10) in terms of H.

Let $p_1, \dots, p_s$, $(s \geq 0)$ be distinct primes which do not divide m. Mahler showed in 1933 that the so-called *Thue–Mahler equation*

$$F(x, y) = m p_1^{z_1} \cdots p_s^{z_s} \quad \text{in } x, y \in \mathbb{Z}, \;\; z_1, \dots, z_s \in \mathbb{Z}_{\geq 0} \tag{12}$$

with $\gcd(x, y, p_1, \dots, p_s) = 1$ has only finitely many solutions. This implies that for $S = \{\infty, p_1, \dots, p_s\}$, (4) has finitely many solutions in S-integers of $\mathbb{Q}$. The theorems of Thue and Mahler were extended by Siegel, Parry and Lang to the case of more general ground rings. All these results were ineffective.

Coates (1970) made effective Mahler's theorem. He established a p-adic analogue of Baker's inequality concerning linear forms in logarithms, and used it to derive an explicit upper bound for the solutions of (12), which depends on n, H, m, s and the maximum of the primes $p_1, \dots, p_s$. Coates' bound was improved by Sprindžuk and others; the best known bounds are due to Bugeaud & Győry (1996b) and Bugeaud (1998). The mentioned improvements imply that

$$P(F(x, y)) > c \log\log \mathrm{X}, \tag{13}$$

where $\mathrm{X} = \max(|x|, |y|)$ with coprime $x, y \in \mathbb{Z}$. Here $P(a)$ denotes the greatest prime factor of a positive integer a (with the convention that $P(1) = 1$), and c is a positive constant depending only on F.

Baker's theorem was extended to the number field case by Baker (1969) (see also Baker & Coates 1970), and to the so-called inhomogeneous case by Sprindžuk, see e.g. Sprindžuk (1993). In the relative case the best known bounds on the solutions of (4) and (12) can be found in Bugeaud & Győry (1996b) and Bugeaud (1998). In Győry (1983) effective finiteness theorems concerning (4) and (12) were established over arbitrary finitely generated ground rings over $\mathbb{Z}$.

The proofs of the above-mentioned effective results involved Baker's theory. Recently Bombieri and Bombieri & Cohen have developed a new effective method in diophantine approximation which is based on a reworked version of the Thue principle, the so-called Dyson lemma and some results from the geometry of numbers. This method provides almost the same bounds for the solutions of Thue equations and Thue–Mahler equations over number fields as those obtained by Baker's theory; see Bombieri's article in this volume.

1.2 Elliptic, hyperelliptic and superelliptic equations

Let $f \in \mathbb{Z}[X]$ be a polynomial of degree $n \geq 3$. Consider the *hyperelliptic equation*

$$y^2 = f(x) \quad \text{in } x, y \in \mathbb{Z}, \tag{14}$$

and the *superelliptic equation*

$$y^m = f(x) \quad \text{in } x, y \in \mathbb{Z} \tag{15}$$

where $m \geq 3$ is a given integer. For $n = 3$, equation (14) is called an *elliptic equation*, while for $f(X) = X^3 + k$ a *Mordell equation*.

Mordell (elliptic case) and later Siegel proved that if f has no multiple zero then equation (14) has only finitely many solutions. Le Veque gave a criterion for equation (15) to have only finitely many solutions. Generalizations were also established to the case of more general ground rings. The proofs depend on the Thue–Siegel method and hence these results are ineffective.

Baker (1968b, 1968c, 1969) was the first to give explicit upper bounds for the solutions of (14) and (15), which depend only on m and the degree and height of f. He used his fundamental inequalities concerning linear forms in logarithms to derive bounds for the solutions in the cases when, in (14), f has at least 3 simple zeros, and, in (15), f has at least 2 simple zeros.

In his 1969 paper Baker reduced equations (14) and (15) to relative Thue equations. Assume, for simplicity that f is monic, and that θ_1, θ_2 and, in case of (14), θ_3 are simple zeros of $f(X)$. Put $K_i = \mathbb{Q}(\theta_i)$ for $i = 1, 2, 3$, and let x, y be a solution of (14) or (15). Then following Siegel's argument one deduces that

$$x - \theta_i = \beta_i \sigma_i^m,$$

where β_i is an integer in K_i with bounded height, and σ_i is an unknown integer in K_i for $i = 1, 2, 3$. This implies that

$$\theta_2 - \theta_1 = \beta_1 \sigma_1^m - \beta_2 \sigma_2^m.$$

For $m \geq 3$, this is a Thue equation over the number field $K_1 K_2$. If $m = 2$, we have also the relation

$$\theta_3 - \theta_1 = \beta_1 \sigma_1^m - \beta_3 \sigma_3^m.$$

Then one more reduction step is needed to arrive at a Thue equation in an appropriate extension field of $K_1 K_2 K_3$. The relative Thue equations so obtained can be reduced in both cases to inequalities concerning linear forms in logarithms to which Baker's theory applies.

Quantitative improvements and generalizations of Baker's theorems concerning (14) and (15) were later obtained by Stark, Sprindžuk, Kotov, Trelina, Turk, Brindza, Poulakis, Voutier, Pintér, Bugeaud, Hajdu, Herendi and others. Brindza (1984) made effective Le Veque's theorem in full generality. Combining Le Veque's arguments with Baker's theory he derived effective upper bounds for the S-integral solutions of (14) and (15), unless m divides the multiplicities of all but one zero of f; or m is even, the multiplicities of all but two zeros of f are divisible by m and the remaining two by $m/2$. In Brindza (1989) he extended his result to the case of arbitrary finitely generated ground rings over $\mathbb{Z}$.

As a remarkable application of Baker's theory, Baker & Coates (1970) gave an explicit upper bound for the sizes of integral points on curves of genus 1. Their bound was later improved by Kotov, Trelina, Schmidt and Bilu.

A major *open problem* is to give a general effective version of Siegel's theorem on integral points of curves of genus greater than 1. In this direction notable partial results were obtained by Kleiman, Bilu, Dvornich, Zannier and Poulakis; see also Bilu's paper in this volume.

Finally, we consider equation (15) with $m \geq 2$ as an unknown. Assume that $|y| > 1$. In 1976 Tijdeman showed by means of Baker's theory that if f has at least two simple rational zeros then in (15) m is bounded by a computable number depending only of f. This was extended by Schinzel & Tijdeman (1976) to the case when f has at least two distinct zeros. Various generalizations and improvements were later obtained by Shorey, van der Poorten, Tijdeman, Schinzel, Sprindžuk, Turk, Brindza, Evertse, Győry, Bérczes, Hajdu and others; for references see, for example, Shorey & Tijdeman (1986), Sprindžuk (1993) and Bérczes, Brindza & Hadju (1998). Important applications were given by Győry, Tijdeman, Voorhoeve, Brindza and others (see Shorey & Tijdeman 1986) to the equation $1^k + 2^k + \ldots + x^k = y^z$ in integers

$x, y, z \geq 2$, and by Tijdeman, Terai and Győry (see Győry 1997) to the equation $\binom{x+u}{u} = y^z$ in integers $x, u, y, z \geq 2$.

1.3 Unit equations and S-unit equations

Let K be an algebraic number field, S_∞ the set of infinite places on K, and S a finite set of places on K with $S \supseteq S_\infty$. Denote by U_K and U_S the unit group and the S-unit group of K, respectively. The equations of the form

$$\lambda_1\mu_1 + \lambda_2\mu_2 = 1 \quad \text{in } \mu_1, \mu_2 \in U_S, \tag{16}$$

where λ_1 and λ_2 are given non-zero elements of K, have a wide range of applications to various areas of number theory. Equation (16) is called an *S-unit equation* and for $S = S_\infty$ a *unit equation*, or more precisely a(n *S*-)*unit equation in two unknowns*. In many cases it is more convenient to consider the S-unit equation in homogeneous form

$$\lambda_1\mu_1 + \lambda_2\mu_2 + \lambda_3\mu_3 = 0 \quad \text{in } \mu_1, \mu_2, \mu_3 \in U_S, \tag{16a}$$

where $\lambda_1, \lambda_2, \lambda_3$ denote fixed elements of $K \setminus \{0\}$.

For a long time these equations were utilized merely in special cases and in an implicit way. It was implicitly proved by Siegel ($S = S_\infty$) and Mahler that equation (16) has only finitely many solutions. This implies the finiteness of the number of solutions of (16a) up to a common proportional factor. In 1960 Lang gave a direct proof for a more general version of this finiteness theorem. All these proofs were ineffective.

As was seen in Section 1.1, the Thue equation (4) can be reduced to equations (8), i.e. to special unit equations of the form (16a). Thue-Mahler equations lead to similar special S-unit equations. In these special situations the first bounds on the solutions of (16a) were implicitly given by Baker (1968b) ($S = S_\infty$) and Coates (1970).

The first general explicit bounds for the solutions of (16) and (16a) were obtained in Győry (1972, 1973, 1974, 1976) in the $S = S_\infty$ case, and Győry (1979) by using Baker's theory. Since the early 1970s we applied these explicit results in a systematic way to irreducible polynomials, arithmetic graphs, algebraic number theory and polynomial diophantine equations. Thereby we extended the applicability of Baker's theory to, amongst others, wide classes of polynomial equations in an arbitrary number of unknowns. We first reduced the diophantine problems under consideration to the study of such systems of unit equations, in which the equations possess certain graph-theoretic connectedness properties. Then we combined our bounds on the solutions of (16) with some combinatorial arguments to derive bounds for the solutions of the initial diophantine problems. In the following sections some remarkable applications of our results on (16) and (16a) will be presented to decomposable form equations and algebraic number theory; for other applications we refer to the survey papers Evertse *et al.* (1988b), Győry (1992, 1996).

We briefly sketch how to apply Baker's theory to derive a bound for the solutions of (16). There are fundamental S-units $\varepsilon_1, \ldots, \varepsilon_{s-1}$ in K with bounded heights. Here s denotes the cardinality of S. Let μ_1, μ_2 be a solution of (16) in S-units. Then

$$\mu_i = \zeta_i \varepsilon_1^{b_{i1}} \cdots \varepsilon_{s-1}^{b_{i,s-1}}$$

where ζ_i is a root of unity in K and $b_{i1}, \ldots, b_{i,s-1}$ are unknown rational integer exponents for $i = 1, 2$. If $H(\mu_1) \geq H(\mu_2)$ then one can deduce that $B \leq c_1 \log H(\mu_1)$ where $B = \max_{i,j} |b_{ij}|$. Further, there is a $v \in S$ such that

$$|\mu_1|_v \leq c_2 \exp\{-c_3 B\},$$

whence

$$0 < |\varepsilon_1^{b_{21}} \cdots \varepsilon_{s-1}^{b_{2,s-1}} - \alpha|_v \leq c_4 \exp\{-c_3 B\} \tag{17}$$

follows with an appropriate $\alpha \in K$ of bounded height. The constants c_1 to c_4 depend at most on K, S and λ_1, λ_2 and can be given explicitly. One can now apply the complex or p-adic version of Baker's theory according as $v \in S_\infty$ or $v \in S \setminus S_\infty$, and this yields an upper bound B_0 for B. Finally, this implies an upper bound for $H(\mu_1)$ and $H(\mu_2)$.

Our upper bounds on the solutions were later improved by Sprindžuk, Schmidt, Bugeaud and myself. The best known bounds are due to Bugeaud & Győry (1996a) and Bugeaud (1998). In these improvements the main new ingradients were among other things the best known estimates for linear forms in logarithms, the utilization of S-regulators and their specific properties as well as the use of fundamental S-units $\varepsilon_1, \ldots, \varepsilon_{s-1}$ having a particularly good upper bound for $\log H(\varepsilon_1) \cdots \log H(\varepsilon_{s-1})$. Further, in Bugeaud (1998) Baker's theory was combined with some recent arguments of Bombieri & Cohen. In the special case $S = S_\infty$, the estimates obtained in Bugeaud & Győry (1996a) and Bugeaud (1998) on the solutions μ_1, μ_2 of (16) are of the form

$$\max_i H(\mu_i) < \exp\{c_1 R_K (\log^* R_K) \log H\} \tag{18}$$

and

$$\max_i H(\mu_i) < \exp\{c_2^2 R_K^2 + c_2 R_K \log H\} \tag{19}$$

respectively, where $H = \max_i H(\lambda_i)$, R_K denotes the regulator of K, and

c_1, c_2 are explicitly given constants which depend only on the degree of K. We note that c_2 in Bugeaud (1998) is much larger than c_1 in Bugeaud & Győry (1996a).

We remark that using their new effective method mentioned in Section 1.1, Bombieri & Cohen derived almost the same bounds for the solutions of (16) as those established in Bugeaud & Győry (1996a) and Bugeaud (1998) by means of Baker's theory; see Bombieri's article in this volume.

The bounds in (18) and (19) can be applied to (16a) to derive a bound for $\max_{i,j} H(\mu_i/\mu_j)$. In most applications of (16a) to polynomial equations there is a subfield L of K such that for some $\mathbb{Q}$-isomorphism σ of L, $\sigma(L) \subset K$ and $\mu_2 = \sigma(\mu_1)$ for each solution under consideration μ_1, μ_2, μ_3 of (16a). Under this assumption we have recently obtained in Győry (1998) such bounds for the solutions of (16a) which give much better quantitative results for polynomial equations than (18) and (19). In particular, for $S = S_\infty$ we obtained

$$\max_{1\le i,j\le 3} H(\mu_i/\mu_j) < \exp\left\{c_3 R_L (\log H) \log\left(\frac{\log H(\mu_1)}{\log H}\right)\right\}, \tag{20}$$

provided that $H(\mu_1) \ge H^{c_4}$, where R_L denotes the regulator of L and c_3, c_4 are explicitly given constants depending only on the degree of K. In (20) the bound depends still on $H(\mu_1)$. However, in the course of applications the polynomial equation in question leads to several other unit equations of the same type which are usually 'connected' in a certain sense; cf. the next sections. After applying (20) to these unit equations and using their connectedness one can ultimately obtain an upper bound for the solutions of the initial polynomial equation.

Baker's theory was also used to derive good upper bounds for the *number of solutions* of (16); see Győry (1979), Evertse *et al.* (1988a), Brindza & Győry (1990) and Bombieri, Mueller & Poe (1997). By a theorem of Evertse the number of solutions of (16) is at most $3 \cdot 7^{4s}$, and this bound is not far from the best possible. The equations (16) and $\lambda_1'\mu_1' + \lambda_2'\mu_2' = 1$ in $\mu_1', \mu_2' \in U_S$ are called S-equivalent if $\lambda_1'/\lambda_1, \lambda_2'/\lambda_2 \in U_S$. In this case they have the same number of solutions. It was proved in Evertse *et al.* (1988a) that apart from finitely many and effectively determinable S-equivalence classes, the number of solutions of (16) is at most $s + 1$. This result was later applied to the Thue–Mahler equations by Evertse, Győry and Stewart. In Evertse *et al.* (1988a) it was also showed that the bound $s + 1$ can be replaced by 2 which is already sharp. However, in this version the method of proof does not make it possible to determine the exceptional S-equivalence classes.

An important generalization of equation (16) is the *S-unit equation in n unknowns*

$$\lambda_1\mu_1 + \cdots + \lambda_n\mu_n = 1 \quad \text{in } \mu_1, \ldots, \mu_n \in U_S, \tag{21}$$

where $\lambda_1, \ldots, \lambda_n$ are given non-zero elements of K. For $n \geq 3$ this equation can have infinitely many solutions. This is the case if the left side has a vanishing subsum and U_S is infinite. Van der Poorten & Schlickewei and independently Evertse proved that (21) has only finitely many solutions for which no proper subsum on the left vanishes. Later this was extended to the case when K is a field of characteristic 0 and U_S is replaced by a finitely generated subgroup of the multiplicative group K^*. The proofs of these results depend on a generalization of Schmidt's subspace theorem and hence are ineffective.

There are explicit upper bounds for the number of solutions of (21), but not for the solutions themselves when $n \geq 3$. A remarkable *unsolved problem* is to make effective for $n \geq 3$ the above finiteness theorem concerning (21). Such an effective version would follow, for example, from an effective variant of the following general result (see Győry 1992) which can be proved by the Thue–Siegel–Roth–Schmidt method. For given $\alpha_1, \ldots, \alpha_{n-1}, \alpha_n \in K \setminus \{0\}$, $v \in S$ and $c > 0$, the inequality

$$0 < \left| \sum_{i=1}^{n-1} \alpha_i \varepsilon_1^{b_{i1}} \cdots \varepsilon_{s-1}^{b_{i,s-1}} - \alpha_n \right|_v < \exp\{-cB\}$$

which is a generalization of (17) has only a finite number of solutions in $b_{ij} \in \mathbb{Z}$ with $\max_{i,j} |b_{ij}| = B$, provided that no subsum on the left vanishes. For $n = 2$ this was established in an effective way by Baker's theory. An effective version for the case $n > 2$ would have major implications for number theory; see for example Section 1.5 and Győry (1992), §7.

1.4 Discriminant form and index form equations

Let K be an algebraic number field of degree $n \geq 3$ with ring of integers O_K and discriminant D_K, and let $1, \alpha_1, \ldots, \alpha_m$ be $\mathbb{Q}$-linearly independent algebraic integers in K. Consider the equation

$$D_{K/\mathbb{Q}}(x_1\alpha_1 + \cdots + x_m\alpha_m) = D \quad \text{in } x_1, \ldots, x_m \in \mathbb{Z}, \tag{22}$$

where D is a given non-zero rational integer and $D_{K/\mathbb{Q}}(\alpha)$ denotes the discriminant of any $\alpha \in K$. Putting $l(\mathbf{X}) = \alpha_1 X_1 + \cdots + \alpha_m X_m$, the form $D_{K/\mathbb{Q}}(l(\mathbf{X}))$ is a decomposable form in $X_1, \ldots, X_m$ with degree $n(n-1)$ and coefficients in $\mathbb{Z}$. It is called a *discriminant form*, and (22) a *discriminant form equation*. If in particular $l(\mathbf{X}) = \alpha_1 X_1 + \cdots + \alpha_{n-1} X_{n-1}$ with a $\mathbb{Z}$-basis $\{1, \alpha_1, \ldots, \alpha_{n-1}\}$ of O_K, then we have

$$D_{K/\mathbb{Q}}(l(\mathbf{X})) = (\mathrm{I}(\mathbf{X}))^2 D_K, \tag{23}$$

where $\mathrm{I}(\mathbf{X}) = \mathrm{I}(X_1, \ldots, X_{n-1})$ is a decomposable form of degree $\frac{n(n-1)}{2}$ with coefficients in $\mathbb{Z}$. Representing a primitive integral element α of K in the form $\alpha = x_0 + x_1\alpha_1 + \cdots + x_{n-1}\alpha_{n-1}$ with $x_0, \ldots, x_{n-1} \in \mathbb{Z}$, we find $|\mathrm{I}(x_1, \ldots, x_{n-1})|$ is precisely the index $\mathrm{I}(\alpha)$ of α, i.e. the index of the subgroup $\mathbb{Z}^+[\alpha]$ in the additive group O_K^+ of O_K. Hence $\mathrm{I}(\mathbf{X})$ is called the *index form* of the $\mathbb{Z}$-basis $\{1, \alpha_1, \ldots, \alpha_{n-1}\}$, and

$$\mathrm{I}(x_1, \ldots, x_{n-1}) = \pm \mathrm{I} \quad \text{in } x_1, \ldots, x_{n-1} \in \mathbb{Z} \tag{24}$$

is an *index form equation*. Here I denotes a given non-zero rational integer. In the case considered in (23) and (24), equations (22) and (24) are obviously equivalent with the choice $D = \mathrm{I}^2 D_K$.

Equations (22) and (24) are of basic importance in algebraic number theory. For instance, for $\mathrm{I} = 1$ equation (24) is equivalent to the equation

$$O_K = \mathbb{Z}[\alpha] \text{ in } \alpha \in O_K \iff \{1, \alpha, \ldots, \alpha^{n-1}\} \ \mathbb{Z}\text{-basis in } O_K. \tag{25}$$

Equations (22), (24) and (25) have been intensively studied by many authors, including Kronecker, Hensel, Hasse, Delone and Nagell. For $n = 3$, Delone and, independently, Nagell, and for $n = 4$, Nagell, showed that these equations have only finitely many solutions. Apart from certain very special cases their results were ineffective.

We gave in Győry (1973, 1974, 1976) explicit upper bounds for the solutions of (22) and (24). In those papers these equations were first reduced to the study of appropriate systems of unit equations. Then our explicit results obtained by Baker's theory on unit equations were utilized. The general effective finiteness theorems so obtained led to many important applications. We mention below some of them in algebraic number theory.

Two elements α, α' of O_K with $\alpha' - \alpha \in \mathbb{Z}$ are called *equivalent*. Such elements have the same discriminant and same index. We proved in quantitative form the following effective finiteness theorems.

- Up to equivalence, there are only finitely many $\alpha \in O_K$ with $D_{K/\mathbb{Q}}(\alpha) = D$. This was first proved by Birch & Merriman (1972) in an ineffective way and, independently, by the author (Győry 1973) in an effective form. The first quantitative version was given in Győry (1974).
- Up to equivalence, there are only finitely many $\alpha \in O_K$ with a given index, and all these can be, at least in principle, determined; see Győry (1976).

- In particular, there are only finitely many pairwise inequivalent $\alpha \in O_K$ satisfying (25), and all these α can be effectively determined; see Győry (1976). This provided the first general algorithm for finding all power integral bases in number fields.
- Apart from translation of the form $f(X) \to f(X+a)$ with $a \in \mathbb{Z}$, there are only finitely many monic polynomials $f \in \mathbb{Z}[X]$ with a given non-zero discriminant and all these polynomials can be, at least in principle, determined; see Győry (1973, 1974, 1976).

For other applications we refer to Győry (1980b, 2000) and the references given there.

The above-mentioned effective results of Győry (1973, 1974, 1976) on equations (22) and (24) were later extended to equations of Mahler type by Trelina and by Győry & Papp, to the relative case by Győry & Papp, and to equations considered over arbitrary finitely generated domains over $\mathbb{Z}$ by Győry. Inhomogeneous versions were established by Gaál. These results led to further applications. Surveys on the subject can be found in Győry (1980b, 1983, 2000).

We now outline how to derive bounds for the solutions of (22) via unit equations. Denote by $K^{(i)}$ the conjugates of K over $\mathbb{Q}$, and by $l^{(i)}(\mathbf{X}) = \alpha_1^{(i)}X_1 + \cdots + \alpha_m^{(i)}X_m$, $i = 1, \ldots, n$, the corresponding conjugates of the linear form $l(\mathbf{X}) = \alpha_1 X_1 + \cdots + \alpha_m X_m$. Put $l_{ij}(\mathbf{X}) = l^{(i)}(\mathbf{X}) - l^{(j)}(\mathbf{X})$. Then equation (22) takes the form

$$\prod_{i \neq j} l_{ij}(\mathbf{x}) = (-1)^{n(n-1)/2} D \quad \text{in } \mathbf{x} \in \mathbb{Z}^m. \tag{26}$$

This implies that for each solution $\mathbf{x}$, $l_{ij}(\mathbf{x}) = \gamma_{ij}\mu_{ij}$, where γ_{ij} is an algebraic integer of bounded height, and μ_{ij} is an unknown unit in $K^{(i)}K^{(j)}$. We have

$$l_{ij}(\mathbf{x}) + l_{jk}(\mathbf{x}) = l_{ik}(\mathbf{x}) \tag{27}$$

for each distinct i, j, k, whence, dividing by $l_{ik}(\mathbf{x})$, we get a unit equation over the number field $K^{(i)}K^{(j)}K^{(k)}$ or, if it is more convenient, over the normal closure, denoted by N, of $K/\mathbb{Q}$. Choosing $i = 1, k = 2$ and representing $\mu_{1j}/\mu_{1,2}, \mu_{j2}/\mu_{1,2}$ by an appropriate system of fundamental units $\varepsilon_1, \ldots, \varepsilon_r$ of $K^{(i)}K^{(j)}K^{(k)}$ or of N, we arrive at a unit equation of the form

$$\lambda_{j1}\varepsilon_1^{b_1} \cdots \varepsilon_r^{b_r} + \lambda_{j2}\varepsilon_1^{b_1'} \cdots \varepsilon_r^{b_r'} = 1 \tag{28}$$

with $\lambda_{j1}, \lambda_{j2} \in K^{(1)}K^{(j)}K^{(2)}$ of bounded height. Using our results presented in the previous section on unit equations, we can derive bounds first for $B = \max_l\{|b_l|, |b_l'|\}$ and then for the height of $l_{j2}(\mathbf{x})/l_{1,2}(\mathbf{x})$, $j = 3, \ldots, n$.

Together with $l_{ij}(\mathbf{x}) = l_{i2}(\mathbf{x}) - l_{j2}(\mathbf{x})$ this implies a bound for the height of $l_{ij}(\mathbf{x})/l_{1,2}(\mathbf{x})$ for any distinct i and j. By virtue of (26) this gives a bound for the height of $l_{1,2}(\mathbf{x})$ and hence for the height of every $l_{ij}(\mathbf{x})$. Finally, one can easily derive a bound for the coordinates $x_1, \ldots, x_m$ of $\mathbf{x}$ by means of Cramer's rule.

In the above sketch of proof it was of crucial importance that for each i, j, the linear forms l_{ij} and $l_{1,2}$ are 'connected' by the relations $l_{ij} = l_{i2} + l_{2j}$ and $l_{1,2} = l_{1j} - l_{2j}$ having l_{2j} as a common term. This is a special case of a more general concept, the triangular connectedness, which will be defined in the next section.

The bounds established in Győry (1973, 1974, 1976) were improved upon in the 1970s and 1980s. All these bounds depend on the parameters (degree, unit rank, discriminant or regulator) of N or on those of the subfields $K^{(i)}K^{(j)}K^{(k)}$ of N. In 1998, we considerably improved (cf. Győry 1998) the previous bounds on the solution of (22) and (24), working in much smaller subfields of N. The subfield $L_{ij} = \mathbb{Q}(\alpha^{(i)} + \alpha^{(j)}, \alpha^{(i)}\alpha^{(j)})$ of $K^{(i)}K^{(j)}$ is independent of the choice of the primitive element α of K. We showed in Győry (1998) that after an appropriate transformation of (22) it is enough to deal with those relations (27) in which $\sigma(l_{ij}) = l_{ik}$ for some $\sigma \in \mathrm{Gal}(N/\mathbb{Q})$. Then we can apply our new improved bound (20) on the solutions of the corresponding unit equations with the choice $K^{(i)}K^{(j)}K^{(k)}$ and L_{ij} for K and L, respectively. Further, as is pointed out in Győry (2000), these unit equations have much fewer unknown exponents than in Győry (1973, 1974, 1976). However, in this restricted sense the system of linear forms involved is *no longer* triangularly connected, and new algebraic number-theoretic and combinatorial arguments are needed to surmount this difficulty. The best known bound so obtained in Győry (1998) for the solutions $(x_1, \ldots, x_m) \in \mathbb{Z}^m$ of (22) is of the form

$$\max_i |x_i| < A^{m-1} \exp\{cR(\log^* R)(R + \log |D|)\}, \tag{29}$$

where A is an upper bound for the sizes of the α_i, c is an explicitly given constant which depends only on n, and R denotes the regulator of N or the maximum of the regulators of L_{ij}, according as N is 'small' (i.e. $[N : K] \leq \frac{n-1}{2}$ and $N = KK^{(i)}$ for some i, see Győry 1998) or not.

As will be pointed out in Section 2.3, the above-mentioned refinement of the earlier method of proof plays an important rôle in the resolution of concrete equations of the type (22), (24) and (25).

Finally, we make a mention of a related result concerning binary forms of given discriminant. The binary forms $F, G \in \mathbb{Z}[X, Y] = \mathbb{Z}[\mathbf{X}]$ are called equivalent if $F(\mathbf{X}) = G(U\mathbf{X})$ for some $U \in SL_2(\mathbb{Z})$. In this case they have the same discriminant. It is a classical theorem that for given $n \geq 2$ and $D \neq 0$, there are only finitely many equivalence classes of binary forms $F \in \mathbb{Z}[\mathbf{X}]$ with degree n and discriminant D. This theorem was proved for $n = 2$ by Lagrange and for $n = 3$ by Hermite in an effective form, and for $n \geq 4$ by Birch & Marriman (1972) in an ineffective way. An effective version was given by Evertse & Győry (1991) in a quantitative form. Later we extended with Evertse our result to decomposable forms of given degree and given discriminant. The main tool in our proofs was the explicit result of Győry (1979) on S-unit equations whose proof, as was seen above, is based on Baker's theory.

1.5 Decomposable form equations of general type

Consider now the *decomposable form equation*

$$F(\mathbf{x}) = a \qquad \text{in } \mathbf{x} \in \mathbb{Z}^m, \tag{30}$$

where a denotes a given non-zero integer, and $F(\mathbf{X}) = F(X_1, \ldots, X_m) \in \mathbb{Z}[\mathbf{X}]$ is an arbitrary *decomposable form*, i.e. a homogeneous polynomial which factorizes into linear factors over $\overline{\mathbb{Q}}$. The most important classes of such equations are the Thue equations, norm form equations, discriminant form and index form equations.

Schmidt (1971) obtained a general finiteness criterion for (30) in the case when $F(\mathbf{X})$ is a norm form. Further, in Schmidt (1972) he gave a description of the structure of the set of solutions of norm form equations. These results were extended to norm form equations of Mahler type by Schlickewei, and to the relative case by Laurent. In 1988 Evertse & Győry generalized these finiteness criteria for arbitrary decomposable form equations. They pointed out that decomposable form equations and general S-unit equations of the form (21) are in fact equivalent. In 1993 Győry described in full generality the structure of the set of solutions of decomposable form equations. All these results depend on Schmidt's subspace theorem or its generalizations and hence are ineffective. For a recent survey on the subject see, for example, Győry (1999).

Győry & Papp (1978) derived explicit bounds for the solutions of (30), subject to the condition that (30) is *triangularly connected*, more precisely that the linear factors of F, denoted by $l_1, \ldots, l_n$, are *triangularly connected*. This means that the graph $\mathcal{G}(F)$ with vertex set $\{l_1, \ldots, l_n\}$ is connected, where $\{l_i, l_j\}$ is an edge whenever there is a linear factor l_k of F such that

$$\lambda_i l_i + \lambda_j l_j + \lambda_k l_k = 0 \tag{31}$$

with non-zero constants $\lambda_i, \lambda_j, \lambda_k$. In Győry (1978, 1980a) this was generalized for equations of Mahler type, and in Győry (1981) for even wider classes of decomposable forms F, when $\mathcal{G}(F)$ is not necessarily connected but its connected components possess certain connectedness properties. The linear factors of binary forms, discriminant forms and index forms are triangularly connected. Hence the results of Győry (1978, 1980a, 1981) include as special cases (13) and the above-presented effective finiteness theorems on Thue equations, Thue–Mahler equations, discriminant form and index form equations. Further, the main results of the papers mentioned can also be applied to certain important classes of norm form equations. In particular, in Győry (1981) an explicit bound is given for the solutions of the norm form equation

$$N_{K/\mathbb{Q}}(x_1\alpha_1 + \cdots + x_m\alpha_m) = a \quad \text{in } x_1, \ldots, x_m \in \mathbb{Z}, \tag{32}$$

subject to the condition that $x_m \neq 0$, $\alpha_1 = 1, \alpha_2, \ldots, \alpha_m$ are $\mathbb{Q}$-linearly independent elements of $K = \mathbb{Q}(\alpha_1, \ldots, \alpha_m)$ and K is of degree ≥ 3 over $\mathbb{Q}(\alpha_1, \ldots, \alpha_{m-1})$. For (32) a slightly weaker theorem was proved independently by Kotov. The assumptions concerning x_m and $\alpha_1, \ldots, \alpha_m$ are in general necessary. The condition $x_m \neq 0$ can be removed if we assume that α_i is of degree ≥ 3 over $\mathbb{Q}(\alpha_1, \ldots, \alpha_{i-1})$ for $i = 2, \ldots, m$.

In Győry & Papp (1978), Győry (1978, 1980a, 1981) the results were established in more general form, over algebraic number fields. Further generalization was obtained in Győry (1983) to the case of arbitrary finitely generated ground rings over $\mathbb{Z}$. An inhomogeneous version concerning norm form equations was proved by Gaál. For a survey, see Győry (1980b) or Evertse & Győry (1988).

To derive bounds for the solutions $\mathbf{x}$ of (30) we used in Győry & Papp (1978), Győry (1978, 1980a, 1981) the same basic idea as earlier in Győry (1973, 1974, 1976) for discriminant form and index form equations; cf. Section 1.4. Using the relations (31), we first reduced equation (30) to an appropriate system of unit equations of the form (16a). Then we applied the explicit bounds of Győry (1979) obtained by Baker's theory on the solutions of unit equations. In the next step the connectedness of the linear factors l_i of F enabled us to derive an upper bound for the heights of the $l_i(\mathbf{x})/l_1(\mathbf{x})$, $i = 2, \ldots, n$. Finally, the height of $l_1(\mathbf{x})$ was bounded above from (30), and then Cramer's rule was used to get a bound for the coordinates of $\mathbf{x}$.

In Győry (1998) the previous bounds on the solutions of (30) were considerably improved, and the general algorithm given in Győry (1981) for solving (30) via unit equations has been significantly refined. Equation (30) was reduced to special unit equations arising from such relations (31) in which at least two linear factors are conjugate to each other. Then our recent estimate (20) was applied to the corresponding unit equations to derive improved bounds for the solutions of (30).

Although the known effective results concerning (30) cover almost all important classes of decomposable form equations, they do not apply to all such equations having only finitely many solutions. It is a major *open problem* to make effective in full generality the general ineffective theorems of Schmidt, Evertse and Győry, and Győry on equation (30). A combination of the proofs of Evertse & Győry with an effective version of the finiteness theorem on the general S-unit equation (21) would yield effective variants of the above-mentioned ineffective results concerning (30).

2 Explicit determination of the solutions

In applications of Baker's theory to general classes of equations the *main steps* are as follows.

1. Reduce the equation to inequalities of the shape

$$0 < |b_1 \log \alpha_1 + \cdots + b_r \log \alpha_r - \log \alpha_{r+1}| \leq c_1 \exp\{-c_2 B\} \quad \text{(3a)}$$

where $\alpha_1, \ldots, \alpha_{r+1}$ are non-zero algebraic numbers, and $b_1, \ldots, b_r$ are unknown rational integers with $B = \max_i |b_i|$.

2. Apply Baker's theory to derive an explicit upper bound B_0 for B.

This implies an upper bound for the initial unknowns. However, in case of concrete equations both B_0 and this bound are too large for practical use.

In their pioneering work Baker & Davenport (1969) initiated the following *general strategy* for the numerical resolution of concrete equations.

3. Reduce B_0 to a much smaller bound B_1 for which $B \leq B_1$.

4. Determine the solutions under this bound B_1, using some search techniques and specific properties of the initial equation.

Baker & Davenport illustrated this strategy by solving completely the system of equations

$$3x^2 - 2 = y^2, \quad 8x^2 - 7 = z^2 \quad \text{in } x, y, z \in \mathbb{Z}. \quad (33)$$

This was first reduced to an inequality of the form (3a) with $r = 2$, and then Baker's (1968a) estimate was used to yield the bound $B_0 = 10^{487}$ for B. After dividing by $\log \alpha_2$ the authors arrived at an inequality of the form

$$0 < |b_1\delta_1 + b_2\delta_2 + \delta| < c_3 \exp\{-c_2 B\} \quad (34)$$

with $\delta_2 = -1$ and appropriate real numbers δ_1, δ. Let $C > 2$ be a number which is not very large. It follows from Dirichlet's theorem of diophantine approximation that there exists a positive integer q such that $q \leq CB_0$ and

$$||q\delta_1|| \leq (CB_0)^{-1},$$

where $||\alpha||$ denotes the distance of a real number α from the nearest integer. Such a q can be quickly found from the continued fraction expansion of δ_1. If $||q\delta|| \geq 2C^{-1}$ which is heuristically plausible when C is large enough, it follows that

$$B \leq c_2^{-1}(\log B_0 + \log(c_3C^2)).$$

With the choice $C = 10^{33}$, the above procedure yielded $B \leq B_1 = 500$. The remaining cases were directly treated by the authors to show that $x = \pm 1$ and ± 11 are the only solutions of (33).

Since 1969, considerable progress has been made in the following directions:

- much *sharper estimates* have been established in Baker's theory on linear forms in logarithms;
- a *computational theory* of algebraic number fields has been developed which produces algorithms and new techniques for computing fundamental/independent units, integral elements of given norm, class numbers, class groups and so on; there are currently computer packages, e.g. KANT, PARI, SIMATH, MAPLE, MATHEMATICA, for performing various number-theoretic calculations;
- the applications of the LLL-basis reduction algorithm of Lenstra, Lenstra & Lovász and other computational techniques led to the development of the *computational diophantine approximation*;
- *supercomputers* and *computer technology* have been greatly developed.

All these played an important rôle in the creation of a new and quickly developing area of number theory, the *constructive* or *computational theory* of diophantine equations.

After the 1969 paper of Baker & Davenport, many people, including (in alphabetical order) Bilu, Ellison, Gaál, Gebel, Hanrot, Herrmann, Heuberger, Lettl, Mignotte, Pethő, Pohst, Smart, Stroecker, Thomas, Tichy, Tzanakis, Voutier, Wakabayashi, de Weger, Wildanger and Zimmer have contributed with significant results to this new area. For a comprehensive account of the present stage of the theory we refer to the recent book Smart (1998) and the references given there.

An important breakthrough was the multidimensional generalization of the Baker–Davenport reduction algorithm. Several authors realized in various special cases that the LLL-basis reduction algorithm can be used to reduce the

Baker-type bound B_0. However, Pethő and Schulenberg (1987) and de Weger (1987) were the first to apply in full generality the LLL-algorithm for developing reduction algorithms for the case $r \geq 2$. In Pethő and Schulenberg (1987) the simultaneous version of the LLL-algorithm was applied to Thue equations with constant term 1; see the next section. De Weger (1987, 1989) proposed the systematic use of the linear form version of the LLL-algorithm for solving diophantine equations. He elaborated the real, complex and p-adic versions of his reduction process which has revolutionized the subject.

In the real case the essence of de Weger's reduction algorithm can be outlined as follows. On applying Baker's theory to the linear form occurring in (3) one can obtain an upper bound B_0 for B. The inequality (3) can be written in the form

$$0 < |b_1\delta_1 + \cdots + b_r\delta_r + \delta| < c_3 \exp\{-c_2 B\} \tag{35}$$

with appropriate real $\delta_1, \ldots, \delta_r, \delta$. Denote by $\mathcal{L}$ the lattice in $\mathbb{R}^{r+1}$ spanned by the column vectors of the matrix

$$\begin{pmatrix} 1 & 0 & \ldots & 0 & 0 \\ 0 & 1 & & & 0 \\ \vdots & & \ddots & & \vdots \\ 0 & & & 1 & 0 \\ C\delta_1 & & \ldots & C\delta_r & C\delta \end{pmatrix},$$

where C is a suitably chosen positive number. Let $\underline{\mathbf{l}}_1$ denote the first basis vector of the LLL-reduced basis of $\mathcal{L}$. Then $\underline{\mathbf{l}}_1$ can be easily computed and

$$|\underline{\mathbf{l}}_1|^2 \leq 2^r |\underline{\mathbf{x}}|^2 \quad \text{for every } \underline{\mathbf{x}} \in \mathcal{L}.$$

Choose C so that

$$|\underline{\mathbf{l}}_1| \geq \sqrt{(r+2)2^r}\, B_0.$$

A suitable choice is $C \sim B_0^{r+1}$. Then using (35), we infer that

$$B \leq c_2^{-1}(\log(c_3 C) - \log B_0) = B_1.$$

This process reduces the bound B_0 approximately to its logarithm. After repeated application of this reduction with B_1 in place of B_0 and so on, we get a relatively small bound B_R for B. Usually the reduction procedure ends in 4 or 5 steps, and the reduced bound lies in the range $100 \leq B_R \leq 1000$.

In the following sections we shall deal with the numerical resolution of the equations treated in Part 1. As was seen above, all these equations can be reduced directly or via other equations to inequalities of the form (3) to which Baker's theory applies. In case of concrete equations Baker's theory excludes the existence of 'large' solutions $b_1, \ldots, b_r$ of the corresponding inequalities (3), and the reduction algorithms can be used to show that no 'medium' solution exists. The final step of solving the original equation is to locate the 'small' integral tuples $b_1, \ldots, b_r$ for which $B \leq B_R$ and which provide the solutions of the initial equation. Even if B_R is moderate (< 100), a direct enumeration of the $(2B_R + 1)^r$ possible tuples $b_1, \ldots, b_r$ is hopeless whenever r is large. Further search techniques are needed, using the specific properties of the initial diophantine equation.

2.1 Thue equations

Let $F \in \mathbb{Z}[X, Y]$ denote an irreducible binary form of degree ≥ 3. If $x, y \in \mathbb{Z}$ is a solution of the Thue inequality $0 < |F(x, y)| \leq m$ with $|y|$ large, then x/y approximates well one of the real roots of $F(X, 1)$. Using the continued fraction expansions of these roots, Pethő (1987) gave an efficient algorithm for the computation of 'small' solutions of Thue inequalities.

For solving concrete Thue equations of the form

$$F(x, y) = m \qquad \text{in } x, y \in \mathbb{Z}, \tag{4}$$

general practical methods were developed by Pethő & Schulenberg (1987) for the case $m = 1$, and by Tzanakis & de Weger (1989) for arbitrary m. The main steps of these methods are as follows. Equation (4) leads to inequalities of the form (3) via those of the form (9). Then Baker's theory yields a bound B_0 for B. The application of the reduction algorithms discussed above gives a much smaller bound B_R for B. Finally, a combination of the continued fraction method with direct enumeration makes it possible to determine all the solutions of (4). By means of this general approach a great number of Thue equations of degree ≤ 5 were completely solved; for references see for example Pethő (1990) and Tzanakis & de Weger (1989). The method was later extended to inhomogeneous Thue equations by Gaál (1988).

A general practical algorithm was given by Tzanakis & de Weger (1992) for solving Thue–Mahler equations by combining Baker's theory with de Weger's version of the LLL-reduction algorithm and with a sieving technique. An extension to the relative case was worked out by Smart (1997a).

A significant advance was made by Bilu & Hanrot (1996, 1999). They developed a more efficient algorithm for solving Thue equations which is also applicable to equations of high degree. To this end they modified the method of Tzanakis & de Weger (1989) as follows:

(i) Tzanakis & de Weger reduced (4) to a single inequality of the form (3) with unknown rational integer coefficients $b_1, \ldots, b_r$. Bilu & Hanrot

(1996) utilized the fact that (4) implies (9) for $n-2$ values of i, whence they deduced $r-1$ linearly independent inequalities of the form (3) in $b_1, \ldots, b_r$.

(ii) After eliminating $r-2$ unknowns b_i, they arrived at an inhomogeneous inequality in two unknowns.

(iii) Instead of the multidimensional reduction algorithm they used the much faster Baker–Davenport reduction method to get a reduced bound B_R for B.

(iv) They showed that for fixed i, the tuple $b_1, \ldots, b_r$ is in fact defined uniquely as soon as b_i is given. Therefore one has to check only $2B_R+1$ possibilities for the tuple $b_1, \ldots, b_r$.

(v) Bilu & Hanrot (1999) made the method even more efficient when $K = \mathbb{Q}(\theta)$, where $F(\theta, 1) = 0$, has a small subfield of degree ≥ 3.

By means of this method the Thue equation (4) can be solved in practice in reasonable time as soon as a system of fundamental units and a complete system of non-associate elements of norm m/a_0 of the ideal $(1, \theta)$ are determined in K. Here a_0 denotes the leading coefficient of $F(X, 1)$. To illustrate the method, consider the real cyclotomic equation

$$F_p(x, y) = \prod_{k=1}^{(p-1)/2} \left(y - x \cdot 2\cos\frac{2k\pi}{p} \right) = \pm 1 \text{ in } x, y \in \mathbb{Z} \qquad (36)$$

where $p > 12$ is a prime number. This equation, where $F_p(X, Y)$ is of degree $(p-1)/2$ and has its coefficients in $\mathbb{Z}$, occurs in the study of primitive divisors of Lucas and Lehmer numbers. It was shown in Bilu & Hanrot (1999) that for $p = 67, 311, 977, 997$ and 5011, the solutions of (36) are

$$(0, \pm 1),\ (\pm 1, 0),\ (\pm 1, \pm 1),\ (\pm 1, \mp 1),\ (\pm 1, \mp 2).$$

As a remarkable application of their method Bilu, Hanrot & Voutier (2001) proved an old conjecture which asserts that for $n > 30$, the nth term of any Lucas or Lehmer sequence has a primitive divisor.

It should be noted that no generalization of the above method is known for Thue–Mahler equations and for the relative case.

2.2 *Unit equations and S-unit equations*

Let K be an algebraic number field, and S a finite set of places on K containing the set of infinite places S_∞. The S-unit equation (16) can be written in the form

$$\lambda_1' \varepsilon_1^{b_{1,1}} \cdots \varepsilon_{s-1}^{b_{1,s-1}} + \lambda_2' \varepsilon_1^{b_{2,1}} \cdots \varepsilon_{s-1}^{b_{2,s-1}} = 1, \qquad (37)$$

where λ_1', λ_2' are given non-zero elements of K, s denotes the cardinality of S, $\varepsilon_1, \ldots, \varepsilon_{s-1}$ are fundamental/independent S-units in K and b_{ij} are rational integer unknowns. Győry (1979) reduced the equations of the shape (37) to inequalities of the form (17), and using Baker's theory he gave an explicit bound B_0 for $B = \max_{i,j} |b_{ij}|$. Then, in case of a concrete equation (37) de Weger's reduction process discussed above can be applied to reduce B_0 to a much smaller bound B_R. A direct enumeration of the possible values of the exponents with $B \leq B_R$ becomes hopeless if s is large, even if B_R is small. For $K = \mathbb{Q}$, de Weger (1987) used an algorithm of Fincke & Pohst for calculating lattice points in ellipsoids to find the 'small' solutions of (37), and as an illustration he resolved, over $\mathbb{Q}$, an S-unit equation with $s = 7$. In the general case Smart (1995, see also Smart 1998) applied a sieving technique to locate the 'small' solutions of (37). Using parallel sieve with suitable prime ideals, he determined all the solutions of several concrete S-unit equations with $s \leq 7$.

In the important special case $S = S_\infty$, Wildanger (1997)† has recently introduced an efficient method for finding the 'small' unknown exponents in equations of the form (37). Using some ideas of de Weger (1987) he reduced the problem for searching lattice points in appropriate ellipsoids of the logarithmic space, and then applied the algorithm of Fincke & Pohst for the enumeration of the lattice points in question. Wildanger's method enabled him to solve completely unit equations in normal extensions of $\mathbb{Q}$ with unit ranks ≤ 10. Some applications of the method to equations (22), (24) and (30) will be discussed in the next sections. A further application is given in Gaál & Pohst (2001) to relative Thue equations. Very recently Wildanger's method has been extended by Smart (1999) to general S-unit equations.

2.3 Discriminant form and index form equations

Keeping the notation of Section 1.4, let again K be an algebraic number field of degree $n \geq 3$, $\{1, \alpha_1, \ldots, \alpha_{n-1}\}$ an integral basis of K, and consider the corresponding discriminant form equation

$$D_{K/\mathbb{Q}}(x_1\alpha_1 + \ldots + x_{n-1}\alpha_{n-1}) = D \quad \text{in } x_1, \ldots, x_{n-1} \in \mathbb{Z} \tag{22a}$$

and index form equation

$$\mathrm{I}(x_1, \cdots, x_{n-1}) = \pm\mathrm{I} \quad \text{in } x_1, \ldots, x_{n-1} \in \mathbb{Z}. \tag{24}$$

For $D = \mathrm{I}^2 D_K$ these equations are equivalent. Even the best known bound (29) on the solutions is too large for practical use. As described in Section 1.4,

† Added in proof. See also Wildanger's paper in *J. Number Theory* **82** (2000), 188–224.

Győry (1973, 1974, 1976) developed a general method for the solution of (22a) and (24) via unit equations by first reducing equations (22a), (24) to unit equations of the form (28) over N or $K^{(i)}K^{(j)}K^{(k)}$, and then deriving by Baker's theory an explicit but large bound B_0 for the maximum absolute value B of the unknown exponents. We have seen above that in the 1980s such reduction algorithms were developed which can be used in concrete cases to reduce B_0 to a much smaller bound B_R. There remained the problem of checking the possible $(2B_R+1)^{2r}$ exponents under the bound B_R in the equations (28) involved, where r denotes the unit rank of the field N or $K^{(i)}K^{(j)}K^{(k)}$. Here r can be large compared to n, attaining the values $n!-1$ and $n(n-1)(n-2)-1$, respectively, when K is totally real and $\mathrm{Gal}(N/\mathbb{Q}) = S_n$. Therefore there are in general too many cases for the exponents to be checked and too many unit equations to be solved. Hence it was generally believed for a time that this general method involving unit equations is only of theoretical interest.

For $n = 3$, when (24) is in fact a cubic Thue equation, Gaál & Schulte (1989) solved (24) with I $= 1$ for cubic number fields K with discriminants $-300 \leq D_K \leq 3137$. The solutions gave all power integral bases in the fields in question.

For $n = 4$, the equation (24) was studied by Gaál, Pethő & Pohst (1993, 1996, and references therein). They developed an efficient algorithm for the resolution of (24) in arbitrary quartic number fields, reducing the problem to solving a cubic Thue equation and several quartic Thue equations. By means of their method the authors made extensive computations and published several numerical tables, providing the complete lists of solutions of a great number of concrete equations (24). They computed *minimal indices* and *all elements of minimal index* in the following number fields:

- in all totally real quartic fields with Galois group A_4 and discriminants not exceeding 10^6 (31 fields);
- in the 50 totally real quartic fields with smallest discriminants and Galois group S_4;
- in the 50 quartic fields with mixed signature and smallest absolute discriminants;
- in all totally complex quartic fields with discriminants not exceeding 10^6 and Galois group A_4 (90 fields) or S_4 (44122 fields).

The method cannot be applied to the case $n > 4$, except for $n = 6, 8, 9$ when K has a quadratic or cubic subfield; then (24) leads to relative cubic or relative quartic Thue equations.

Smart (1993, 1995, 1996) was the first to solve discriminant form and index form equations by the method involving unit equations. In his 1996 paper

Smart worked in the normal closure N and diminished the number of unit equations (28) to be solved by using the action of $\mathrm{Gal}(N/\mathbb{Q})$ on these equations. Further, he applied his sieving process mentioned in the preceding section to find the 'small' exponents in the unit equations involved. To illustrate his algorithm he solved equation (24) for $\mathrm{I} = 1$ in some sextic fields having an imaginary quadratic subfield.

As was discussed in Section 2.2, Wildanger (1997) has recently worked out an efficient method which can be used to determine the 'small' exponents in unit equations of the form (28). Combining his algorithm with the general method of Győry (1973, 1974, 1976) and with reduction algorithms, Wildanger (1997) made it possible to solve equations (22a), (24) for normal extensions K of $\mathbb{Q}$ with unit ranks not exceeding 10. In particular, he completely solved equation (24) for $\mathrm{I} = 1$ in cyclotomic fields of degree at most 12.

In the approaches of Smart and Wildanger the equations (28) were considered over N and $K^{(i)}K^{(j)}K^{(k)}$, respectively. Then for given degree n, the number of unknown exponents in (28) can attain or exceed the value $2(n(n-1)(n-2)+1)$, i.e. for $n = 4, 5, 6$ the values $46, 118, 238$, respectively. This shows that for $n \geq 4$ there are in general too many unknowns to utilize the algorithms of Smart or Wildanger.

In Section 1.4 it was mentioned that recently we have considerably refined in Győry (1998, 2000) the general approach of Győry (1973, 1974, 1976) by reducing (22a), (24) to unit equations over much smaller number fields, having much fewer unknown exponents. Namely, for given degree n, the total number of unknown exponents of the unit equations involved in our refined version is at most $\frac{n(n-1)}{2} - 1$, i.e. for $n = 4, 5, 6$ at most $5, 9, 14$, respectively. The combination of this refined version of the general approach with a variant of Wildanger's enumeration method given by Gaál & Pohst makes it feasible to solve equations (22a), (24) in any number field of degree $n \leq 5$. For quintic fields such an algorithm was described in Gaál & Győry (1999). As an illustration we present the following *example* from there.

Consider the quintic field $K = \mathbb{Q}(\theta)$ where θ is a zero of the polynomial $X^5 - 6X^3 + X^2 + 4X + 1$. Then K is totally real with discriminant $D_K = 36497$, Galois group S_5 (most difficult case) and integral basis $\{1, \theta, \theta^2, \theta^3, \theta^4\}$. Then up to translation by elements of $\mathbb{Z}$, all the $\alpha \in O_K$ for which $\{1, \alpha, \ldots, \alpha^4\}$ is an integral basis in K are given by $\alpha = \pm(x_1\theta + x_2\theta^2 + x_3\theta^3 + x_4\theta^4)$ where

$$\begin{aligned}&(x_1, x_2, x_3, x_4)\\ &\quad = (1, -6, 0, 1), (1, 0, 0, 0), (2, -6, 0, 1), (2, -5, 0, 1), (3, -11, 0, 2),\\ &\qquad (3, -5, 0, 1), (3, 0, -5, 2), (4, -5, -1, 1), (4, 0, -3, -1),\end{aligned}$$

$$(4, 5, -1, -1), (6, -6, -1, 1), (6, 15, -2, -3), (7, -12, -1, 2),$$
$$(7, -11, -1, 2), (8, -12, -1, 2), (9, -18, -1, 3), (9, -17, -1, 3),$$
$$(11, -23, -1, 4), (13, -18, -2, 3), (15, -24, -2, 4),$$
$$(16, -23, -2, 4), (19, -41, -2, 7), (31, -46, -4, 8),$$
$$(53, 62, -14, -13), (80, -159, -9, 27), (115, -166, -15, 29).$$

Very recently I gave a further refinement of the general method by reducing the corresponding unit equations to $n-2$ inequalities concerning linear forms in logarithms with coefficients b_l. Hence $n-3$ exponents b_l can be eliminated, and the number of the remaining exponents to be determined is at most $\frac{n(n-1)}{2} - (n-2)$. This recent refinement will yield further numerical applications to equations (22a), (24) and (25).

Finally, we mention a related result concerning binary forms. The effective finiteness theorem of Evertse & Győry (1991) presented in Section 1.4 on binary forms was proved in a more general form, for binary forms with given degree and with discriminants divisible by a given finite set of primes. On applying his sieving process concerning S-unit equations, recently Smart (1997b) turned Evertse & Győry's proof into a practical algorithm, and used it to determine all hyperelliptic curves of genus 2 with good reduction away from 2.

2.4 Decomposable form equations of general type

Under the assumptions formulated in Section 1.5, a general algorithm was given in Győry (1978, 1980a, 1980b, 1981) for solving decomposable form equations and their generalizations of Mahler type. The decomposable form equations under consideration were first reduced to S-unit equations of the form (37), and then the results of Győry (1979) obtained by Baker's theory were used to derive an explicit bound B_0 for the unknown exponents in the S-unit equations involved. However, this bound B_0 is too large for practical use. In case of triangularly connected decomposable form equations of Mahler type, Smart (1995) made the general method more practical. Following the arguments of Győry (1978, 1980a 1980b, 1981), he arrived first at the explicit bound B_0 for the unknown exponents. Then he used the reduction techniques developed in de Weger (1989) to reduce the bound B_0. Finally, he applied his sieving process mentioned in Section 2.2 to locate the 'small' solutions. An application was also given in Smart (1995) to finding curves of genus 2 with good reduction outside a given finite set of primes and Weierstrass points in given number fields.

In Győry (1998) we have recently made more efficient for practical use the general method given in Győry (1981) for solving equation (30) and its generalization of Mahler type. The decomposable form equation was reduced to special S-unit equations having properties specified in Section 1.5, which have much fewer unknown exponents than the corresponding S-unit equations in Győry (1981). This refinement of the general method can be combined with reduction techniques and with recent algorithms of Wildanger and Smart concerning unit equations and S-unit equations, respectively, to develop an efficient algorithm for the resolution of concrete decomposable form equations satisfying the conditions made in Győry (1998), and also specified above. Such an algorithm has been recently described by Gaál & Győry (1999) for discriminant form and index form equations in quintic fields (cf. Section 2.3), and by Gaál (2000) for norm form equations of the form (32) which satisfy the assumptions formulated in Section 1.5 and for which the relative unit rank of $K/\mathbb{Q}(\alpha_1, \ldots, \alpha_{m-1})$ is at most 11. In both of those papers numerical examples were also presented.

2.5 Elliptic equations

Consider the elliptic curve defined by the equation

$$E: \quad y^2 = x^3 + ax + b \tag{38}$$

where a, b are given rational integers with $4a^3 + 27b^2 \neq 0$. Let $E(\mathbb{Q})$ denote the set of points $(x, y) \in \mathbb{Q}^2$ satisfying (38) and the infinite point. There exist classical methods which reduce the computation of integral points $(x, y) \in E(\mathbb{Z})$ to finitely many Thue equations or unit equations in two unknowns; see e.g. Mordell (1969), Smart (1998) and Pethő & Zimmer (2000). In concrete cases the Thue equations and unit equations involved can be solved by combining Baker's theory with reduction techniques and with the methods of Bilu & Hanrot (1996) and Wildanger (1997), respectively. These classical approaches require, however, a lot of computations in number fields.

A completely different method for proving the finiteness of $E(\mathbb{Z})$ was proposed by Lang which makes use of the group structure of the elliptic curve, and transforms the search of integral points on E to an inequality concerning linear form in elliptic logarithms. Combining this idea with an explicit lower bound of David (1995) for linear forms in elliptic logarithms, Gebel, Pethő & Zimmer (1994) and, independently, Stroecker & Tzanakis (1994) worked out an efficient algorithm for the determination of integral points of E. The main steps of this method are as follows.

Let $P_1, \ldots, P_r$ be a basis of the infinite part of the Mordell–Weil group of $E(\mathbb{Q})$. Then every point $P = (x, y) \in E(\mathbb{Q})$ has the representation

$$P = b_1 P_1 + \cdots + b_r P_r + T \tag{39}$$

where $b_1, \ldots, b_r \in \mathbb{Z}$ and $T \in E_{\text{tors}}(\mathbb{Q})$ is a torsion point. The elements of $E_{\text{tors}}(\mathbb{Q})$ can be computed easily. To find the integral points on E, we need to determine which values of the variables b_i can take to make the point P integral. Assume that in (39) $P \in E(\mathbb{Z})$. Put $B(P) = \max_i |b_i|$. One can show that

$$\left| \sum_{i=1}^{r} b_i \delta_i + \delta \right| \leq c_0 \exp\{-c_1 B^2(P) + c_2\} \tag{40}$$

with appropriate explicit constants c_0, c_1, c_2. Here δ_i denotes the normalized elliptic logarithm of P_i for $i = 1, \ldots, r$, and $\delta \in \mathbb{Z}$. Using David's explicit lower bound for the left-hand side of (40) and comparing it with the upper bound, one obtains an explicit upper bound B_0 for $B(P)$. Then, in concrete cases, de Weger's variant of the reduction algorithm can be applied to reduce B_0 to a much smaller bound B_R. Finally, all points $P \in E(\mathbb{Q})$ with $B(P) \leq B_R$ can be tested for integrality when the rank r is not large ($r \leq 6$), and so all $P \in E(\mathbb{Z})$ can be determined.

Gebel, Pethő & Zimmer (1998) applied their method to the Mordell equation

$$y^2 = x^3 + k \quad \text{in } x, y \in \mathbb{Z} \tag{41}$$

where k is a given non-zero integer, and made extensive computations. They found all solutions for each k with $0 < |k| \leq 10^4$, and the computations were partially extended to the range $0 < |k| \leq 10^5$. Several numerical tables were established which enabled the authors to make some interesting observations on the distribution of the solutions of (41). For example, for $0 < |k| \leq 10^5$ their numerical results include all integral points $P = (x, y)$ on $E: \ y^2 = x^3 + k$ with $x \geq 10^9$:

k	rank	x	$\pm y$
28024	4	3 790 689 201	233 387 325 399 875
–64432	4	3 171 881 612	178 638 660 622 364
91017	3	1 979 757 358	88 088 243 191 777
99207	2	1 303 201 029	47 045 395 221 186
–88688	3	1 053 831 624	34 210 296 678 956

Recently Pethő *et al.* (1999) have extended their method to computing all S-integral points on the curve E defined by (38). Let $S = \{q_1 = \infty, q_2, \ldots, q_s\}$ where $q_2, \ldots, q_s$ are distinct primes, and let $\mathbb{Z}_S$ denote the ring of S-integers in $\mathbb{Q}$. For each point $P = (x, y) \in E(\mathbb{Z}_S)$, consider the representation of P in the form (39). Then, for some $q \in S$, one can deduce an inequality of the form (40) with the left side replaced by $|\sum_{i=1}^{r} b_i \delta_{iq} + \delta_q|_q$, where δ_{iq} denotes the q-adic elliptic logarithm of P_i, $i = 1, \ldots, r$, and $\delta_q \in \mathbb{Z}$ if $q = \infty$ and $\delta_q = 0$ otherwise. An explicit lower bound for linear forms in q-adic elliptic logarithms would make it possible to derive an upper bound for $B(P)$. However, no q-adic analogue of David's result is known for $r > 2$. To overcome the absence of such a lower bound, it is shown in Pethő *et al.* (1999) that $B^2(P) \leq c_3 \log H(x) + c_4$, where c_3, c_4 are constants. In the next step an explicit bound B_0 is derived for $B(P)$ by using a recent bound of Hajdu & Herendi (1998) for $H(x)$, which was obtained by a combination of the complex and p-adic versions of Baker's theory. Then, in concrete cases, de Weger's reduction process can be used to reduce the value of B_0. Finally, a similar argument works as for the points in $E(\mathbb{Z})$ to test the S-integrality and compute all S-integral points on E. To illustrate the method in practice, all 144 S-integral points on

$$E: \quad y^2 = x^3 - 172x + 505$$

are determined for $S = \{\infty, 3, 5, 7\}$. In this case the rank $r = 4$.

The elliptic logarithmic method was successfully applied in many other papers, written by Bremner, Gebel, Hajdu, Herrmann, Pethő, Smart, Stroecker, Tzanakis, de Weger, Zimmer and others. Recently, the method has been extended by Smart & Stephens (1997) to elliptic curves over algebraic number fields, by Tzanakis (1996) to the case of certain quartic equations, and by Stroecker & de Weger (1999) to general cubic curves having at least one rational point.

The elliptic method requires the knowledge of a basis of the Mordell–Weil group of $E(\mathbb{Q})$. Although the Mordell–Weil theorem is in general ineffective, there are certain processes for computing a basis which work fine in practice. Further, as 'most' elliptic curves do have very small ranks, the absence of efficient techniques for testing the S-integrality for large $r(\geq 8)$ is no problem in practice. The elliptic method is by *experience* very efficient. For a more detailed exposition of the method and for its comparison with the classical methods, see Smart (1998) and Pethő & Zimmer (2000).

Finally, we note that recently Bilu & Hanrot (1998) described a practical method for solving superelliptic equations without intermediate use of Thue equations or unit equations. They reduced the initial equation directly to linear forms in logarithms of algebraic numbers, which the complex version of Baker's theory can be applied to. Then they used for concrete cases the method of Baker & Davenport to reduce the Baker-type bound, and developed an efficient algorithm to find the solutions under the reduced bound.

References

Baker, A. (1966), Linear forms in the logarithms of algebraic numbers I, *Mathematika*, **13**, 204–216.

Baker, A. (1967a), Linear forms in the logarithms of algebraic numbers II, *Mathematika*, **14**, 102–107.

Baker, A. (1967b), Linear forms in the logarithms of algebraic numbers III, *Mathematika*, **14**, 220–228.

Baker, A. (1968a), Linear forms in the logarithms of algebraic numbers IV, *Mathematika*, **15**, 204–216.

Baker, A. (1968b), Contributions to the theory of Diophantine equations, *Philos. Trans. Roy. Soc. London Ser. A*, **263**, 173–208.

Baker, A. (1968c), The Diophantine equation $y^2 = ax^3 + bx^2 + cx + d$, *J. London Math. Soc.*, **43**, 1–9.

Baker, A. (1969), Bounds for the solutions of the hyperelliptic equation, *Proc. Camb. Philos. Soc.*, **65**, 439–444.

Baker, A. (1975), *Transcendental Number Theory*, Cambridge University Press; 2nd edition with additional material (1979); 3rd edition with updated material (1990).

Baker, A. (1988), ed., *New Advances in Transcendence Theory*, Cambridge University Press.

Baker, A. (1998), Logarithmic forms and the *abc*-conjecture, in *Number Theory*, K. Győry, A. Pethő, V.T. Sós (eds.), de Gruyter, 37–44.

Baker, A. & J. Coates (1970), Integer points on curves of genus 1, *Proc. Camb. Philos. Soc.*, **67**, 595–602.

Baker, A. & H. Davenport (1969), The equations $3x^2 - 2 = y^2$ and $8x^2 - 7 = z^2$, *Quart. J. Math. Oxford Ser. (2)*, **20**, 129–137.

Baker, A. & D.W. Masser (1977), eds., *Transcendence Theory: Advances and Applications*, Academic Press.

Baker, A. & G. Wüstholz (1993), Logarithmic forms and group varieties, *J. Reine Angew. Math.*, **442**, 19–62.

Bérczes, A., B. Brindza & L. Hajdu (1998), On the power values of polynomials, *Publ. Math. Debrecen* **53**, 375–381.

Bilu, Y. & G. Hanrot (1996), Solving Thue equations of high degree, *J. Number Theory*, **60**, 373–392.

Bilu, Y. & G. Hanrot (1998), Solving superelliptic Diophantine equations by Baker's method, *Compositio Math.*, **112**, 273–312.

Bilu, Y. & G. Hanrot (1999), Thue equations with composite fields, *Acta Arith.*, **88**, 311–326.

Bilu, Y., G. Hanrot & P.M. Voutier (2001), Existence of primitive divisors of Lucas and Lehmer numbers (with an appendix by M. Mignotte), to appear.

Birch, B.J. & J.R. Merriman (1972), Finiteness theorems for binary forms with given discriminant, *Proc. London Math. Soc. (3)*, **24**, 385–394.

Bombieri, E., J. Mueller & M. Poe (1997), The unit equation and the cluster principle, *Acta Arith.*, **79**, 361–389.

Brindza, B. (1984), On S-integral solutions of the equation $y^m = f(x)$, *Acta Math. Hungar.*, **44**, 133–139.

Brindza, B. (1989), On the equation $f(x) = y^m$ over finitely generated domains, *Acta Math. Hungar.*, **53**, 377–383.

Brindza, B. & K. Győry (1990), On unit equations with rational coefficients, *Acta Arith.*, **53**, 367–388.

Brindza, B., J.-H. Evertse & K. Győry (1991), Bounds for the solutions of some Diophantine equations in terms of discriminants, *J. Austral. Math. Soc. Ser. A*, **51**, 8–26.

Bugeaud, Y. (1998), Bornes effectives pour les solutions des équations en S-unités et des équations de Thue–Mahler, *J. Number Theory*, **71**, 227–244.

Bugeaud, Y. & K. Győry (1996a), Bounds for the solutions of unit equations, *Acta Arith.*, **74**, 67–80.

Bugeaud, Y. & K. Győry (1996b), Bounds for the solutions of Thue–Mahler equations and norm form equations, *Acta Arith.*, **74**, 273–292.

Coates, J. (1970), An effective p-adic analogue of a theorem of Thue II. The greatest prime factor of a binary form, *Acta Arith.*, **16**, 399–412.

David, S. (1995), Minorations de formes linéaires de logarithmes elliptiques, *Mém. Soc. Math. France (N.S.)*, 143pp.

Evertse, J.-H. & K. Győry (1988), Decomposable form equations, in *New Advances in Transcendence Theory*, A. Baker (ed.), Cambridge University Press, 175–202.

Evertse, J.-H. & K. Győry (1991), Effective finiteness results for binary forms with given discriminant, *Compositio Math.*, **79**, 169–204.

Evertse, J.-H., K. Győry, C.L. Stewart & R. Tijdeman (1988a), On S-unit equations in two unknowns, *Invent. Math.*, **92**, 461–477.

Evertse, J.-H., K. Győry, C.L. Stewart & R. Tijdeman (1988b), S-unit equations and their applications, in *New Advances in Transcendence Theory*, A. Baker (ed.), Cambridge University Press, 110–174.

Fel'dman, N.I. & Y.V. Nesterenko (1998), *Transcendental Numbers*, Springer.

Gaál, I. (1988), On the resolution of inhomogeneous norm form equations in two dominating variables, *Math. Comp.*, **51**, 359–373.

Gaál, I. (2000), An efficient algorithm for the explicit resolution of norm form equations, *Publ. Math. Debrecen* **56**, 375–390.

Gaál, I. & K. Győry (1999), Index form equations in quintic fields, *Acta Arith.*, **89**, 379–396.

Gaál, I., A. Pethő & M. Pohst (1993), On the resolution of index form equations in quartic number fields, *J. Symbolic Comput.*, **16**, 563–584.

Gaál, I., A. Pethő & M. Pohst (1996), Simultaneous representation of integers by a pair of ternary quadratic forms—with an application to index form equations in quartic number fields, *J. Number Theory*, **57**, 90–104.

Gaál, I. & M. Pohst (2001), On the resolution of relative Thue equations, to appear.

Gaál, I. & N. Schulte (1989), Computing all power integral bases of cubic fields, *Math. Comp.*, **53**, 689–696.

Gebel, J., A. Pethő & H.G. Zimmer (1994), Computing integral points on elliptic curves, *Acta Arith.*, **68**, 171–192.

Gebel, J., A. Pethő & H. G. Zimmer (1998), On Mordell's equation, *Compositio Math.*, **110**, 335–367.

Győry, K. (1972), Sur l'irréductibilité d'une classe des polynômes II, *Publ. Math. Debrecen*, **19**, 293–326.

Győry, K. (1973), Sur les polynômes à coefficients entiers et de discriminant donné I, *Acta Arith.*, **23**, 419–426.

Győry, K. (1974), Sur les polynômes à coefficients entiers et de discriminant donné II, *Publ. Math. Debrecen*, **21**, 125–144.

Győry, K. (1976), Sur les polynômes à coefficients entiers et de discriminant donné III, *Publ. Math. Debrecen*, **23**, 141–165.

Győry, K. (1978), On the greatest prime factors of decomposable forms at integer points, *Ann. Acad. Sci. Fenn. Ser. A. I.*, **4**, 341–355.

Győry, K. (1979), On the number of solutions of linear equations in units of an algebraic number field, *Comment. Math. Helv.*, **54**, 583–600.

Győry, K. (1980a), Explicit upper bounds for the solutions of some diophantine equations, *Ann. Acad. Sci. Fenn. Ser. A. I.*, **5**, 3–12.

Győry, K. (1980b), *Résultats effectifs sur la représentation des entiers par des formes décomposables*, Queen's Papers in Pure and Applied Math., **56**, A.J. Coleman and P. Ribenboim (eds.), Kingston, Canada.

Győry, K. (1981), On the representation of integers by decomposable forms in several variables, *Publ. Math. Debrecen*, **28**, 89–98.

Győry, K. (1983), Bounds for the solutions of norm form, discriminant form and index form equations in finitely generated integral domains, *Acta Math. Hungar.*, **42**, 45–80.

Győry, K. (1992), Some recent applications of S-unit equations, *Astérisque*, **209**, 17–38.

Győry, K. (1996), Applications of unit equations, in *Analytic Number Theory*, Kyoto, Y. Motohashi (ed.), 62–78.

Győry, K. (1997), On the diophantine equation $\binom{n}{k} = x^l$, *Acta Arith.* **80**, 289–295.

Győry, K. (1998), Bounds for the solutions of decomposable form equations, *Publ. Math. Debrecen*, **52**, 1–31.

Győry, K. (1999), On the distribution of solutions of decomposable form equations, in *Number Theory in Progress*, K. Győry, H. Iwaniec and J. Urbanowicz (eds.), de Gruyter, 237–265.

Győry, K. (2000), Discriminant form and index form equations, in *Algebraic Number Theory and Diophantine Analysis*, F. Halter-Koch and R.F. Tichy (eds.), de Gruyter, 191–214.

Győry, K. & Z.Z. Papp (1978), Effective estimates for the integer solutions of norm form and discriminant form equations, *Publ. Math. Debrecen*, **25**, 311–325.

Hajdu, L. & T. Herendi (1998), Explicit bounds for the solutions of elliptic equations with rational coefficients, *J. Symbolic Computation*, **25**, 361–366.

Matveev, E.M. (1998), An explicit lower bound for a homogeneous rational linear form in logarithms of algebraic numbers, *Izvestiya: Mathematics*, **62**, 723–772.

Mordell, L.J. (1969), *Diophantine Equations*, Academic Press.

Pethő, A. (1987), On the resolution of Thue inequalities, *J. Symbolic Computation*, **4**, 103–109.

Pethő, A. (1990), Computational methods for the resolution of Diophantine equations, in *Number Theory*, R.A. Mollin (ed.), de Gruyter, 479–492.

Pethő, A. & R. Schulenberg (1987), Effektives Lösen von Thue Gleichungen, *Publ. Math. Debrecen*, **34**, 189–196.

Pethő, A. & H.G. Zimmer (2000), S-integer points on elliptic curves, theory and practice, in *Algebraic Number Theory and Diophantine Analysis*, F. Halter-Koch and R.F. Tichy (eds.), de Gruyter, 351–363.

Pethő, A., H. G. Zimmer, J. Gebel & E. Herrmann (1999), Computing all S-integral points on elliptic curves, *Math. Proc. Cambridge Philos. Soc.*, **127**, 383–402.

Schinzel, A. & R. Tijdeman (1976), On the equation $y^m = P(x)$, *Acta Arith.* **31**, 199–204.

Schmidt, W.M. (1971), Linearformen mit algebraishen Koeffizienten II, *Math. Ann.*, **191**, 1–20.

Schmidt, W.M. (1972), Norm form equations, *Ann. of Math.*, **96**, 525–551.

Shorey, T.N. & R. Tijdeman (1986), *Exponential Diophantine Equations*, Cambridge University Press.

Siegel, C.L. (1929), Über einige Anwendungen Diophantischer Approximationen, *Abh. Preuss. Akad. Wiss.*, 1–41.

Smart, N.P. (1993), Solving a quartic discriminant form equation, *Publ. Math. Debrecen*, **43**, 29–39.

Smart, N.P. (1995), The solution of triangularly connected decomposable form equations, *Math. Comp.*, **64**, 819–840.

Smart, N.P. (1996), Solving discriminant form equations via unit equations, *J. Symbolic Comput.*, **21**, 367–374.

Smart, N.P. (1997a), Thue and Thue–Mahler equations over rings of integers, *J. London Math. Soc. (2)*, **56**, 455–462.

Smart, N.P. (1997b), S-unit equations, binary forms and curves of genus 2, *Proc. London Math. Soc.*, **75**, 271–307.

Smart, N.P. & N.M. Stephens (1997), Integral points on elliptic curves over number fields, *Proc. Camb. Phil. Soc.*, **122**, 9–16.

Smart, N.P. (1998), *The Algorithmic Resolution of Diophantine Equations*, Cambridge University Press.

Smart, N.P. (1999), Determining the small solutions to S-unit equations, *Math. Comp.*, **68**, 1687–1699.

Sprindžuk, V.G. (1993), *Classical Diophantine Equations*, (Lecture Notes in Math. **1559**), Springer.

Stroeker, R.J. & N. Tzanakis (1994), Solving elliptic diophantine equations by estimating linear forms in elliptic logarithms, *Acta Arith.*, **67**, 177–196.

Stroeker, R.J. & B.M.M. de Weger (1999), Solving elliptic diophantine equations: the general cubic case, *Acta Arith.*, **87**, 339–365.

Tzanakis, N. (1996), Solving elliptic diophantine equations by estimating linear forms in elliptic logarithms. The case of quartic equations, *Acta Arith.*, **75**, 165–190.

Tzanakis, N. & B.M.M. de Weger (1989), On the practical solution of the Thue equation, *J. Number Theory*, **31**, 99–132.

Tzanakis, N. & B.M.M. de Weger (1992), How to explicitly solve a Thue-Mahler equation, *Compositio Math.*, **84**, 223–288.

Waldschmidt, M. (1993), Minorations de combinaisons linéaires de logarithmes de nombres algébriques, *Canad. J. Math.*, **45**, 176–224.

de Weger, B.M.M. (1987), Solving exponential diophantine equations using lattice basis reduction algorithms, *J. Number Theory*, **26**, 325–367.

de Weger, B.M.M. (1989), *Algorithms for Diophantine Equations*, Center for Mathematics and Computer Science, Amsterdam .

Wildanger, K. (1997), *Über das Lösen von Einheiten- und Indexformgleichungen in algebraishen Zahlenkörpern mit einer Anwendung auf die Bestimmung aller ganzen Punkte einer Mordellschen Kurve*, PhD thesis, Technishen Universität Berlin.

Yu, K. (1994), Linear forms in p-adic logarithms III, *Compositio Math.*, **91**, 241–276.

Yu, K. (1999), p-adic logarithmic forms and group varieties II, *Acta Arith.*, **89**, 337–378.

5

Baker's Method and Modular Curves

Yuri F. Bilu

Abstract

We review Baker's method for effective analysis of S-integral points on curves. As an example, we show that S-integral points on the modular curve $X_0(N)$ are effectively bounded; here N is a positive integer, distinct from 1, 2, 3, 5, 7 and 13.

1 Introduction

Let K be a number field, and S a finite set of places of K, including all Archimedean places. Denote by $\mathcal{O}_S$ the ring of S-integers of the field K.

Let C be a smooth projective curve over K of genus $\mathbf{g}$ and $x \in K(C)$ a non-constant rational function on C. We denote by $C(K)$ the set of K-rational points and by $C(\mathcal{O}_S, x)$ the set of S-integral points with respect to x:

$$C(\mathcal{O}_S, x) := \{P \in C(K) : x(P) \in \mathcal{O}_S\}. \tag{1}$$

According to the classical *theorem of Siegel* (see Siegel 1929, Lang 1983, Serre 1989), *the set $C(\mathcal{O}_S, x)$ is finite if $\mathbf{g} \geq 1$ or if x has at least three distinct poles.* In 1983 Faltings (see Falting 1983, 1984; Cornell & Silverman 1986) confirmed Mordell's conjecture, by proving that the very set $C(K)$ of rational points is finite if $\mathbf{g} \geq 2$.

The results of Siegel and Faltings are both ineffective in the sense that they imply no explicit bound for the size of S-integral or rational points. In spite of numerous efforts of many mathematicians, no effective approach to the study of rational points is known. On the other hand, there is a general method for effective analysis of integral points, developed by Alan Baker (1966, 1967a,b, 1968a,b,c, 1969). Using Baker's method, one may obtain effective versions of Siegel's theorem for curves of genus 0 and 1 (Baker and Coates 1970) and for certain particular curves of higher genus.

In Section 3 we give a general overview of Baker's method, as presented in Bilu (1993, 1995). In Section 4 we apply it to S-integral points on modular curves (with respect to the rational function defined by the j-invariant). In particular, we obtain the following result.

Theorem 10 *Let N be a positive integer, distinct from* 1, 2, 3, 5, 7 *and* 13. *Then S-integral points on on the modular curve $X_0(N)$ are effectively bounded in terms of K, S and N.*

Notation We denote by $\overline{K}$ the algebraic closure of the field K.

Acknowledgements I am pleased to thank David Masser for many interesting discussions, and Elina Wojciechowska for a helpful advice. I also thank Yann Bugeaud, Dale Brownawell, Damien Roy and the referee for pointing out several inaccuracies.

2 Heights

In this section we recall the definition and simplest properties of the Weil height.

Let α be an algebraic number, and let K be any number field containing α. We normalize the valuations $|\cdot|_v$ of K so that their restrictions to $\mathbb{Q}$ coincide with the usual infinite or p-adic valuations of $\mathbb{Q}$. Now put

$$\|\alpha\|_v = \max\left(1, |\alpha_v|^{[K_v:\mathbb{Q}_v]}\right), \qquad H(\alpha) = \left(\prod_v \|\alpha\|_v\right)^{1/[K:\mathbb{Q}]}, \qquad (2)$$

where the product extends to all places of K. It is well-known that $H(\alpha)$ is independent of the choice of the field K.

It follows from the definition that, for $\alpha, \beta \in \overline{\mathbb{Q}}$ and $n \in Z$,

$$H(\alpha \pm \beta) \le 2H(\alpha)H(\beta), \quad H(\alpha\beta) \le H(\alpha)H(\beta), \quad H(\alpha^n) = H(\alpha)^{|n|}. \qquad (3)$$

It also follows from the definition that $H(\alpha) \ge 1$. A more precise statement is given by the classical *theorem of Kronecker*: $H(\alpha) = 1$ *if and only if $\alpha = 0$ or α is a root of unity; moreover, if $H(\alpha) > 1$ and $\deg\alpha = d$, then $H(\alpha) \ge 1 + c(d)$, where $c(d)$ is a positive effective constant.* (Lehmer's famous conjecture '$c(d) = c/d$ with an absolute constant $c > 0$' is still open. See Smyth (1971) and Dobrowolski (1979) for the best results in this direction.)

If C is a curve over $\overline{\mathbb{Q}}$ and $x \in \overline{\mathbb{Q}}(C)$ is a non-constant rational function, we define the Weil height on $C(\overline{\mathbb{Q}})$ with respect to x as $H_x(P) = H(x(P))$, where we put, by definition, $H(\infty) = 1$. The index x will be omitted if this does not confuse.

If $y \in \overline{\mathbb{Q}}(C)$ is a different non-constant rational function, then

$$\log H_y(P) \leq c_1 \log H_x(P) + c_2, \tag{4}$$

where c_1 and c_2 effectively depend on C, x and y. In fact, a stronger assertion,

$$\lim_{H_x(P)\to\infty} \frac{\log H_y(P)}{\log H_x(P)} = \frac{[\overline{\mathbb{Q}}(C) : \overline{\mathbb{Q}}(y)]}{[\overline{\mathbb{Q}}(C) : \overline{\mathbb{Q}}(x)]}$$

(called *quasi-equivalence of heights*), holds, but (4) is sufficient for our purposes.

It is very easy to prove (4). If $y \in \overline{\mathbb{Q}}(x)$ then (4) is a consequence of relations (3). In the general case, y satisfies an algebraic equation of the form $y^m + z_1 y^{m-1} + \cdots + z_m = 0$, where $z_1, \ldots, z_m \in \overline{\mathbb{Q}}(x)$. Then for any place v

$$\log \|y(P)\|_v \leq \log \max \{\|z_1(P)\|_v, \ldots, \|z_m(P)\|_v\} + \log \|m\|_v.$$

Summing over v, we obtain $\log H_y(P) \leq \log H_{z_1}(P) + \cdots + \log H_{z_m}(P) + \log m$. Since an inequality of the form (4) holds for every z_i, it holds for y as well.

We say that an algebraic number is *effectively bounded* (in terms of certain parameters) if its height is bounded from above by a constant, effectively depending on these parameters. Further, if C and x are as above, then a point $P \in C(\overline{\mathbb{Q}})$ is effectively bounded if the algebraic number $x(P)$ is effectively bounded.

3 Baker's method

Hier stehe ich. Ich kann nicht anders.
Luther (Melancthon 1548)

In this section we briefly review Baker's method, in the form developed in Bilu (1993, 1995). The main feature of the our approach is that it avoids the linear units equations, which had been indispensable in all existing expositions of the method. (See, for instance, the books of Sprindžuk (1982) and Serre (1989).) Instead, we systematically use *functional units*, see below.

The main justification of our approach is that it has a wider range of applications. For example, it does not seem that Theorem 10 can be proved using linear unit equations. See Bilu (1995), Section 6, for another example where unit equations, are, most probably, non-applicable.

Even in cases when unit equations are applicable, the present approach provides neater and more conceptual proofs, see Bilu (1997, 1998), Bugeaud (1998). This is especially important for the numerical solution of Diophantine equations using Baker's method, because the technology of functional units helps one to choose various auxiliary number fields in the most economical way, see Bilu & Hanrot (1998, 1999). Using functional units, Bilu, Hanrot & Voutier (2001) managed to solve the long-standing problem of primitive divisors.

It is not my intention to write a comprehensive survey of Baker's method. Therefore, many trends and results are not discussed, and many important references are left out. One can find extensive bibliographies in the monographs of Sprindžuk (1982), Shorey & Tijdeman (1986), surveys of Evertse *et al.* (1988) and Győry (1992, 1996).

Baker's theorem

The principal basis of Baker's method is the following theorem.

Theorem 1 *Let* $\beta_1, \ldots, \beta_m$ *be algebraic numbers,* $b_1, \ldots, b_m$ *rational integers,* v *a place of the number field* $\mathbb{Q}(\beta_1, \ldots, \beta_m)$*, and* $\varepsilon > 0$*. Assume that*

$$0 < \left|\beta_1^{b_1} \cdots \beta_m^{b_m} - 1\right|_v < e^{-\varepsilon B}, \tag{5}$$

where $B = \max\{b_1, \ldots, b_m\}$*. Then* $B \le B_0$*, where* B_0 *is effectively computable in terms of* $\beta_1, \ldots, \beta_m$, ε *and* v.

Several comments are to be made here. A similar statement without effectiveness can be easily deduced from the classical Diophantine approximations result of Thue–Siegel–Roth (or even Thue–Siegel) type, see Gelfond (1952), Theorem I.IV, or Lang (1983), Corollary 7.1.2. On the other hand, the *effective* inequality†

$$\left|\beta_1^{b_1} \cdots \beta_m^{b_m} - 1\right|_v \ge e^{-cB}, \tag{6}$$

with a positive constant c easily follows from the product formula. Baker's contribution was in replacing c by an arbitrarily small ε without loosing the effectiveness.

† provided the left-hand side is non-zero, which will be always assumed here

At present, there exist three methods for proving Theorem 1. Baker (1966, 1967a,b, 1968a) himself used his theory of logarithmic forms. He proved that

$$\left|\beta_1^{b_1}\cdots\beta_m^{b_m}-1\right|_v \geq e^{-c\log^\kappa B}, \tag{7}$$

where $\kappa > m+1$ and c is effective. (Later, κ was reduced to 1.) This estimate clearly implies Theorem 1. Baker himself considered only Archimedean case; the non-Archimedean theory was developed by different authors, notably van der Poorten and Yu.

Here is not the proper place for a detailed historical survey of Baker's theory. We juust mention that best known forms of (7) are due to Baker & Wüstholz (1993), Waldschmidt (1993) and Matveev (1998) in the Archimedean case, and Yu (1999) in the non-Archimedean case.

Quite recently, it has been discovered (Bugeaud 1998, Bilu & Bugeaud 2001) that one does not need the full strength of Baker's theory to obtain Theorem 1: it can be easily deduced from an estimate for linear forms in just two logarithms (of Gelfond–Feldman type).

Bombieri (1993) (see also Bombieri & Cohen 1997, 2001) suggested an approach completely independent of the theory of logarithmic forms. It is based on Dyson's lemma and some geometry of numbers. This method inspired the argument of Bugeaud (1998), Bilu & Bugeaud (2001); in particular, the key geometric lemma used in the latter is due to Bombieri & Cohen (1997).

We shall apply Theorem 1 in the following equivalent form.

Theorem 1′ *Let G be a finitely generated multiplicative group of algebraic numbers, v a place of the number field generated by G, and $\varepsilon > 0$. Let $\alpha \in G$ satisfy $0 < |\alpha - 1|_v < H(\alpha)^{-\varepsilon}$. Then $H(\alpha)$ is effectively bounded in terms of G, v and ε.*

For the reader's convenience, we indicate the proof of equivalence of Theorems 1 and 1′. It is sufficient to establish the following statement (essentially, due to Dirichlet): *let $\beta_1, \ldots, \beta_m$ be multiplicatively independent non-zero algebraic numbers, $b_1, \ldots, b_m \in \mathbb{Z}$ and $\alpha = \beta_1^{b_1}\cdots\beta_m^{b_m}$; then $B \ll \log H(\alpha)$, the implicit constant effectively depending on $\beta_1, \ldots, \beta_m$.*

To prove this, denote by S the set of all places v of the number field $\mathbb{Q}(\beta_1, \ldots, \beta_m)$ with the following property: there exists a β_i with $|\beta_i|_v \neq 1$. Let Γ be the subgroup of $\mathbb{R}^{|S|}$ generated by the m vectors $\left(\log|\beta_i|_v\right)_{v\in S}$. By the theorem of Kronecker (see Section 2), the rank of Γ is m, and the norms of its non-zero elements are bounded from below by an effective positive constant. Hence, the $(|S| \times m)$-matrix $\left(\log|\beta_i|_v\right)_{\substack{v\in S\\ 1\leq i\leq m}}$ is of rank m, and it has an $(m \times m)$-submatrix of determinant effectively bounded from below. Inverting this submatrix, we express $b_1, \ldots, b_m$ as linear combinations of $\log |\alpha|_v \quad (v \in S)$ with effectively bounded coefficients. This proves that $B \ll \log H(\alpha)$.

Functional units

In this subsection we show how Theorem $1'$ applies to the effective study of S-integral points on curves.

Let C be a projective curve over a field K, and Σ a finite subset of $C(\overline{K})$. A $\overline{K}$-rational function on C is called Σ-unit if its poles and zeros belong to Σ. The multiplicative group of Σ-units is isomorphic to $\overline{K}^* \times \mathbb{Z}^\rho$, where the rank $\rho = \rho(\Sigma)$ satisfies $0 \le \rho(\Sigma) \le |\Sigma| - 1$.

It should be mentioned that, for a given set Σ, the existence of a non-constant Σ-unit can be effectively verified, and, if existing, such a unit can be explicitly constructed. Indeed, fix $Q \in \Sigma$ and consider the Jacobian embedding $C \hookrightarrow J(C)$ with Q going to the origin. Then $\rho(\Sigma) \ge 1$ if and only if the set $\Sigma \setminus \{Q\}$ is linearly dependent on $J(C)$. The work of Masser (1988) implies that if this set is linearly dependent, then there exists a non-trivial linear relation with effectively bounded coefficients. Hence the linear dependence can be effectively verified.

Alternatively speaking, if a non-constant Σ-unit exists, then there exists a unit with effectively bounded orders of poles and zeros†. Such a unit can be constructed using the effective Riemann–Roch theorem due to Coates (1970) or Schmidt (1991). See Bilu (1993) Chapter 1, where this construction is described in the full detail.

Applying this procedure to proper subsets of Σ, one can compute $\rho(\Sigma)$ and construct a full rank system of independent Σ-units.

The principal property of a Σ-unit is

Proposition 2 *Let K, S, C and x be as in Section 1. Denote by Σ the set of poles of x and let $y \in \overline{K}(C)$ be a Σ-unit. Then, after replacing K be a finite extension, and adding to S finitely many new places, we have the following: for any $P \in C(\mathcal{O}_S, x)$, the specialization $y(P)$ is an S-unit of the field K.*

The proof is very simple. By extending the base field, we may assume that $y \in K(C)$. Since all poles of y are among the poles of x, the function y is

† It must be pointed out that Masser's bounds, while universal, are rather huge. In concrete cases much sharper bounds are usually available.

integral over the ring $K[x]$. Expanding the set S, we may assume that y is integral over the ring $\mathcal{O}_S[x]$. Similarly, y^{-1} is integral over $\mathcal{O}_S[x]$ after a further expansion of S. Hence, for $P \in C(\mathcal{O}_S, x)$, both $y(P)$ and $y^{-1}(P)$ are integral over $\mathcal{O}_S$, which means that $y(P)$ is an S-unit.

In what follows, C stands for an algebraic curve defined over $\overline{\mathbb{Q}}$ and $x \in \overline{\mathbb{Q}}(C)$ is non-constant. Given a number field K, a K-model of the pair (C, x) is a K-model of C such that $x \in K(C)$. We shall say that *Siegel's theorem is effective for the pair* (C, x) if, for any number field K, for any K-model of the (C, x) and for any finite set S of places of K, the set $C(\mathcal{O}_S, x)$ is effectively bounded.

The heart of (our version of) Baker's method is the following simple theorem.

Theorem 3 (Bilu 1993, 1995) *Denote by Σ the set of poles of x and assume that $\rho(\Sigma) \geq 2$. Then Siegel's theorem is effective for the pair (C, x).*

We sketch a proof. (See Bilu 1995 for a detailed argument, and Bilu 1993, Chapter 1, for a quantitative statement.) Since $\rho(\Sigma) \geq 2$, for every $Q \in \Sigma$ there is an (effectively constructable) Σ-unit y_Q with $y_Q(Q) = 1$. Fix a number field K such that C is definable over K and $x \in K(C)$, and a finite set S of places of K. Extending K and expanding S as in Proposition 2, we may assume that $y_Q(P)$ is an S-unit for any S-integral point P and any $Q \in \Sigma$.

Fix an S-integral point P. Since $x(P)$ is an S-integer, there exists $v \in S$ such that

$$\|x(P)\|_v \geq H_x(P)^{1/|S|}. \tag{8}$$

Further, there exists $Q \in \Sigma$ such that the v-adic distance $d_v(Q, P)$ (see Remark 4) satisfies

$$d_v(Q, P)^{[K_v:\mathbb{Q}_v]} \ll \|x(P)\|_v^{-1/t}, \tag{9}$$

where $t = -\mathrm{ord}_Q(x)$ and the implicit constant is effective (as well as everywhere below). Notice that $t > 0$, because Q is a pole of x.

Writing y instead of y_Q, and using (4), (8) and (9), we obtain

$$|y(P) - 1|_v = |y(P) - y(Q)|_v \ll d_v(Q, P) \ll H_y(P)^{-\varepsilon}, \tag{10}$$

where ε is an effective positive constant. Since the points P with $y(P) = 1$ are effectively bounded, we may assume that $y(P) \neq 1$. In other words, $\alpha = y(P)$ satisfies $0 < |\alpha - 1|_v \ll H(\alpha)^{-\varepsilon}$.

Since α is an S-unit (that is, belongs to a finitely-generated multiplicative group), Theorem 1$'$ yields an effective upper bound for $H(\alpha) = H_y(P)$. Again using (4), but with x and y interchanged, we conclude that $H_x(P)$ is effectively bounded. This proves the theorem.

Remark 4 How to define the v-adic distance $d_v(Q,\cdot)$ is up to the reader's taste, background and fantasy. One can, for instance, fix a rational function on $z \in K(C)$ with the single pole Q of multiplicity m (which exists for sufficiently large m by the Riemann–Roch theorem) and define the distance from $d_v(Q,P)^{[K_v:\mathbb{Q}_v]} = \|z(P)\|_v^{-1/m}$. Or, one can use the language of Weil functions Lang (1983), Chapter 10. Or, one can apply the canonical local heights defined through Arakelov theory (Lang 1998, Gross 1986). Finally, one can avoid the notion of the v-adic distance at all, using instead Puiseux expansions, as in Bilu (1993, 1995).

Coverings

Theorem 3 becomes much more powerful when enriched with the technique of coverings. The main fact about coverings is the following *Chevalley–Weil principle*, which formalizes standard arguments of the kind 'if $(x-a)(x-b)$ is a square, then each of $x-a$ and $x-b$ is almost a square'.

Proposition 5 (Chevalley–Weil principle) *Let K, S, C and x be as in Section 1. Let $\widetilde{C}\to C$ be a finite covering étale outside the poles of x. Then there is an effectively constructable number field L so that every point of $\widetilde{C}$ above $C(\mathcal{O}_S,x)$ is L-rational.*

The Chevalley–Weil principle holds not only for curves, but for varieties of arbitrary dimension as well. See Lang (1983), Section 2.8, Serre (1989), Section 8.1, and Vojta (1987), Section 5.1, for different proofs. A very explicit form of the Chevalley–Weil principle for curves can be found in Bilu (1993), Chapter 4.

Combined with the Chevalley–Weil principle, Theorem 3 implies the following.

Proposition 6 *Let C and x be as in the previous subsection. Let $\widetilde{C}\to C$ be a finite covering étale outside the poles of x, and let $\widetilde{\Sigma}$ be the set of points of $\widetilde{C}$ above the poles of x. Assume that $\rho(\widetilde{\Sigma})\ge 2$. Then Siegel's theorem is effective for the pair (C,x).*

This result is quite general: it contains all known (to the author) cases of the effective version of Siegel's theorem.

As the simplest example, we will show how Proposition 6 implies the effective Siegel theorem for curves of genus 0 or 1 (due to Baker & Coates 1970). See Bilu (1995) and Bilu (1993), Chapter 5, for many further examples.

Theorem 7 *Let C and x be as above. Assume that either $\mathbf{g}(C) = 0$ and x has at least 3 poles, or $\mathbf{g}(C) = 1$. Then Siegel's theorem is effective for the pair (C, x).*

Proof If $\mathbf{g}(C) = 0$ then $\rho(\Sigma) = |\Sigma| - 1$ for any finite set $\Sigma \subset C(\overline{K})$. If Σ is the set of poles of x then $|\Sigma| \geq 3$ by the assumption, which means that $\rho(\Sigma) \geq 2$ and one can apply Theorem 3.

In the genus 1 case one has to use coverings. We endow C with a group structure and consider an isogeny $\widetilde{C} \to C$ of degree $m \geq 3$. Let $\mathcal{F}$ be the fiber above a pole of x. Then $mP - mQ$ is a principal divisor for any $P, Q \in \mathcal{F}$. Hence, $\rho(\mathcal{F}) = |\mathcal{F}| - 1 = m - 1 \geq 2$, and we can apply Proposition 6. □

4 Modular curves

Preliminaries

In this section we apply Theorem 3 and Proposition 6 to modular curves.

Recall the definitions; for all details see Shimura (1971), Chapter 1, and Lang (1976), Chapter 3. The group $\mathrm{SL}_2(\mathbb{Z})$ acts on the extended upper half-plane $\overline{\mathcal{H}} := \mathcal{H} \cup \mathbb{Q} \cup \{\infty\}$ by fractional linear transformations. This defines an action of $\mathrm{PSL}_2(\mathbb{Z}) = \mathrm{SL}_2(\mathbb{Z})/\{\pm I\}$, because the matrix $-I = \begin{bmatrix} -1 & 0 \\ 0 & -1 \end{bmatrix}$ acts trivially. Here and below, Γ will always stand for a finite index subgroup of $\mathrm{SL}_2(\mathbb{Z})$, and $\overline{\Gamma}$ will denote the image of Γ in $\mathrm{PSL}_2(\mathbb{Z})$.

For any Γ denote by X_Γ the quotient $\Gamma\backslash\overline{\mathcal{H}} = \overline{\Gamma}\backslash\overline{\mathcal{H}}$, endowed with the natural structure of a compact Riemann surface. The images of $\mathbb{Q} \cup \{i\infty\}$ in X_Γ are called the *cusps of* X_Γ. The number of the cusps is usually denoted by $\nu_\infty(\Gamma)$. The modular invariant j defines a rational function on X_Γ, its poles being exactly the cusps.

Non-identical elements of $\mathrm{PSL}_2(\mathbb{Z})$ of finite order are called *elliptic*; they can have orders 2 or 3 only. Pre-images of elliptic elements in $\mathrm{SL}_2(\mathbb{Z})$ are also called elliptic. An element of $\mathrm{SL}_2(\mathbb{Z})$ is elliptic if and only if its trace is equal to 0, -1 or 1, the order being 4, 3 or 6, respectively.

For $z \in \overline{\mathcal{H}}$ denote by Γ_z (respectively, $\overline{\Gamma}_z$) the stabilizer of z in Γ (respectively, $\overline{\Gamma}$). If $z \in \mathcal{H}$ then group $\overline{\Gamma}_z$ can be either trivial, or cyclic of order 2 or 3.

Table 1.

N	2	4	other
$\nu_\infty(\Gamma(N))$	3		$\frac{1}{2}N^2\prod_{p\mid N}\left(1-p^{-2}\right)$
$\nu_\infty(\Gamma_1(N))$	2	3	$\frac{1}{2}\sum_{d\mid N}\varphi(d)\varphi(N/d)$
$\nu_\infty(\Gamma_0(N))$			$\sum_{d\mid N}\varphi\,(\gcd(d, N/d))$

Let Γ' be a finite index subgroup of Γ. Then we have a finite covering $X_{\Gamma'}\to X_\Gamma$ of degree $[\overline{\Gamma}:\overline{\Gamma}']$. For a point P of $X_{\Gamma'}$ fix a pre-image $z\in\overline{\mathcal{H}}$. Then $[\overline{\Gamma}_z:\overline{\Gamma}'_z]$ does not depend on the choice of z and is equal to the ramification index of P over X_Γ. It follows that *the covering $X_{\Gamma'}\to X_\Gamma$ is étale outside the cusps if and only if Γ' contains all elliptic elements of Γ.*

Three important classes of subgroups of $\mathrm{SL}_2(\mathbb{Z})$ are $\Gamma(N)$, $\Gamma_1(N)$ and $\Gamma_0(N)$, where N is a positive integer. They consist of matrices A satisfying

$$A\equiv I \bmod N,\qquad A\equiv\begin{bmatrix}1&*\\0&1\end{bmatrix}\bmod N\quad\text{and}\quad A\equiv\begin{bmatrix}*&*\\0&*\end{bmatrix}\bmod N,$$

respectively. The corresponding modular curves are denoted by $X(N)$, $X_1(N)$ and $X_0(N)$, respectively. The numbers of cusps of these curves are given in Table 1 (cf. Shimura 1971, Section 1.6).

One calls Γ a *congruence subgroup* if it contains $\Gamma(N)$ for some N.

Effective Siegel theorem for modular curves

Defined originally as a compact Riemann surface, X_Γ has a $\overline{\mathbb{Q}}$-model such that $j\in\overline{\mathbb{Q}}(X_\Gamma)$. (This is the 'easy' part of Belyi's theorem, cf. Serre 1989, Section 5.4.) Hence, one can study Diophantine properties of modular curves. In particular, one may wonder if Siegel's theorem is effective for the pair (X_Γ, j). The first result in this direction is due to Kubert & Lang (1981), Theorem 8.1.2. They proved that Siegel's theorem is effective for $(X(N), j)$, when $N\geq 7$. (Kubert & Lang did not explicitly refer to the effectiveness, but their argument is certainly effective.)

The following was observed in Bilu (1995).

Proposition 8 (Bilu 1995, Proposition 5.1(a)) *Let Γ be a congruence subgroup such that X_Γ has at least* 3 *cusps. Then Siegel's theorem is effective for (X_Γ, j).*

Indeed, if Γ is a congruence subgroup, then the set Σ of cusps satisfies $\rho(\Sigma) = |\Sigma| - 1$, by the classical theorem of Manin–Drinfeld: Lang (1976), Section 4.2. Hence Proposition 8 is a consequence of Theorem 3.

Corollary 9 *Siegel's theorem is effective for* $(X(N), j)$ *when* $N \geq 2$, *and for* $(X_1(N), j)$ *when* $N \geq 4$

Indeed, according to Table 1, each of these curves has at least 3 cusps.

When N is a composite number, $X_0(N)$ has at least 3 cusps as well. Therefore, Siegel's theorem is effective for $(X_0(N), j)$ when N is composite. It turns out that this can be strengthened.

Theorem 10 *Siegel's theorem is effective for* $(X_0(N), j)$ *when* $N \notin \{1, 2, 3, 5, 7, 13\}$.

Notice that the curves $X_0(1) = X_1(1) = X(1)$, $X_0(2) = X_1(2)$, $X_0(3) = X_1(3)$, $X_0(5)$, $X_0(7)$ and $X_0(13)$ must be excluded from consideration, because they are of genus 0 and have at most 2 cusps, which means that they do not satisfy the assumption of Siegel's theorem.

The proof of Theorem 10 will be given below.

It is proved in Bilu (2001) that Siegel's theorem is effective for the pair (X_Γ, j), for all congruence subgroups Γ with finitely many exceptions. The proof of this general result is similar to that of Theorem 10, but involves more technicalities.

One may ask about conditions for effective Siegel's theorem which hold for some non-congruence subgroups as well. The only reasonable example I know is the following:

Proposition 11 *Siegel's theorem is effective for* (X_Γ, j) *if* Γ *has no elliptic elements.*

See Bilu (1995), Section 5, for two simple proofs of this result.

In connection with Proposition 11, mention that any curve definable over $\overline{\mathbb{Q}}$ is realizable as X_Γ where Γ is a free group with two generators (and, in particular, has no elliptic elements). This is a version of Belyi's theorem, see Serre (1989), page 71, the proof of '(3)⇒(2)'.

Notice also that $\Gamma(N)$ and $\Gamma_1(N)$ have no elliptic elements for $N \geq 2$, respectively $N \geq 4$. Hence Proposition 11 gives an alternative proof for Corollary 9.

Proof of Theorem 10

We start with the following general assertion.

Proposition 12 *Let* Γ *have a subgroup* Γ' *satisfying the following three conditions:* Γ' *is a congruence subgroup;* $X_{\Gamma'}$ *has at least three cusps;* Γ' *contains all elliptic elements from* Γ. *Then Siegel's theorem is effective for* (X_Γ, j).

Proof Since Γ' contains all elliptic elements from Γ, the covering $X_{\Gamma'} \to X_\Gamma$ is étale outside the cusps. By the Manin–Drinfeld Theorem, the set Σ' of cusps of $X_{\Gamma'}$ satisfies $\rho(\Sigma') = |\Sigma'| - 1 \geq 2$. Hence the result follows from Proposition 8. □

Proposition 13 *Let* p *be a prime number and let* G *be a proper subgroup of* $\mathbb{F}_p^*$ *containing* -1. *Put*

$$\Gamma' = \left\{ \begin{bmatrix} a & b \\ c & d \end{bmatrix} \in \Gamma_0(p) : a \in G (\operatorname{mod} p) \right\}. \tag{11}$$

Then $X_{\Gamma'}$ *has at least three cusps.*

Proof If Γ is a subgroup of $\mathrm{SL}_2(\mathbb{Z})$ containing $-I$, then the cusps of X_Γ stand in a one-to-one correspondence with the Γ-orbits of $\mathbb{Q} \cup \{\infty\}$. Since $X_0(p)$ has exactly two cusps, it is sufficient to show that the Γ'-orbit of some element of $\mathbb{Q} \cup \{\infty\}$ is a proper subset of its $\Gamma_0(p)$-orbit. We will do this for the Γ'-orbit of ∞. The $\Gamma_0(p)$-orbit of ∞ includes, besides ∞ itself, all rational numbers with denominator divisible by p. However, the Γ'-orbit includes (besides ∞) only the rational numbers with denominator divisible by p and numerator belonging $\operatorname{mod} p$ to G. Since G is a proper subgroup of $\mathbb{F}_p^*$, the Γ'-orbit is indeed a proper subset of the $\Gamma_0(p)$-orbit. The proposition is proved. □

Proof of Theorem 10 We have already mentioned that, for composite N, the curve $X_0(N)$ has at least three cusps, and one may use Proposition 8. We are left with $X_0(p)$, where $p > 7$ is a prime number distinct from 13. Let G be the subgroup of $\mathbb{F}_p^*$ consisting of elements of order dividing 12, and let Γ' be defined from (11). Since $p > 7$ and $p \neq 13$, the group G is a proper subgroup of $\mathbb{F}_p^*$. By Proposition 13, the curve $X_{\Gamma'}$ has at least three cusps.

Let $A = \begin{bmatrix} a & b \\ c & d \end{bmatrix}$ be an elliptic element of $\Gamma_0(p)$. Then

$$I = A^{12} \equiv \begin{bmatrix} a^{12} & * \\ 0 & d^{12} \end{bmatrix} \operatorname{mod} p.$$

Therefore $a^{12} = 1 \bmod p$, which means that $A \in \Gamma'$. Thus, Γ' meets the assumption of Proposition 12 (with $\Gamma = \Gamma_0(p)$). Hence, Siegel's theorem is effective for $(X_0(p), j)$. The theorem is proved. □

Consequences for elliptic curves

Let E be an elliptic curve over a number field K and S a finite set of places of K. Recall that, by the classical result of Deuring (see Husemöller 1987, Subsections 5.7.5 and 5.7.6), the j-invariant $j(E)$ is an S-integer if and only if E has potentially good reduction outside S. Hence the above results about $X_1(N)$ and $X_0(N)$ can be reformulated as follows.

Proposition 14 *Let K, S and E be as above, and assume that E has a potentially good reduction outside S. Then we have the following.*

(a) *If E has a K-torsion point of order $N \geq 4$, then $j(E)$ is effectively bounded in terms of K, S and N.*

(b) *If E has a cyclic subgroup, defined over K, of order $N \notin \{1, 2, 3, 5, 7, 13\}$, then $j(E)$ is effectively bounded in terms of K, S and N.*

Recall that a subset of $E(\overline{K})$ is *defined over K* if it is invariant under the action of the Galois group $\mathrm{Gal}(\overline{K}/K)$.

By the celebrated theorem of Merel (1996), the order of a K-torsion point in effectively bounded in terms of K (and even in terms of the degree $[K:\mathbb{Q}]$). (See the work of Parent (1999) for an explicit version of Merel's result.) Merel's theorem implies the following 'uniform version' of Proposition 14(a).

Proposition 15 *Let K, S and E be as above. Assume that E has a potentially good reduction outside S, and a K-torsion point of order at least* 4. *Then $j(E)$ is effectively bounded in terms of K and S.*

References

Baker, A. (1966), Linear forms in the logarithms of algebraic numbers I, *Mathematika* **13**, 204–216.

Baker, A. (1967a), Linear forms in the logarithms of algebraic numbers II, *Mathematika* **14**, 102–107.

Baker, A. (1967b), Linear forms in the logarithms of algebraic numbers III, *Mathematika* **14**, 220–224.

Baker, A. (1968), Linear forms in the logarithms of algebraic numbers IV, *Mathematika* **15**, 204–216.

Baker, A. (1968b), Contribution to the theory of Diophantine equations, *Phil. Trans. Roy. Soc. London* **A263**, 173–208.

Baker, A. (1968c), The Diophantine equation $y^2 = ax^3 + bx^2 + cx + d$. *J. London Math. Soc.* **43**, 1–9.

Baker, A. (1969), Bounds for solutions of hyperelliptic equations, *Proc. Cambridge Phil. Soc.* **65**, 439–444.

Baker, A. (ed.) (1988), *New Advances in Transcendence Theory (Durham, 1986)*, Cambridge University Press.

Baker, A. & J. Coates (1970), Integer points on curves of genus 1, *Math. Proc. Camb. Phil. Soc.* **67**, 592–602.

Baker, A. & G. Wüstholz (1993), Logarithmic forms and group varieties, *J. Reine Angew. Math.* **442**, 19–62.

Bilu, Yu. (1993), *Effective analysis of integral points on algebraic curves*, PhD Thesis, Beer-Sheva.

Bilu, Yu. (1995), Effective analysis of integral points on algebraic curves, *Israel J. Math* **90**, 235–252.

Bilu, Yu. (1997), Quantitative Siegel's theorem for Galois coverings, *Compositio Math.*, **106**, 125–158.

Bilu, Yu. (1998), Integral points and Galois covers, *Math. Contemp.*, **14**, 1–11.

Bilu, Yu. (2001), Effective Siegel's theorem for modular curves, in preparation.

Bilu, Yu. & Y. Bugeaud (2001), Démonstration du théorèm de Baker-Feldman via les formes linéaires en deux logarithmes, *J. Th. Nombres Bordeaux*, to appear.

Bilu, Yu. & G. Hanrot (1998), Solving superelliptic Diophantine equations by Baker's method, *Compositio Math.*, **112**, 273–312.

Bilu, Yu. & G. Hanrot (1999), Thue equations with composite fields, *Acta Arith.*, **88**, 311–326.

Bilu, Yu., G. Hanrot & P.M. Voutier, (2001) Existence of primitive divisors of Lucas and Lehmer numbers (with and appendix by M. Mignotte), *J. Reine Angew. Math.*, to appear; available for downloading from the electronic preprint server of the Mathematical Institute of the University of Basel `ftp://www.math.unibas.ch/pub/bilu/` .

Bombieri, E. (1993), Effective Diophantine approximation on $\mathbb{G}_m$, *Ann. Scuola Norm. Sup. Pisa Cl. Sci.* **20**, 61–89.

Bombieri, E. & P.B. Cohen (1997), Effective Diophantine approximation on $\mathbb{G}_M$ II, *Ann. Scuola Norm. Sup. Pisa Cl. Sci.* **24**, 205–225.

Bombieri, E. & P.B. Cohen (2001), Effective Diophantine approximation on $\mathbb{G}_M$ III, preprint.

Bugeaud, Y. (1998), Bornes effectives pour les solutions des équations en S-unités et des équations de Thue–Mahler, *J. Number Theory* **71**, 227–244.

Bugeaud, Y. (2000), On the greatest prime factor of $ax^m + by^n$. II., *Bull. London Math. Soc.*, **32**, 673–678.

Coates, J. (1970), Construction of rational functions on a curve, *Math. Proc. Camb. Phil. Soc.* **68**, 105–123.

Cornell, G. & J.H. Silverman (eds.) (1986), *Arithmetic Geometry*, Springer.

Dobrowolski, E. (1979), On a question of Lehmer and the number of irreducible factors of a polynomial, *Acta Arith.* **34**, 391–401.

Evertse, J.-H., K. Győry, C.L. Stewart & R. Tijdeman, (1988) S-unit equations and their applications, in *New Advances in Transcendence Theory (Durham, 1986)*, A. Baker (ed.) Cambridge University Press, 110–174.

Faltings, G. (1983), Endlichkeitssätze für abelsche Varietäten über Zahlkörpern, *Invent. Math.* **73**, 349–366; Erratum: **75** (1984), 381.

Gelfond, A.O. (1952), *Transcendent and Algebraic Numbers* (Russian), GITTL, Moscow; English translation, Dover, 1960.

Gross, B.H. (1986), Local heights on curves, in *Arithmetic Geometry*, G. Cornell & J.H. Silverman (eds.), Springer, 327–339.

Győry, K. (1992), Some recent applications of S-unit equations *Astérisque* **209**, 17–38.

Győry, K. (1996), Applications of unit equations, in *Analytic Number Theory (Kyoto, 1994)*, Sūrikaisekikenkyūsho Kōkyūroku **958**, 62–78.

Husemöller, D. (1987), *Elliptic Curves*, Springer.

Kubert, K. & S. Lang (1981), *Modular Units*, Springer.

Lang, S. (1976), *Introduction to Modular Forms*, Springer.

Lang, S. (1983), *Fundamentals of Diophantine Geometry*, Springer.

Lang, S. (1988), *Introduction to Arakelov Theory*, Springer.

Masser, D.W. (1986), Linear relations on algebraic groups, in *New Advances in Transcendence Theory (Durham, 1986)*, A. Baker (ed.) Cambridge University Press, 248–262.

Matveev, E. (1998), An explicit lower bound for a homogeneous rational linear form in logarithms of algebraic numbers (Russian), *Izv. Ross. Akad. Nauk Ser. Mat.* **62**, 81–136.

Melancthon, Ph. (1548), *Historia de vita et actis Lutheri*, Heidelberg.

Merel, L. (1996), Bornes pour la torsion des courbes elliptiques sur les corps de nombres, *Invent. Math.* **124**, 437–449.

Parent, P. (1999), Bornes effectives pour la torsion des courbes elliptiques sur les corps de nombres, *J. Reine Angew. Math.* **506**, 85–116.

Schmidt, W.M. (1991), Construction and Estimation of Bases in Function Fields, *J. Number Th.* **39**, 181–224.

Serre, J.-P. (1989), *Lectures on the Mordell–Weil Theorem*, Vieweg.

Shimura, G. (1971), *Introduction to the Arithmetic Theory of Automorphic Functions*, Iwanami Shoten and Princeton University Press.

Shorey, T.N. & R. Tijdeman (1986), *Exponential Diophantine equations*, Cambridge University Press.

Siegel, C.L. (1929), Über einige Anwendungen Diophantischer Approximationen, *Abh. Preuss Akad. Wiss. Phys.-Math. Kl.*, Nr. 1; *Ges. Abh.*, Band 1, 209–266, Springer, 1966.

Smyth, C.J. (1971), On the product of the conjugates outside the unit circle of an algebraic integer, *Bull. London Math. Soc.* **3**, 169–175.

Sprindžuk, V.G. (1982), *Classical Diophantine Equations in Two Unknowns* (Russian), Nauka, Moscow; English translation: Lecture Notes in Math., Vol. 1559, Springer, 1994.

Vojta, P. (1987), *Diophantine Approximations and Value Distribution Theory*, Lecture Notes in Math. **1239**, Springer.

Waldschmidt, M. (1993), Minorations de combinaisons linéaires de logarithmes de nombres algébriques, *Canadian J. Math.* **45**, 176–224.

Yu, Kunrui (1999), p-adic logarithmic forms and group varieties II, *Acta Arith.* **89**, 337–378.

6

Application of the André–Oort Conjecture to some Questions in Transcendence

Paula B. Cohen and Gisbert Wüstholz

Abstract

We show how a problem concerning the transcendence of values of the classical hypergeometric function, and originating in work of Siegel on G-functions, can be solved using a special case of a conjecture of André–Oort on the distribution of complex multiplication (or special) points on algebraic curves in Shimura varieties. The special case in question has recently been proven, at our suggestion, by Edixhoven & Yafaev (2001); see also Yafaev (2001b). This settles the question of which classical hypergeometric functions with rational parameters, satisfying certain natural assumptions, take only finitely many algebraic values at algebraic points. The fact that such a function cannot have an arithmetic monodromy group goes back to work of Wolfart (1988). We introduce a number of related problems.

Note added in revision In the original version of this article, we introduced a number of open problems motivated by transcendence questions on the classical hypergeometric function. These are summarised in Problems 1, 2, 3 and 4 of §1. One of the main points of this article is to show how Problems 1 and 2 follow from Problem 4, which is in turn related to the André–Oort Conjecture, Oort (1997) concerning the distribution of complex multiplication points on subvarieties of Shimura varieties. Some special cases of Problem 4 had been previously solved by Edixhoven, but the general solution to Problem 4 was announced by Edixhoven & Yafaev (2001) subsequent to our original manuscript, so settling Problems 1 and 2 and leading to the present revised version of our original article. Problem 3 still remains open in general, although related special cases have been solved by André and, under GRH, by Edixhoven (1998, 2001) and by Yafaev (2001a), see also Moonen (1995, 1998). As the general case of the André–Oort Conjecture remains open, the problems raised in §4 are still largely unsolved.

1 Some problems on hypergeometric functions

The hypergeometric functions in one variable we study are the classical Gauss functions arising from integrals of differential forms on algebraic curves varying in a family parametrised by $Q = \mathbb{P}_1(\mathbb{C}) \setminus \{0, 1, \infty\}$. More precisely, let A, B, C, N be positive coprime integers and consider the topologically trivial fibre bundle $\mathcal{X}$ over Q given by

$$\mathcal{X} = \{(y, u, x) \in \mathbb{P}_1 \times \mathbb{P}_1 \times Q \mid y^N = u^A(u-1)^B(u-x)^C\}.$$

The fibre $\mathcal{X}_x$ of $\mathcal{X}$ above $x \in Q$ is an irreducible abelian cover of $\mathbb{P}_1$ ramified at $0, 1, \infty, x$ with covering group $G = (\mathbb{Z}/N\mathbb{Z})$. To the family of algebraic curves $\mathcal{X}$, we can associate the family of abelian varieties $\mathrm{Jac}(\mathcal{X})$ with fibre $\mathrm{Jac}(\mathcal{X}_x)$ above $x \in Q$, studied also in Wolfart (1988), Cohen & Wolfart (1990) and de Jong & Noot (1991). The curve $\mathcal{X}_x$ has an automorphism given by

$$\kappa : (u, y) \mapsto (u, \zeta_N^{-1} y)$$

where ζ_N is a primitive Nth root of unity, which we fix from now on. This automorphism induces an action of ζ_N on the space $H^0(\mathrm{Jac}(\mathcal{X}_x), \Omega)$ of differential forms of the first kind on $\mathcal{X}_x$, and $\mathrm{Jac}(\mathcal{X}_x)$ has up to isogeny a decomposition of the form

$$\mathrm{Jac}(\mathcal{X}_x) \mathrel{\hat{=}} T_x \oplus \sum_{f|N} \mathrm{Jac}(\mathcal{X}_{x,d}).$$

Here T_x is a principally polarised abelian variety of dimension $\varphi(N)$, where φ is Euler's function, with lattice isomorphic to $\mathbb{Z}[\zeta_N]^2$. The space $H^0(T_x, \Omega)$ is generated by the eigenforms for the action of ζ_N by its images ζ_N^s, $s \in (\mathbb{Z}/N\mathbb{Z})^*$, $1 \le s \le N$, by $\mathrm{Gal}(\mathbb{Q}(\zeta_N)/\mathbb{Q})$. The abelian varieties $\mathrm{Jac}(\mathcal{X}_{x,d})$ for $d|N$ are the Jacobians of the curves

$$y^d = u^A(u-1)^B(u-x)^C.$$

Therefore T_x can be thought of as the 'new part' of the Jacobian $\mathrm{Jac}(\mathcal{X}_x)$.

The differential form $\omega(x) = du/y$, which is of the first or of the second kind if $x \neq 0, 1, \infty$, is an eigendifferential for the action of ζ_N induced by κ. If γ is an integration cycle on $\mathcal{X}_x$ then $\int_\gamma \omega(x)$ is called a period of $\omega(x)$. For the integration cycle on $\mathcal{X}_x$, we can take a Pochhammer cycle around 0 and x, or around 1 and ∞: in fact the induced action of $(\mathbb{Z}/N\mathbb{Z})$ via κ on these cycles generates a subgroup of finite index in $H_1(\mathcal{X}_x, \mathbb{Z}[\zeta_N])$.

It is convenient at this stage to associate to A, B, C, N the two sets of parameters

$$\mu_1 = \frac{A}{N}, \quad \mu_2 = \frac{B}{N}, \quad \mu_3 = \frac{C}{N}, \quad \mu_4 = 2 - \left(\frac{A}{N} + \frac{B}{N} + \frac{C}{N}\right),$$

and

$$a = \frac{A}{N} + \frac{B}{N} + \frac{C}{N} - 1, \quad b = \frac{C}{N}, \quad c = \frac{A}{N} + \frac{C}{N}.$$

These parameters are related by

$$a = 1 - \mu_4, \quad b = \mu_3, \quad c = \mu_1 + \mu_3,$$

$$\mu_1 = c - b, \quad \mu_2 = a + 1 - c, \quad \mu_3 = b, \quad \mu_4 = 1 - a,$$

and also

$$\sum_{i=1}^{4} \mu_i = 2.$$

We have

$$\omega(x) = \frac{du}{y} = u^{-\mu_1}(u-1)^{-\mu_2}(u-x)^{-\mu_3} du.$$

In this article, we restrict our attention to the case where $\omega(x)$ is a differential form of the first kind. We therefore suppose that

$$\mu_i < 1, \qquad i = 1, \ldots, 4 \tag{H1}$$

or equivalently

$$0 < a < c, \quad b < 1, \quad c - b < 1. \tag{H1a}$$

We also require that

$$\omega(0) := u^{-(\mu_1+\mu_3)}(u-1)^{-\mu_2} du$$

be of the first kind, so that

$$\mu_1 + \mu_3 < 1, \tag{H2}$$

that is

$$c < 1. \tag{H2b}$$

Under the above assumptions, the classical Gauss hypergeometric function associated to the parameters $\mu = (\mu_1, \ldots, \mu_4)$ can be written as

$$\begin{aligned} F(x) = F_\mu(x) &= \frac{\int_1^\infty \omega(x)}{\int_1^\infty u^{-\mu_1-\mu_3}(u-1)^{-\mu_2}du} \\ &= \frac{1}{B(1-\mu_4, 1-\mu_2)} \\ &\quad \times \int_1^\infty u^{-\mu_1}(u-1)^{-\mu_2}(u-x)^{-\mu_3}du. \end{aligned} \tag{1.1}$$

The function $\int_1^\infty \omega(x)$ for $x \in Q$ is, up to multiplication by a non-zero algebraic number, the same as $\int_\gamma \omega(x)$ where γ is a Pochhammer cycle around 1 and ∞ (see Yoshida 1987, p. 11). Recall that one can characterise the Gauss differential equation by its solution space which is the $\mathbb{C}$-vector space of multivalent functions of one variable having two linearly independent branches and prescribed ramification

$$\begin{aligned} \nu_0^{-1} &= |1-c|^{-1} = |1-\mu_1-\mu_3|^{-1} && \text{at} \quad 0 \\ \nu_1^{-1} &= |c-a-b|^{-1} = |1-\mu_2-\mu_3|^{-1} && \text{at} \quad 1 \\ \nu_\infty^{-1} &= |a-b|^{-1} = |1-\mu_4-\mu_3|^{-1} && \text{at} \quad \infty. \end{aligned} \tag{1.2}$$

The function $F(x)$ above is the element of this space which is holomorphic at $x = 0$ and which has $F(0) = 1$. One has the following series development around the origin, usually expressed in terms of a, b, and c,

$$F(x) = F(a, b, c; x) = \sum_{n=0}^{\infty} \frac{(a,n)(b,n)}{(c,n)} \cdot \frac{x^n}{n!}, \qquad |x| < 1,$$

where for $w \in \mathbb{C}$ we have $(w, n) = \prod_{j=1}^{n}(w + n - j)$.

The projectivised monodromy group of the Gauss hypergeometric differential equation is realisable as a subgroup of $\mathrm{PSL}(2, \mathbb{R})$ when the ramifications in (1.2) satisfy the hyperbolicity condition

$$\nu_0 + \nu_1 + \nu_\infty = |1-c| + |c-a-b| + |a-b| < 1. \tag{HYP}$$

Suppose that

$$0 < \mu_i, \qquad i = 1, \ldots, 4. \tag{H3}$$

In terms of a, b, c this gives

$$0 < b < c, \quad a < 1, \quad c - a < 1. \tag{H3a}$$

This is just (H1a) with a replaced by b. Notice that (H1a) and (H3a) together ensure the hyperbolicity condition (HYP). The equivalent conditions (H1) and (H3)

$$0 < \mu_i < 1, \quad i = 1, \ldots 4, \quad \sum_{i=1}^{4} \mu_i = 2$$

are the so-called ball 4-tuple conditions of Deligne–Mostow (Deligne & Mostow 1986). We collect all the above assumptions as

$$0 < \mu_i < 1, \quad i = 1, \ldots 4, \quad \sum_{i=1}^{4} \mu_i = 2, \quad \mu_1 + \mu_3 < 1 \tag{A}$$

or equivalently

$$c < 1, \quad 0 < a < c, \quad 0 < b < c. \tag{A1}$$

For the action of ζ_N, the dimension of the eigenspace of $H^0(T_x, \Omega)$ with eigenvalue ζ_N^s, $s \in (\mathbb{Z}/N\mathbb{Z})^*$ is

$$r_s = \{s\mu_1\} + \{s\mu_2\} + \{s\mu_3\} + \{s\mu_4\} - 1,$$

where $\{\rho\}$ denotes the fractional part of a real number ρ. For a proof of this formula see Chevalley & Weil (1934) or Deligne & Mostow (1986), where it is also shown that r_s equals 0,1, or 2 and $r_s + r_{-s} = 2$. We let S' be the set of $s \in (\mathbb{Z}/N\mathbb{Z})^*$, $1 \le s \le N$, satisfying

$$\{s\mu_1\} + \{s\mu_2\} + \{s\mu_3\} + \{s\mu_4\} = 2$$

If $s \in S'$, then $-s \in S'$. We let S be a set of representatives of $S'/\{\pm 1\}$ and M be the cardinality of S. There is a CM-type $\{\xi_j, j = 1, \ldots, \frac{1}{2}\varphi(N)\}$ for the field $\mathbb{Q}(\zeta_N)$ such that T_x has the generalised CM-type for $\mathbb{Q}(\zeta_N)$ of the form

$$\Phi = \sum_{j=1}^{M} (\xi_j + \xi_{-j}) + 2 \sum_{j=M+1}^{\frac{1}{2}\varphi(N)} \xi_j,$$

encoding the action of $\mathbb{Z}[\zeta_N]$ on the complex vector space of holomorphic 1-forms of T_x. This action determines $2M$ one-dimensional eigenspaces on which $\xi_j \mathbb{Q}(\zeta_N)$ and $\xi_{-j}\mathbb{Q}(\zeta_N)$ act by scalar multiplication for $j = 1, \ldots, M$ and $\frac{1}{2}\varphi(N) - M$ two-dimensional eigenspaces on which $\xi_{M+1}\mathbb{Q}(\zeta_N), \ldots, \xi_{\frac{1}{2}\varphi(N)}\mathbb{Q}(\zeta_N)$ act by scalar multiplication. By work of Albert (1934), Siegel (1963) and Shimura (1963, 1979) the family of principally polarised abelian varieties of dimension $\varphi(N)$, with lattices isomorphic to $\mathbb{Z}[\zeta_N]^2$ and with an action of $\mathbb{Q}(\zeta_N)$ via the generalised CM-type Φ are parametrised by H^M, where H is the upper half plane. The $T_x, x \in Q$ form a 1-dimensional subfamily of this family. Two members of the family of abelian varieties are isomorphic exactly when their corresponding parameters in H^M lie in the same orbit of a certain arithmetic group Γ acting discontinuously on H^M. We have a corresponding morphism of quasi-projective varieties defined over $\overline{\mathbb{Q}}$

$$\phi : Q \to V,$$

where the Shimura variety $V(\mathbb{C})$ is isomorphic to the orbit space H^M/Γ and $\phi(x) \in V(\mathbb{C})$ is the point corresponding to the isomorphism class of T_x, $x \in Q$. For details see Cohen & Wolfart (1990).

Recall that a complex multiplication (CM) point has its class consisting of abelian varieties with complex multiplication in the sense of Shimura and Taniyama. A better terminology is 'special point', since an abelian variety A may have complex multiplications (that is a non-trivial endomorphism ring) without having complex multiplication in the sense of Shimura & Taniyama. An abelian variety A has complex multiplication in the sense of Shimura & Taniyama when it is isogeneous to a product of powers of mutually non-isogeneous simple abelian varieties A_i:

$$A \simeq A_1^{n_1} \times \cdots \times A_m^{n_m}$$

with the endomorphism algebra of each A_i a field K_i with $[K_i : \mathbb{Q}] = 2\dim(A_i)$. By abuse of terminology, we sometimes say an element in a given universal cover of a Shimura variety above a CM point, is also a CM point.

The following problem has its origins in Siegel's work on G-functions Siegel (1929).

Problem 1 *Let* $\mu = \{\mu_j\}_{j=1}^4$ *be a* 4*-tuple of rational numbers satisfying* (A) *and*

$$E = \{x \in \bar{\mathbb{Q}} \mid F_\mu(x) \in \bar{\mathbb{Q}}\}.$$

Let W *be the Zariski closure of the set* $\phi(E \cap Q)$ *in* $V(\mathbb{C})$. *Prove that* W *is a finite union of subvarieties of* $V(\mathbb{C})$ *of Hodge type.*

The cases where the rational 4-tuple μ does not satisfy (A) are largely covered by the discussions in Wolfart (1988), so we make here only a few comments. By imposing in (A) the condition (HYP), we exclude the elliptic and spherical monodromy groups. In the latter case, the function $F_\mu(x)$ is algebraic and so takes algebraic values at all algebraic points. The case $c \in \mathbb{Z}$, also excluded by (A) is treated in §3 of Wolfart (1988). In particular, when $a, b, c-a, c-b \notin \mathbb{Z}$, then (1.1) is, up to multiplication by a non-zero algebraic number, a quotient by π of a period of the second kind, and so for $x \in \bar{\mathbb{Q}}$ is either zero or transcendental (Wolfart & Wüstholz 1985, Satz 2). An example is

$$(\nu_0, \nu_1, \nu_\infty) = (0,0,0), \quad (a,b,c) = \left(\frac{1}{2}, \frac{1}{2}, 1\right), \quad \mu = \left(\frac{1}{2}, \frac{1}{2}, \frac{1}{2}, \frac{1}{2}\right).$$

The associated monodromy group is conjugate to $\Gamma[2]$, the principal congruence subgroup of level 2 in the arithmetic group $\mathrm{PSL}(2, \mathbb{Z})$. On the other hand, for the latter group, we have

$$\begin{aligned}(\nu_0, \nu_1, \nu_\infty) &= \left(\frac{1}{2}, 0, \frac{1}{3}\right), \quad (a, b, c) = \left(\frac{1}{12}, \frac{5}{12}, \frac{1}{2}\right), \\ \mu &= \left(\frac{1}{12}, \frac{7}{12}, \frac{5}{12}, \frac{11}{12}\right),\end{aligned}$$

so that $c < 1$. Moreover E is infinite and has been studied extensively in Beukers & Wolfart (1988). When $c > 1$, one remarks that (a, b, c) can be replaced by $(b + 1 - c, a + 1 - c, 2 - c)$ and affects a permutation

$$\begin{aligned}\mu_1' &= 1 - a = \mu_4, \\ \mu_2' &= b = \mu_3, \\ \mu_3' &= a + 1 - c = \mu_2, \\ \mu_4' &= c - b = \mu_1.\end{aligned}$$

The resulting monodromy group for the 4-tuple μ' is isomorphic to that for μ and the ramifications ν_0, ν_1, ν_∞ are unchanged. Therefore, from the point of view of the monodromy group, the assumption $c < 1$ is not too restrictive. It is nonetheless crucial when we compare later on $\int_1^\infty \omega(0)$ to $\int_1^\infty \omega(x)$ as we will use the fact that both $\omega(0)$ and $\omega(x)$ are of the first kind. If $c > 1$ then by (HYP) $\omega(x)$ is of the first kind whereas $\omega(0)$ is of the second kind and by the results of Wüstholz (1986), for $x \in \bar{\mathbb{Q}}$ the quotient (1.1) is either zero or transcendental.

In §3 of this article, we show that Problem 1 can be solved using a weakened form of the André–Oort Conjecture given as Problem 4 below. That Problem 4 solves Problem 1 involves showing that $\phi(E \cap Q)$ is contained in the CM points of $V(\mathbb{C})$. Moreover, one can fix a CM point P on the Zariski closure of $\phi(Q)$ in $V(\mathbb{C})$ such that every $R \in \phi(Q)$ whose corresponding class of abelian varieties is isogenous to those in the class of P corresponds to a point of E, and reciprocally. The following can therefore be solved by first treating Problem 1.

Problem 2 *The set E is of finite cardinality if and only if the Zariski closure Z of $\phi(Q)$ is not of Hodge type.*

We note that this reflects a fairly typical situation in transcendence theory. If a transcendental function like $F_\mu(x)$ takes an algebraic value at an algebraic point, there must be some special arithmetic reason and, if it happens often, then that must be reflected in some arithmetic geometric way.

These problems can be solved using the following special case, as yet unsolved in general, of the André–Oort Conjecture (for some particular cases, see André (1998), Edixhoven (1998, 2001)).

Problem 3 *Let Z be an irreducible algebraic curve in $V(\mathbb{C})$ such that there are infinitely many complex multiplication points on Z. Then Z is a subcurve of $V(\mathbb{C})$ of Hodge type.*

For Problem 1, we shall see in §3 that it is sufficient to solve the following weaker form of Problem 3.

Problem 4 *Let Z be an irreducible algebraic curve in $V(\mathbb{C})$ containing a complex multiplication point P. If there are infinitely many $Q \in Z$ whose corresponding class of abelian varieties is isogenous to those in the class of P, then Z is of Hodge type.*

A solution to Problem 4 (with V replaced by an arbitrary Shimura variety) has been proved in Edixhoven & Yafaev (2001).

2 Fuchsian triangle groups

The papers Wolfart (1988) and Cohen & Wolfart (1990) mainly treat the case where the Gauss hypergeometric differential equation gives rise to a monodromy representation of $\pi_1(Q)$ with image a Fuchsian triangle group. This corresponds to the situation where, in addition to the assumption (A) of §1, one supposes that the ramifications at 0, 1, ∞ in (1.2) are positive integers p, q, r with $\frac{1}{p}+\frac{1}{q}+\frac{1}{r}<1$. The triangle group $\Delta=\Delta(p,q,r)$ of signature (p,q,r) is defined up to conjugation in $\mathrm{PSL}(2,\mathbb{R})$ by the presentation

$$\langle M_1, M_2, M_3 \mid M_1^p = M_2^q = M_3^r = M_1 M_2 M_3 = 1\rangle,$$

where for infinite p,q or r, say $p=\infty$, the corresponding relation, say $M_1^p = 1$, corresponds to a transformation of infinite order and can be omitted. The group Δ acts discontinuously on the upper half plane $\mathcal{H}$ and has as fundamental domain two copies of a hyperbolic triangle with vertex angles $\pi/p, \pi/q, \pi/r$. If t of the three integers p, q, r are infinite, then the quotient space $\mathcal{H}/\Delta$ is isomorphic to $\mathbb{P}_1(\mathbb{C}) \setminus \{t \text{ points}\}$.

Let Δ be a Fuchsian triangle group, and denote also by Δ a lift to $\mathrm{SL}_2(\mathbb{R})$. Let k be the trace field of Δ, that is the field generated over $\mathbb{Q}$ by the set

$$\{\mathrm{tr}(\gamma) \mid \gamma \in \Delta\}.$$

If $\Delta = \Delta(p,q,r)$, then k is the totally real field $\mathbb{Q}(\cos(\pi/p), \cos(\pi/q),$

$\cos(\pi/r))$. In Takeuchi (1977), it is shown that the algebra $\mathcal{B} = k[\Delta]$ in $M_2(\mathbb{R} \cap \bar{\mathbb{Q}})$ (not the formal group algebra, but the algebra obtained by using the relations in the presentation of the group Δ) is a quaternion algebra over k. Moreover, the ring $\mathcal{O} = \mathcal{O}_k[\Delta]$ is an order in $\mathcal{B}$ with norm unit group

$$\Gamma = \Gamma(\mathcal{B}; \mathcal{O}) = \{\varepsilon \in \mathcal{O} \mid \varepsilon\mathcal{O} = \mathcal{O}, \det(\varepsilon) = 1\}$$

containing Δ. By definition, the group $\Delta \subset \mathrm{SL}_2(\mathbb{R})$ is arithmetic if and only if it is commensurable with the norm unit group of a quaternion algebra over a totally real number field (for a comparison with other definitions of arithmeticity see Mochizuki (1998)). Takeuchi (1977) computes explicitly the list of signatures (p, q, r) giving rise to arithmetic Fuchsian triangle groups. Up to permutation of p, q, r, there are 85 such signatures. Therefore, there are infinitely many signatures giving rise to non-arithmetic Fuchsian triangle groups.

An example of an arithmetic signature is of course $(2, 3, \infty)$ as, up to conjugacy, the corresponding group is $\mathrm{PSL}(2, \mathbb{Z})$ with generating transformations

$$z \mapsto z + 1, \qquad z \mapsto -1/z, \qquad z \in \mathcal{H}.$$

The signature $(2, 5, \infty)$ is non-arithmetic and, up to conjugacy, the corresponding group is generated by the transformations

$$z \mapsto z + \frac{1}{2}(3 + \sqrt{5}), \qquad z \mapsto -\frac{1}{2}\frac{(3 + \sqrt{5})}{z}, \qquad z \in \mathcal{H}.$$

As explained in Shimura (1979), the group Γ above acts in a natural way on $\mathcal{H}^m$ where the integer m is the number of infinite places at which $\mathcal{B}$ is unramified. Indeed, there exists an $\mathbb{R}$-linear isomorphism

$$\iota : \mathcal{B} \otimes_{\mathbb{Q}} \mathbb{R} \to M_2(\mathbb{R})^m \times \mathbb{H}^{n-m}.$$

Here $\mathbb{H}$ is the algebra of Hamiltonians. Let π_ν, $\nu = 1, \ldots, n$ be the composition of the map ι with projection onto the νth factor of its image. We can suppose that π_1 is the identity map on $\mathcal{B}$. Then the maps π_ν, $\nu = 1, \ldots, n$ on $\mathcal{B}$ extend the Galois embeddings of k into $\mathbb{R}$. For $\gamma \in \mathcal{B}$ we write $\gamma_j = \pi_j(\gamma)$. There is an induced action of $\gamma \in \Gamma$ on $\mathcal{H}^m$ given by the action of $(\gamma_1, \ldots, \gamma_m)$. This action is discontinuous on $\mathcal{H}^m$ and so we may form the quotient $V = \mathcal{H}^m/\Gamma$, which is isomorphic to the complex points of a Shimura variety V having the structure of a quasi-projective variety defined over $\bar{\mathbb{Q}}$. There is a natural embedding of Δ into Γ. The induced action on $\mathcal{H}^m$ is given essentially by the restriction of scalars map, where we exclude $n - m$ Galois embeddings. In other words, the coefficients of the matrices of Δ are defined over a real, and at most quadratic, extension L of k. There are m real Galois embeddings $\xi_1, \ldots, \xi_m$ of L into $\mathbb{R}$, extending m Galois embeddings of k into $\mathbb{R}$ such that any matrix M in Γ, and hence any matrix in Δ, acts on $\mathcal{H}^m$ by $(M^{\xi_1}, \ldots, M^{\xi_m})$, where M^{ξ_i} denotes the matrix obtained by the action of ξ_i on the matrix coefficients of M. We have $m = 1$ precisely when Δ is arithmetic: it will then be of finite index in Γ.

In Cohen & Wolfart (1990), we constructed explicitly a complex analytic embedding,

$$F : \mathcal{H} \hookrightarrow \mathcal{H}^m$$

compatible with the embedding of Δ acting on $\mathcal{H}$ into Γ acting on $\mathcal{H}^m$. Furthermore, the induced quotient map

$$\mathcal{H}/\Delta \to \mathcal{H}^m/\Gamma$$

can be taken as providing a morphism

$$\phi : \mathcal{C} \to V$$

defined over $\bar{\mathbb{Q}}$ between the corresponding underlying quasi-projective or projective varieties over $\bar{\mathbb{Q}}$, with $\mathcal{C}(\mathbb{C})$ isomorphic to $\mathcal{H}/\Delta$. It is useful to have such an explicit construction, especially for understanding the analytic properties of ϕ and its behaviour at the points which are images of the fixed points of the action of Δ on $\mathcal{H}$. Nonetheless, outside the images of such fixed points, the existence of ϕ comes naturally from a sort of 'period map', associating to a family of curves a family of abelian varieties coming from the 'new part' of their Jacobian. This was made clear in §1.

A remark is in order here: returning to the family of abelian varieties T_x, $x \in \mathbb{P}_1(\mathbb{C})\backslash\{0, 1, \infty\}$ of §1 and the associated Shimura variety H^M/Γ_1, it may happen that $F_\mu(x)$ has monodromy group a Fuchsian triangle group Δ with associated Shimura variety as above H^m/Γ_2 but that m is strictly smaller than (in fact, in practice divides) M. This corresponds to the fact that there may be two distinct integers s_1, s_2 in the set S of §1 with $(\{s_1\mu_j\})_{j=1}^4$ and $(\{s_2\mu_j\})_{j=1}^4$ being equivalent up to permutation. In this case, the abelian varieties T_x will split up to isogeny into powers of elements of the family parametrised by H^m/Γ_1. For example, in Cohen & Wolfart (1990), Example 1, we consider the case of the Hecke triangle group of signature $(2, 5, \infty)$. (There is an error in the dimension counting of the abelian varieties for this example in Cohen & Wolfart (1990), which we correct here.) The least common denominator of the μ_is is 20, so that we are lead to considering the cyclotomic field $\mathbb{Q}(\zeta_{20})$ which has associated non-reduced set

$$S = \{1, -3, 7, -9\}$$

giving rise to four 4-tuples $(\{s\mu_j\})_{j=1}^4$, $s \in S$. However, only two of these

4-tuples are inequivalent under permutation, so that in fact $m = 2$ and we are reduced to considering $\mathcal{H}^2$ with the action of Γ, the Hilbert modular group for the field $\mathbb{Q}(\sqrt{5})$ with $V = V_5$ being the corresponding Hilbert modular surface. The abelian variety T_x has dimension $\varphi(20) = 8$ and endomorphism algebra containing $\mathbb{Q}(\zeta_{20})$. Moreover, for each $x \in \mathbb{P}_1(\mathbb{C}) \setminus \{0, 1, \infty\}$, it factors into 2 abelian varieties each of dimension 4 and with endomorphism algebra containing $\mathbb{Q}(\zeta_5)$. Each factor is an element of the family $\mathcal{T}$. These elements in turn factor into 2 abelian surfaces with real multiplication by $\mathbb{Q}(\sqrt{5})$. Up to isogeny T_x decomposes as the 4th power of any of these abelian surfaces.

We end this section by noting the following.

Proposition 1 *The variety $Z = \phi(\mathcal{C}) \subset V$ is an irreducible algebraic variety defined over $\bar{\mathbb{Q}}$ and is of Hodge type if and only if Δ is arithmetic.*

Proof The group $\iota(\Delta)$ is contained in the group

$$\Delta' = \{\gamma \in \Gamma \mid \gamma(F(\mathcal{H})) = F(\mathcal{H})\}$$

with finite index. (One can replace Δ by the maximal triangle group containing it.) The variety Z is of Hodge type if and only if Δ' is arithmetic, therefore if and only if Δ is arithmetic. □

The solution of the following problem is therefore contained in that of Problem 2 of §1.

Problem 5 *Let Δ be a Fuchsian triangle group and F the corresponding Gauss hypergeometric function. Consider the set*

$$E = \{x \in \bar{\mathbb{Q}} \mid F(x) \in \bar{\mathbb{Q}}\}.$$

Prove that the set E has finite cardinality if and only if Δ is non-arithmetic.

Problem 5 was first formulated as a theorem in Wolfart (1988). Unfortunately, that paper contains a serious error (as noticed by Walter Gubler). Nonetheless, the ideas in it strongly influenced both Cohen & Wolfart (1990) and the present article. That Δ is arithmetic implies E is infinite follows from the fact that $\mathcal{C}(\mathbb{C}) = \mathcal{H}/\Delta$ is the set of complex points of a Shimura curve. The hard part of Problem 5 is to show that E is finite for Δ a non-arithmetic Fuchsian triangle group.

3 Solving Problem 4 solves Problem 1

To relate Problems 4 and 1 we apply the Analytic Subgroup Theorem of Wüstholz (1986, 1989). As in §1, let $\mu = (\mu_i)_{i=1}^4$ be a rational 4-tuple satisfying (A) and recall that the classical Gauss hypergeometric function $F = F_\mu(x)$ can be expressed for $x \in Q$ as the quotient of

$$\lambda(x) := \int_1^\infty u^{-\mu_1}(u-1)^{-\mu_2}(u-x)^{-\mu_3}du$$

and

$$\lambda_0 := B(1-\mu_4, 1-\mu_2) = \int_0^1 u^{-\mu_4}(1-u)^{-\mu_2}du.$$

As we saw in §1, up to multiplication by an algebraic number, we can view $\lambda(x)$ as a period of the first kind on T_x. If $x \in Q \cap \bar{\mathbb{Q}}$, then $\mathcal{X}_x$, its Jacobian and T_x are all defined over $\bar{\mathbb{Q}}$. By Koblitz & Rohrlich (1978), up to multiplication by a non-zero algebraic number (and as we've assumed $\mu_2 + \mu_4 > 1$), we can view λ_0 as a period of the first kind on a (non-uniquely defined) abelian variety A_0, defined over $\bar{\mathbb{Q}}$, and with complex multiplication, in the sense of Shimura & Taniyama, by $\mathbb{Q}(\zeta_N)$ of a certain CM type. The abelian variety A_0 need not be simple, but it is isogenous to a power of a simple abelian variety. We can assume that $F_\mu(x) \in \bar{\mathbb{Q}}^*$ for $x \in \bar{\mathbb{Q}}$, i.e. that x is not a zero of $F_\mu(x)$ (see Wolfart 1988, §10). The two 'periods' $\lambda(x)$ and λ_0 then differ only up to multiplication by a non-zero algebraic number. We have the following consequence of the Analytic Subgroup Theorem: see Wüstholz (1986), Theorem 5.

Proposition 2 *Let A and B be abelian varieties defined over $\bar{\mathbb{Q}}$ and denote by V_A the $\bar{\mathbb{Q}}$-vector subspace of $\mathbb{C}$ generated by all the periods $\int_\gamma \omega$ of A with $\gamma \in H_1(A, \mathbb{Z})$ and $\omega \in H^0(A, \Omega_{\bar{\mathbb{Q}}})$ and similarly for B. Then $V_A \cap V_B \neq \{0\}$ if and only if there exist simple abelian subvarieties A' of A and B' of B with A' isogeneous to B'.*

As we saw in §1, for $x \in Q$, the abelian varieties T_x are of dimension $\varphi(N)$ and have endomorphism algebras containing $\mathbb{Q}(\zeta_N)$. If T_x has a factor with CM by the field $\mathbb{Q}(\zeta_N)$ and isogenous to A_0 then it must have a complementary factor isogenous to a fixed abelian variety A_1 with CM by the same field. Moreover, in this situation, as $\omega(x)$ generates the eigenspace of $H^0(T_x, \Omega)$ for the action of ζ_N corresponding to $s = 1$, the periods of $\omega(x)$ generate V_{A_0} over $\bar{\mathbb{Q}}$ since $r_1 = r_{-1} = 1$. We deduce the following.

Proposition 3 *Let μ satisfy (A). There are two fixed abelian varieties A_0 and A_1 with complex multiplication by $\mathbb{Q}(\zeta_N)$ such that if $x \in Q \cap \bar{\mathbb{Q}}$ and $F_\mu(x) \in \bar{\mathbb{Q}}^*$, then T_x is isogeneous to the product $A_0 \times A_1$.*

Using Proposition , we deduce at once from the above discussion that Problem 4 implies Problem 1. We now turn to an example.

The triangle groups with signature (5, 5, 5), (7, 7, 7)

These triangle groups are related to some examples of de Jong & Noot (1991). They found counterexamples to a conjecture of Coleman that the number of complex isomorphism classes of algebraic curves of genus g whose Jacobians have complex multiplication in the sense of Shimura and Taniyama is finite once $g \geq 4$. Indeed, counterexamples are given by the families of curves with affine models,

$$Y_5(x): \qquad y^5 = u(u-1)(u-x), \qquad x \neq 0, 1, \infty, \qquad \text{genus} = 4,$$

and

$$Y_7(x): \qquad y^7 = u(u-1)(u-x), \qquad x \neq 0, 1, \infty, \qquad \text{genus} = 6.$$

In de Jong & Noot (1991) it was shown that, in both the above cases, there are an infinite number of $x \neq 0, 1, \infty$ such that the corresponding Jacobians have CM in the sense of Shimura & Taniyama. We can see this also as a direct consequence of the discussion of this article. Consider the curve $Y_5(x)$. Then with $\mu_j = \frac{2}{5}$, $j = 1, \ldots, 4$ we have $\mathrm{Jac}(Y_5(x)) = T_x$ for $x \neq 0, 1, \infty$. The corresponding triangle group Δ has signature (5, 5, 5) and is arithmetic (Takeuchi 1977). Therefore the orbits of the action of Δ on $\mathcal{H}$ correspond to the points of a Shimura curve and to the isomorphism classes of the $\mathrm{Jac}(Y_5(x))$. Therefore, an infinite number of these Jacobians have CM in the sense of Shimura & Taniyama. A similar discussion can be carried out for the family $Y_7(x)$.

4 The André–Oort Conjecture for the Siegel moduli space

Let $\mathcal{A}_g(\mathbb{C})$ denote the moduli space of principally polarised complex abelian varieties of dimension g (see for example Faltings, Wüstholz *et al.* 1984). We have the following more general form of the André–Oort conjecture.

Problem 6 *Let Z be an irreducible algebraic subvariety of $\mathcal{A}_g(\mathbb{C})$ such that the complex multiplication points on Z are dense for the Zariski topology. Show that Z is a subvariety of $\mathcal{A}_g(\mathbb{C})$ of Hodge type.*

This conjecture in the case $\dim(Z) = 1$ was first stated by André in André (1989) p. 216, Problem 3 and the general conjecture is due independently to Oort (1997). A thorough discussion of the conjecture in a more general form not appealing to abelian varieties is given in Edixhoven (1999). An earlier detailed account can be found in Moonen (1995, 1998a,b) where some special cases are proved. Foundational material on varieties of Hodge type is given in Deligne (1971, 1979) and Deligne *et al.* (1989). In the case where $\mathcal{A}_g(\mathbb{C})$ is replaced by the variety $\mathbb{P}_1(\mathbb{C}) \times \mathbb{P}_1(\mathbb{C})$, the corresponding conjecture was proved assuming the generalised Riemann hypothesis in Edixhoven (1998) and unconditionally in André (1998). It says that if C is an irreducible algebraic curve in $\mathbb{P}_1(\mathbb{C}) \times \mathbb{P}_1(\mathbb{C})$, with neither projection to $\mathbb{P}_1(\mathbb{C})$ being constant, and carrying infinitely many points of the form (j_1, j_2) with $j_1, j_2 \in \mathbb{P}_1(\mathbb{C})$ complex multiplication moduli, then C is necessarily a modular curve given by points of the form $(j(z), j((az+b)/d))$ with $a, b, d \in \mathbb{Z}$, $a, d \neq 0$ and $z \in \mathcal{H}$, where $\mathcal{H}$ is the upper half plane. Yafaev (2001a) has generalised this result to the case of products of two Shimura curves associated to quaternion algebras over $\mathbb{Q}$. Edixhoven (2001) has proved, again assuming the generalised Riemann hypothesis, the André–Oort Conjecture with $\mathcal{A}_g(\mathbb{C})$ replaced by a Hilbert modular surface.

The points of $\mathcal{A}_g(\mathbb{C})$ correspond to isomorphism classes of polarised abelian varieties, and a complex multiplication point has its class consisting of abelian varieties with complex multiplication in the sense of Shimura and Taniyama. We studied in the previous sections a special case of the conjecture arising from the theory of hypergeometric functions. In this section, we mention another application of the André–Oort Conjecture.

Recall that the elliptic modular function can be defined as the holomorphic function

$$j : \mathcal{H} \to \mathbb{C}$$

which is invariant under the action of $\mathrm{SL}(2, \mathbb{Z})$, so that

$$j((az+b)/(cz+d)) = j(z), \qquad z \in \mathcal{H}, \qquad \begin{pmatrix} a & b \\ c & d \end{pmatrix} \in \mathrm{SL}(2, \mathbb{Z}),$$

and whose Fourier expansion is of the form

$$j(z) = e^{-2\pi i z} + 744 + \sum_{n=1}^{\infty} a_n e^{2\pi i n z}, \qquad a_n \in \mathbb{C}, \quad z \in \mathcal{H}.$$

The following theorem is proved in Schneider (1937).

Theorem 1 *We have $z \in \mathcal{H} \cap \bar{\mathbb{Q}}$ and $j(z) \in \bar{\mathbb{Q}}$ if and only if z is a complex multiplication point, or equivalently, if and only if the field $\mathbb{Q}(z)$ is an imaginary quadratic extension of $\mathbb{Q}$.*

The Siegel moduli space $\mathcal{A}_g$ can be obtained as the quotient of a complex symmetric domain by an arithmetic group in the following way. We let, for $g \geq 1$,

$$\mathcal{H}_g = \{z \in M_g(\mathbb{C}) \mid z = z^t, \quad \mathrm{Im}(z) >> 0\}$$

and

$$\mathrm{Sp}_{2g} = \left\{\gamma \in \mathrm{GL}_{2g} \mid \gamma \begin{pmatrix} 0 & -1_g \\ 1_g & 0 \end{pmatrix} \gamma^t = \begin{pmatrix} 0 & -1_g \\ 1_g & 0 \end{pmatrix}\right\}.$$

Then,

$$\mathcal{A}_g(\mathbb{C}) = \mathrm{Sp}_{2g}(\mathbb{Z})\backslash\mathcal{H}_g$$

for the action of $\mathrm{Sp}_{2g}(\mathbb{Z})$ on $\mathcal{H}_g$ given by

$$z \mapsto (Az + B)(Cz + D)^{-1}, \qquad z \in \mathcal{H}_g, \quad \begin{pmatrix} A & B \\ C & D \end{pmatrix} \in \mathrm{Sp}_{2g}(\mathbb{Z}),$$

where A, B,C, D are integral $g \times g$ matrices. Of course, the case $g = 1$ reduces to the action of SL(2, $\mathbb{Z}$) on $\mathcal{H}$ and we have $\mathcal{A}_1(\mathbb{C}) = \mathbb{C}$. By general results on the existence of canonical models, one knows that $\mathcal{A}_g$ has an underlying structure of a quasi-projective variety defined over $\bar{\mathbb{Q}}$. There exists a holomorphic $\mathrm{Sp}_{2g}(\mathbb{Z})$ invariant map

$$J : \mathcal{H}_g \to \mathcal{A}_g(\mathbb{C})$$

carrying complex multiplication points to points of $\mathcal{A}_g(\bar{\mathbb{Q}})$ and we can, and will, fix from now on a choice of such a J. The following theorem due jointly to Cohen, Shiga & Wolfart can de found in Shiga & Wolfart (1995) and Cohen (1996).

Theorem 2 *We have $z \in \mathcal{H}_g \cap M_g(\bar{\mathbb{Q}})$ and $J(z) \in \mathcal{A}_g(\bar{\mathbb{Q}})$ if and only if z is a complex multiplication point.*

We deduce the following equivalent formulation of the André–Oort Conjecture of §1.

Problem 7 *Let Z be the Zariski closure of any set of moduli $J \in \mathcal{A}_g(\bar{\mathbb{Q}})$ which are values of the J-map at some $z \in \mathcal{H}_g \cap M_g(\bar{\mathbb{Q}})$. Then Z is a finite union of subvarieties of $\mathcal{A}_g(\mathbb{C})$ of Hodge type.*

In the case where Z is an irreducible algebraic curve defined over $\bar{\mathbb{Q}}$ not of Hodge type in a Siegel modular variety, Problem 7 says that $J(z) \notin Z(\bar{\mathbb{Q}})$ for $z \in \mathcal{H}_g \cap M_g(\bar{\mathbb{Q}})$ unless z is in one of finitely many special $\mathrm{Sp}_{2g}(\mathbb{Z})$-orbits of complex multiplication points.

Acknowledgements The first author thanks the Ellentuck Fund and the School of Mathematics of the Institute for Advanced Study, Princeton, for their support during the preparation of this article. The authors thank Pierre-Antoine Desrousseaux for suggesting several improvements to the original version of this manuscript.

References

Albert, A.A. (1934), A solution of the principal problem of Riemann matrices, *Ann. Math.* **35**, 500–515.

André, Y. (1989), *G-functions and Geometry*, Vieweg.

André, Y. (1998), Finitude des couples d'invariants modulaires singuliers sur une courbe algébrique plane non modulaire donnée, *Journ. Reine Angew. Math.* **505**, 203–208.

Beukers, F. & J. Wolfart (1988), Algebraic values of hypergeometric functions. In *New Advances in Transcendence Theory*, A. Baker (ed.), Cambridge University Press.

Borel, A. (1969), *Introduction aux Groupes Arithmétiques*, Hermann.

Chevalley, Cl. & A. Weil (1934), Über das Verhalten der Integrale 1. Gattung bei Automorphismen des Fonktionenkörpers, *Abh. Hamburger Math. Sem.* **10**, 358–361.

Cohen, P.B. (1996), Humbert surfaces and transcendence properties of automorphic functions, *Rocky Mount. J. of Mathematics* **26** (3), 987–1001.

Cohen, P. & J. Wolfart (1990), Modular embeddings for some non-arithmetic Fuchsian groups, *Acta Arithmetica* **LVI**, 93–110.

Deligne, P. (1971), Travaux de Shimura, *Sém. Bourbaki, 23ème année (1970/71)*, LNM 398, Springer, 123–165.

Deligne, P. (1979), Variétés de Shimura: interprétation modulaire et techniques de construction de modèles canoniques, *Proc. Symp. Pure Math.* **33**, 247–290.

Deligne, P., J.S. Milne, A. Ogus & K.-Y. Shen (1989), *Hodge Cycles, Motives and Shimura Varieties*, LNM 900, Springer.

Deligne, P. & G.D. Mostow (1986), Monodromy of hypergeometric functions, *Publ. Math. IHES* **63**, 5–90.

Edixhoven, S. (1998), Special points on the product of two modular curves, *Compos. Math.* **114**, 315–328.

Edixhoven, S. (2001), On the André–Oort Conjecture for Hilbert modular surfaces, *Progress in Mathematics* **195**, Birkhäuser, 133–155, `http://www.maths.univ-rennes1.fr/~edix`.

Edixhoven, S. and Yafaev, A. (2001), Subvarieties of Shimura varieties, to appear in *Annals of Math.*

Faltings, G., G. Wüstholz *et al.* (1984), *Rational Points*, Vieweg.

de Jong, J. & R. Noot (1991), Jacobians with complex multiplication. In *Arithmetic Algebraic Geometry*, G. van der Geer, F. Oort & J. Steendbrink (eds.), Birkhäuser, 177–192.

Koblitz, N. & D. Rohrlich (1978), Simple factors in the Jacobian of a Fermat curve, *Can. J. Math.* **XXX** (6), 1183–1205.

Mochizuki, S. (1998), Correspondences on hyperbolic curves, *J. Pure Appl. Algebra* **131**, 227–244.

Moonen, B. (1995), *Special points and linearity properties of Shimura varieties*, PhD thesis, Utrecht.

Moonen, B. (1998a), Linearity properties of Shimura varieties, I, *J. Algebr. Geom.* **7**, 539–567.

Moonen, B. (1998b), Linearity properties of Shimura varieties, II, *Compos. Math.* **114**, 3–35.

Oort, F. (1997), Canonical liftings and dense sets of CM-points. In *Arithmetic Geometry, Cortona 1994*, F. Catanese (ed.), Ist. Naz. Mat. F. Severi, Cambridge University Press, 228–234.

Schmidt, Th., Klein's cubic surface and a 'non-arithmetic' curve, *Math. Ann.* **309** (4), 533–539.

Schneider, Th. (1937), Arithmetische Untersuchungen elliptischer Integrale, *Math. Ann.* **113**, 1–13.

Shiga, H. & J. Wolfart (1995a), Criteria for complex multiplication and transcendence properties of automorphic functions, *J. Reine Angew. Math.* **463**, 1–25.

Shimura, G. (1963), On analytic families of polarized abelian varieties and automorphic functions, *Ann. Math.* **78**, 149–192.

Shimura, G. (1966), Moduli of abelian varieties and number theory, *Proc. Sympos. Pure Math.* **9**, 312–332.

Shimura, G. (1979), Automorphic forms and the periods of abelian varieties, *J. Math. Soc. Japan* **31**, 561–592.

Siegel, C.L. (1929), Über einige Anwendungen Diophantischer Approximationen, *Abh. Preuss Akad. Wiss. Phys.-Math. Kl.*, Nr. 1; *Ges. Abh.*, Band 1, 209–266, Springer, 1966.

Siegel, C.L. (1963), *Lectures on Riemann Matrices*, Tata Institute, Bombay.

Takeuchi, K. (1977), Arithmetic triangle groups, *J. Math. Soc. Japan* **29**, 91–106.

Wolfart, J. (1988), Werte hypergeometrische Funktionen, *Invent. Math.* **92**, 187–216.

Wolfart, J. & G. Wüstholz (1985), Der Überlagerungsradius gewisser algebraischer Kurven une die Werte der Betafunktion an rationalen Stellen, *Math. Ann.* **273**, 1–15.

Wüstholz, G. (1986), Algebraic groups, Hodge theory, and transcendence, in *Proc. ICM, Berkeley*.

Wüstholz, G. (1989), Algebraische Punkte auf analytischen Untergruppen algebraischer Gruppen, *Annals of Math.* **129**, 501–517.

Yafaev, A. (2001a), Special points on products of two Shimura curves, *Manuscripta Mathematica*, to appear.

Yafaev, A. (2001b), *Sous-variétés des variétés de Shimura*, Thèse doctorale, Université de Rennes 1, `http://www.maths.univ-rennes1.fr/~edix`.

Yoshida, M. (1987), *Fuchsian Differential Equations*, Vieweg.

7

Regular Dessins, Endomorphisms of Jacobians, and Transcendence

Jürgen Wolfart

Which algebraic curves have Jacobians of CM type? The present article tries to answer this question using on the one hand Grothendieck's dessins d'enfants and on the other hand Wüstholz' analytic subgroup theorem which generalizes Alan Baker's fundamental work on linear forms in logarithms.

The first section shows that it is reasonable to restrict the question to algebraic curves or compact Riemann surfaces with many automorphisms, i.e. corresponding to a regular dessin. The second section applies transcendence via the canonical representation of the automorphism group on the differentials using its effect on the periods. The final section works with transcendence results about period quotients in the Siegel upper half space, mainly for low genus curves with regular dessins.

1 Triangle groups and dessins

Let Y denote a compact Riemann surface or equivalently a complex nonsingular projective algebraic curve. We are looking for properties of $\mathrm{Jac}\, Y$ and may therefore suppose that the genus of Y is $g > 0$. A simple complex polarized abelian variety A has *complex multiplication* if its endomorphism algebra

$$\mathrm{End}_0 A \;:=\; \mathbb{Q} \otimes_{\mathbb{Z}} \mathrm{End}\, A$$

is a number field $\mathbb{K}$ of degree

$$[\mathbb{K} : \mathbb{Q}] \;=\; 2 \dim A\ .$$

Then $\mathbb{K}$ is necessarily a *CM field*, i.e. a totally imaginary quadratic extension of a totally real field of degree $\dim A$. If the polarized complex abelian variety A is not simple, it is isogenous to a direct product of simple abelian varieties. Then, A is said to be of *CM type* if the simple factors have complex multiplication. Abelian varieties with complex multiplication and hence abelian varieties of CM type are $\mathbb{C}$-isomorphic to abelian varieties defined over number fields (Shimura & Taniyama 1961). In short: they *may be defined over number fields*. Since curves and their Jacobians may be defined over the same field (Milne 1986), we can restrict to curves which may be defined over number fields. The corresponding Riemann surfaces given by their complex points can be characterized in a reformulation of Belyĭ's theorem (Belyĭ 1980) as quotients $\Gamma\backslash\mathcal{H}$ of the upper half plane $\mathcal{H}$ by a subgroup Γ of finite index in some cocompact Fuchsian triangle group Δ (Cohen, Itzykson & Wolfart 1996). Therefore one has

Theorem 1 *Let Y be a compact Riemann surface of genus $g > 0$ with a Jacobian* Jac Y *of CM type. Then Y is isomorphic to $\Gamma\backslash\mathcal{H}$ for some subgroup Γ of finite index in a cocompact Fuchsian triangle group Δ.*

In this theorem, Δ and Γ are not uniquely defined. They are constructed by means of a Belyĭ function $\beta : Y \to \mathbb{P}_1(\mathbb{C})$ ramified at most above $0, 1$ and $\infty \in \mathbb{P}_1$. If we identify Y with $\Gamma\backslash\mathcal{H}$, the Belyĭ function β is the canonical projection

$$\beta : \Gamma\backslash\mathcal{H} \to \Delta\backslash\mathcal{H} \cong \mathbb{P}_1(\mathbb{C}) \tag{1}$$

ramified only over the orbits of the three fixed points of Δ of order p, q and r. The Δ-orbits of these fixed points may be identified with $0, 1$ and $\infty \in \mathbb{P}_1$, and p, q, r must be multiples of the respective ramification orders of β over $0, 1, \infty$. But this is the only condition imposed upon the signature of Δ; we may and will normalize Δ by a minimal choice of the signature, i.e. by assuming that p, q, r are the *least* common multiples of all ramification orders of β above $0, 1, \infty$ respectively. Under this normalization, the groups Δ and Γ are uniquely determined by Y and β up to conjugation in $PSL_2\mathbb{R}$ (assuming they are Fuchsian groups, see below). Such a pair (Γ, Δ) will be called *minimal*. In very special cases these **minimal pairs** are not pairs of Fuchsian, but of euclidean groups acting on $\mathbb{C}$ instead of $\mathcal{H}$. The euclidean triangle groups have signature $\langle 2, 3, 6\rangle$, $\langle 3, 3, 3\rangle$ or $\langle 2, 4, 4\rangle$, therefore the Riemann surfaces Y are elliptic curves $\mathbb{C}/\Gamma$, in these cases isogenous to elliptic curves with fixed points of order 3 or 4 induced by the action of Δ, hence with complex multiplication. They are treated in detail in Singerman & Sydall (1997); in most cases, we will take no notice of them in the present article. The restriction to minimal pairs is useful by the following reason.

Lemma 1 *Let Γ be a cocompact Fuchsian group, contained with finite index in a triangle group Δ. Suppose that the genus of $\Gamma\backslash\mathcal{H}$ is > 0 and that the pair (Γ, Δ) is minimal. There is a maximal subgroup N of Γ which is normal and* *of finite index in Δ. This group N is torsion-free, hence the universal covering group of $X := N\backslash\mathcal{H}$.*

The subgroup N can be constructed by taking the intersection of all Δ-conjugates of Γ. That the index $[\Delta : N]$ is finite, may be seen either by group-theoretic arguments or by the fact that $N\backslash\mathcal{H} \to \Delta\backslash\mathcal{H}$ is the normalization of the covering $\Gamma\backslash\mathcal{H} \to \Delta\backslash\mathcal{H}$. Since we have a normal covering, all ramification orders above the respective Δ-fixed point are the same; on the other hand, they must be common multiples of the respective ramification orders of the covering $\Gamma\backslash\mathcal{H} \to \Delta\backslash\mathcal{H}$. (It will turn out that they are even the least common multiples, but we do not need this fact.) Now, by minimality, these ramification orders are multiples of the orders occurring in the signature of Δ. If N had an elliptic element γ, its fixed point would be a Δ-fixed point of order p, say; but in this fixed point – or rather in its image on X –, the normal covering map would be ramified with order $p/\operatorname{ord}\gamma$, a proper divisor instead of a multiple of p in contradiction to our assumption.

The Riemann surfaces X found in this Lemma are of special interest: a Riemann surface of genus $g > 1$ is said to have *many automorphisms* (Rauch 1970, Popp 1972) if it corresponds to a point x of the moduli space M_g of all compact Riemann surfaces of genus g with the following property: there is a neighbourhood $U = U(x)$ in the complex topology of M_g such that to any $z \in U$, $z \neq x$, corresponds a Riemann surface Z with strictly fewer automorphisms than X. Using rigidity properties of triangle groups one easily proves (Wolfart 1997) the

Lemma 2 *The compact Riemann surface X of genus $g > 1$ has many automorphisms if and only if it is isomorphic to a quotient $N\backslash\mathcal{H}$ of the upper half plane by a torsion-free normal subgroup N of a cocompact Fuchsian triangle group Δ.*

It should be noticed that in the present paper we do not only use *clean* Belyĭ functions β, i.e. such β with constant ramification order $q = 2$ above 1. Therefore, we have to use a more general definition of dessins than the usual one given in Grothendieck (1997). We define *dessins* as (oriented) *hypermaps* or *bipartite graphs* on X with white vertices (the points of $\beta^{-1}\{0\}$) and black vertices (the points of $\beta^{-1}\{1\}$); the (open) edges are the connected components of $\beta^{-1}]0,1[$. For other definitions of hypermaps better reflecting the triality between $\beta^{-1}\{0\}$, $\beta^{-1}\{1\}$ and $\beta^{-1}\{\infty\}$, see Jones & Singerman (1996). Finite index subgroups Γ of arbitrary triangle groups Δ correspond via (1) to such more general dessins, and normal subgroups N correspond to *regular dessins*, i.e. those whose automorphism group ($\cong \Delta/N \cong$ the covering group of β – a normal covering in that case) acts transitively on the edges.

We will restrict our considerations of this article mainly to these regular dessins or equivalently, their Riemann surfaces with many automorphisms. A solution of the problem in this special case often gives an answer to the question for arbitrary Riemann surfaces by the following reason.

Lemma 3 *Let Δ be a cocompact Fuchsian triangle group, Γ a subgroup of finite index and genus $g > 0$ and N the maximal subgroup of Γ which is normal in Δ. For the Riemann surfaces*

$$Y := \Gamma\backslash\mathcal{H} \qquad \textit{and} \qquad X := N\backslash\mathcal{H}$$

we have: $\operatorname{Jac} Y$ *is of CM type if* $\operatorname{Jac} X$ *is of CM type, and* $\operatorname{Jac} X$ *has CM factors if* $\operatorname{Jac} Y$ *is of CM type.*

Proof Since the covering map $X \to Y$ induces a surjective morphism of Jacobians, $\operatorname{Jac} Y$ is a homomorphic image, hence isogenous to a factor of $\operatorname{Jac} X$. □

2 Integration on regular dessins

As above, let X be a Riemann surface with many automorphisms of genus $g > 1$, given as a quotient $N\backslash\mathcal{H}$ where N denotes its universal covering group, which, by Lemma 2 is a normal subgroup of a Fuchsian triangle group $\Delta = \langle p, q, r\rangle$. We consider $G = \Delta/N$ as an automorphism group of X (in fact, it is *the* automorphism group of X if Δ is the normalizer of N in $PSL_2\mathbb{R}$, which is true at least if Δ is a maximal triangle group; the possible inclusions among triangle groups are well known from Singerman (1972) and of the regular dessin $\mathcal{D}$ defined by the Belyĭ function (1) on X (with Γ replaced by N). Next we need the *canonical representation*

$$\Phi : G \to GL(H^0(X, \Omega)) \quad ; \quad \Phi(\alpha) : \omega \mapsto \omega \circ \alpha^{-1} \tag{2}$$

for all $\alpha \in G$, $\omega \in H^0(X, \Omega)$. We can identify $H^0(X, \Omega)$ with $H^0(\operatorname{Jac} X, \Omega)$ and G with the group of automorphisms induced on $\operatorname{Jac} X$. Since X is an algebraic curve with dessin, it and its Jacobian may be defined over $\overline{\mathbb{Q}}$. The same is true for the elements of G, so we can restrict our attention to differentials defined over $\overline{\mathbb{Q}}$ and consider Φ as a representation on the $\overline{\mathbb{Q}}$-vector spaces $H^0(X, \Omega_{\overline{\mathbb{Q}}})$ or $H^0(\operatorname{Jac} X, \Omega_{\overline{\mathbb{Q}}})$.

More generally, the endomorphism algebra of an abelian variety A defined over $\overline{\mathbb{Q}}$ acts on the vector space $H^0(A, \Omega_{\overline{\mathbb{Q}}})$ of holomorphic differentials defined over $\overline{\mathbb{Q}}$ and also on the homology $H_1(A, \mathbb{Z})$. These actions imply $\overline{\mathbb{Q}}$-linear relations between the *periods (of the first kind)* $\int_\gamma \omega$, $\omega \in H^0(A, \Omega_{\overline{\mathbb{Q}}})$, $\gamma \in H_1(A, \mathbb{Z})$. By the analytic subgroup theorem (Wüstholz 1986) it is in fact known that *all* $\overline{\mathbb{Q}}$-linear relations between such periods are induced by these actions of End A. In Proposition 3 of Shiga & Wolfart (1995) the 'only if' part of the following Lemma is derived from Wüstholz (1986), but the 'if' part is easily proved taking eigendifferentials of the CM fields $\mathbb{K}_j = \mathrm{End}_0 A_j$ for the simple factors A_j of A.

Lemma 4 *Let A be an abelian variety defined over $\overline{\mathbb{Q}}$.*

(a) *A has a factor with complex multiplication if and only if there exists a nonzero differential $\omega \in H^0(A, \Omega_{\overline{\mathbb{Q}}})$, all of whose periods are algebraic multiples of one another.*
(b) *A is of CM type if and only if there is a basis $\omega_1, \ldots, \omega_n$ of $H^0(A, \Omega_{\overline{\mathbb{Q}}})$ with the property that for all ω_ν, $\nu = 1, \ldots, n$, all periods*

$$\int_\gamma \omega_\nu \, , \quad \gamma \in H_1(A, \mathbb{Z}) \, ,$$

lie in a one-dimensional $\overline{\mathbb{Q}}$-vector space $V_\nu \subset \mathbb{C}$.

This Lemma provides a first application of transcendence to the question if Jac X is of CM type or has CM factors.

Theorem 2 *Let X be a compact Riemann surface of genus $g > 1$ with many automorphisms, let $G \subseteq$ Aut X be the automorphism group of a regular dessin on X, and Φ its canonical representation on the space of differentials of the first kind on X.*

(a) *If Φ has a one-dimensional invariant subspace,* Jac X *has a factor with complex multiplication.*
(b) *If Φ splits into one-dimensional subrepresentations,* Jac X *is of CM type.*
(c) *If G is abelian,* Jac X *is of CM type.*

Proof Let $\omega \in H^0(X, \Omega_{\overline{\mathbb{Q}}})$ generate a one-dimensional G-invariant $\overline{\mathbb{Q}}$-linear subspace U of Φ, and let δ be any edge of the dessin $\mathcal{D}$ on X. Since G acts transitively on the edges of $\mathcal{D}$ and since $\mathcal{D}$ cuts X into simply connected cells, every period of ω is a $\mathbb{Z}$-linear combination of integrals

$$\int_{\alpha(\delta)} \omega = \int_\delta \omega \circ \alpha \, , \quad \alpha \in G \, ,$$

hence all periods of ω lie in the $\overline{\mathbb{Q}}$-vector space generated by the single number $\int_\delta \omega \in \mathbb{C}$. Thus, Lemma 4 implies the Theorem. □

Remark Theorem 2 is only a new look on the fact, well-known from Koblitz & Rohrlich (1978), that Fermat curves of exponent n have Jacobians of CM type. On the one hand, their universal covering groups N are the commutator subgroups $[\Delta, \Delta]$ of $\Delta = \langle n, n, n \rangle$ (Wolfart 1997, Cohen, Itzykson & Wolfart 1996, Jones & Singerman 1996); for the exponent $n = 3$ we obtain an elliptic curve with complex multiplication as already noted following Theorem 1. On the other hand – as pointed out to me by Gareth Jones and David Singerman – every CM factor detected by Theorem 2 is a factor of a Jacobian of some Fermat curve which can be seen as follows. Suppose that Φ has a one-dimensional subrepresentation with invariant subspace generated by some $\omega \neq 0$. It is not the unit representation since otherwise ω would be a differential lifted by the canonical Belyĭ function

$$\beta : X = N \backslash \mathcal{H} \to \Delta \backslash \mathcal{H} \cong \mathbb{P}^1(\mathbb{C})$$

from a differential on $\mathbb{P}^1$; hence $\omega = 0$. Therefore we have a homomorphism of $G = \Delta / N$ onto a nontrivial abelian group. Now suppose $\Delta = \langle p, q, r \rangle$ and let n be the least common multiple of p, q, r. Then we can replace Δ by $\Delta_1 = \langle n, n, n \rangle$ and N by some $N_1 \lhd \Delta_1$ which is no longer torsion-free but still satisfies $X \cong N_1 \backslash \mathcal{H}$; see the remarks after Theorem 1 about the possible choices of the signature of Δ for a given β. On Δ_1, we have a homomorphism onto a nontrivial abelian group whose kernel K contains both N_1 and the commutator subgroup $[\Delta_1, \Delta_1]$. Therefore, the quotient $K \backslash \mathcal{H}$ is a common quotient of X and of the Fermat curve of exponent n, and ω is a lift of a differential on this quotient.

By more sophisticated arguments one can use other canonical representations Φ showing that their decomposition – which can be effectively given with procedures invented by Streit (1996, 2001a) – has interesting consequences for the decomposition and endomorphism structure of Jac X as well (Wolfart 2001). We mention in particular the other extreme case:

Theorem 3 *Under the hypotheses of Theorem 2 let Φ be irreducible. Then* Jac X *is isogenous to a power A^k of a simple abelian variety A, and the endomorphism algebra $D = \mathrm{End}_0 A$ acts irreducibly on $H^0(A, \Omega)$. Moreover:*

(a) *either $g = k$, i.e. A is an elliptic curve;*
(b) *or $g = 2k$, $\dim A = 2$, and $\mathrm{End}_0 A$ is an indefinite quaternion algebra $\mathbb{B}$ over $\mathbb{Q}$.*

In Wolfart (2001) a proof based on transcendence techniques is given. We sketch here another possible argument. If there were non-isogenous simple factors of Jac X we could easily construct proper invariant subspaces for Φ, hence Jac X is of type A^k for a simple factor A. In a similar way we could construct proper invariant subspaces of Φ if the complex representation of D on $H^0(A, \Omega)$ were reducible. Therefore, by standard facts about endomorphism algebras D of simple abelian varieties and their representations (see Shimura 1963 or Runge 1999), the center of D must be $\mathbb{Q}$ or a quadratic imaginary field, and

$$\dim A \;=\; \dim H^0(A, \Omega) \;=\; q\,,$$

where q^2 is the dimension of D over its centre. On the other hand, $\dim_{\mathbb{Q}} D$ ($= q^2$ or $2q^2$) divides $2\dim_{\mathbb{C}} A = 2q$, hence $q = 1$ or 2. Finally, the last statement of the theorem follows again from Albert's classification of endomorphism algebras of simple abelian varieties (Shimura 1963, Runge 1999).

3 Shimura families and the Jacobi locus

To apply another tool coming from transcendence, recall that every principally polarized complex abelian variety A of dimension n is isomorphic to an abelian variety whose underlying complex torus is

$$A_Z \;:=\; \mathbb{C}^n/(\mathbb{Z}^n \oplus Z\mathbb{Z}^n)\,, \qquad Z \in \mathcal{H}_n\,,$$

where $\mathcal{H}_n$ denotes the Siegel upper half space of symmetric complex $n \times n$-matrices with positive definite imaginary part. For principally polarized abelian varieties, Z is uniquely determined by A up to transformations under the Siegel modular group $\Gamma_n := \mathrm{Sp}\,(2n, \mathbb{Z})$. In particular, the property that Z is an *algebraic point* of $\mathcal{H}_n$ or not, i.e. whether or not the matrix Z has algebraic entries, depends only on the complex isomorphism class of A. We call Z a *period quotient* or a *normalized period matrix* of A. In this terminology, we have the following special case of the Main Theorem of Shiga & Wolfart (1995) (again a consequence of Wüstholz (1986); for a more modern version, see Cohen 1996).

Lemma 5 *The complex abelian variety A defined over $\overline{\mathbb{Q}}$ and of dimension n is of CM type if and only if its period quotient Z is an algebraic point of the Siegel upper half space $\mathcal{H}_n$.*

The way from a smooth complex projective curve or compact Riemann surface X of genus $g = n$ to these period quotients is well known: Take a basis $\omega_1, \ldots, \omega_n$ of the holomorphic differentials and a *symplectic* basis $\gamma_1, \ldots, \gamma_{2n}$ of the homology of X, i.e. with intersection matrix

$$J := \begin{pmatrix} 0 & E \\ -E & 0 \end{pmatrix}, \quad E \quad \text{the } n \times n \text{ unit matrix}$$

and form the $(n \times 2n)$-period matrix

$$(\Omega_1, \Omega_2) := \left(\int_{\gamma_j} \omega_i \right), \quad i = 1, \ldots, n \ , \ j = 1, \ldots, 2n$$

with $n \times n$-matrices Ω_1 and Ω_2. Then $Z := \Omega_2^{-1}\Omega_1$ is called a *period quotient for* X since it is a period quotient for Jac X. Note that Z is independent of the choice of the basis of the differentials and that another choice of the symplectic basis of the cycles gives another point of the orbit of Z under the action of Γ_n.

Let X be a compact Riemann surface of genus $g > 1$ with many automorphisms and let G denote its automorphism group. Since every automorphism of X induces a unique automorphism of Jac X, we can consider G as subgroup of the (polarization preserving) automorphism group G_J of Jac X. It is known (see e.g. Baily 1962, p. 375 or Lemma 1 and 2 in Popp 1972) that

Lemma 6

(a) $G_J = G$ *if* X *is hyperelliptic;*
(b) $[G_J : G] = 2$ *if* X *is not hyperelliptic.*

Let Z be a fixed period quotient for X and Jac X in the Siegel upper half space $\mathcal{H}_g$, and let L be the algebra $\mathrm{End}_0\mathrm{Jac}\, X$. Fixing a symplectic basis of the homology of X, we obtain not only a fixed Z but also a rational representation of L. For an explicit version of this representation see Section 2 of Runge (1999). Here we consider therefore L as subalgebra of the matrix algebra $M_{2g}(\mathbb{Q})$. The action of G_J on a symplectic basis of the homology of X gives

Lemma 7 *The automorphism group* G_J *is isomorphic to the stabilizer subgroup*

$$\Sigma_J := \{\gamma \in \Gamma_g \mid \gamma(Z) = Z\}$$

of Z *in the Siegel modular group* $\Gamma_g = \mathrm{Sp}_{2g}\mathbb{Z}$.

A different choice of Z induced by some other choice of the homology basis on X gives a Γ_g-conjugate stabilizer and a Γ_g-conjugate rational representation of the endomorphism algebra as well. If we fix Z, hence Σ_J, we can linearize the analytic subset

$$S_G := \{W \in \mathcal{H}_g \mid \gamma(W) = W \quad \text{for all} \quad \gamma \in \Sigma_J\} \tag{3}$$

of the Siegel upper half space by a generalized Cayley transform (see e.g. Gottschling 1961) to see that S_G is in fact a submanifold. By definition, $Z \in S_G$. Since we consider G as a subgroup of L, S_G contains a complex submanifold $\mathbb{H}(L)$ of $\mathcal{H}_g$ parametrizing a *Shimura family* of principally polarized complex abelian varieties A_W, $W \in \mathbb{H}(L)$, containing L in their endomorphism algebras – always considered as subalgebra of $M_{2g}(\mathbb{Q})$ in a fixed rational representation. For explicit equations defining $\mathbb{H}(L)$ see Sections 2 and 3 of Runge (1999), in particular Runge's Lemma 2 and the following definitions. The submanifold $\mathbb{H}(L)$ is a complex symmetric domain; we will call it the *Shimura domain* for Z or for X, suppressing in this notation the dependence on the chosen rational representation. The dimension of $\mathbb{H}(L)$ is well known and depends on the type of L only (Shimura 1963, Runge 1999). In particular we have

Lemma 8 $\dim \mathbb{H}(L) = 0$, *i. e.* $\mathbb{H}(L) = \{Z\}$ *if and only if* Jac X *is of CM type.*

The period quotients of all Jacobians of compact Riemann surfaces of genus g form another locally analytic set $\mathcal{J}_g \subseteq \mathcal{H}_g$ of complex dimension $3g-3$, the *Jacobi locus*. By definition of 'many automorphisms', X has no deformation preserving the automorphism group G, whence Jac X has no deformation as a Jacobian preserving its automorphism group, with the following exception: it is possible that X is hyperelliptic with an automorphism group $G = G_J$ and has non-hyperelliptic deformations X' whose automorphism group G' is isomorphic to an index-two subgroup of G, such that $G'_J \cong G$ (recall that the hyperelliptic involution gives the matrix $-E \in \mathrm{Sp}_{2g}\mathbb{Z}$ hence the identity in $P\mathrm{Sp}_{2g}$). Then, $S_G = S_{G'}$ – for an example see the side remark in the proof of Theorem 5. Since the hyperelliptic involution generates a cyclic group C_2 central in G, we have $G \cong G' \times C_2$ in these cases. In all other cases, S_G intersects $\mathcal{J}_g$ in isolated points. *A fortiori*, we obtain

Theorem 4 *Let X be a Riemann surface with many automorphisms. If X is hyperelliptic and $g \geq 3$, we suppose further that the hyperelliptic involution does not generate a direct factor of the automorphism group. Then its period quotient Z is an isolated point of the intersection $\mathbb{H}(L) \cap \mathcal{J}_g$ of its Shimura domain and the Jacobi locus.*

Because $\mathcal{J}_2$, $\mathcal{J}_3$ are open and dense in $\mathcal{H}_2$, $\mathcal{H}_3$ respectively, Z is in these cases an isolated point of $\mathbb{H}(L)$, hence $\dim \mathbb{H}(L) = 0$. Lemma 8 implies that Jac X is of CM type proving already the genus 2 part (for three non-isomorphic Riemann surfaces with many automorphisms) and the non-hyperelliptic genus 3 part (four non-isomorphic surfaces) of

Theorem 5 *All compact Riemann surfaces with many automorphisms of genus* ≤ 4 *have Jacobians of CM type with the exception of the hyperelliptic genus* 3 *surface given by*

$$y^2 = x^8 - 14x^4 + 1 \tag{4}$$

and the two genus 4 *surfaces given by the respective equations*

$$\frac{4}{27}\frac{(x^2-x+1)^3}{x^2(x-1)^2} + \frac{4}{27}\frac{(y^2-y+1)^3}{y^2(y-1)^2} = 1 \tag{5}$$

$$x_1^n + \ldots + x_5^n = 0 \quad \text{for } n = 1, 2, 3. \tag{6}$$

For these three exceptional surfaces, the Jacobians are isogenous to powers of elliptic curves without complex multiplication.

Sketch of the proof. In genus 3, there are four non-isomorphic hyperelliptic Riemann surfaces with many automorphisms (For detailed information see Wolfart 2001 and the references quoted there.) With the exception of (4), the hyperelliptic involution does not generate a direct factor of the automorphism group, so we may apply Theorem 4 as in the non-hyperelliptic case; another possibility is the application of Theorem 2 since all these surfaces have a regular dessin with abelian automorphism group – in general a proper subgroup of the automorphism group of the surface. In contrast, the Jacobian of the curve (4) has the elliptic curve factor

$$y^2 = x^4 - 14x^2 + 1$$

with j-invariant $16 \cdot 13^3/9$. Since it is not an integer, the elliptic curve has no complex multiplication.

Side remark The polynomial on the right hand side of (4) is constructed in such a way that the zeros form the vertices of a regular cube or the centres of the faces of a regular octahedron inside the Riemann sphere. The automorphism group of (4) is therefore $G \cong S_4 \times C_2$, with a regular dessin coming from a surjective homomorphism

$$\Delta = \langle 2, 4, 6 \rangle \to G$$

with torsion-free kernel N, the universal covering group of (4). Apparently, (4) is an example for the hyperelliptic exception mentioned in Theorem 4.

With the index 2 factor $G' = S_4$ of G there is a complex 1-dimensional family of curves X_τ of genus 3 with automorphism group $\mathrm{Aut}\, X_\tau \supseteq G'$ and $\mathrm{Aut\,Jac}\, X_\tau \supseteq G'_J \cong G$. The existence of this family can be deduced from the fact that the Fuchsian (genus 0) quadrangle groups of signature $\langle 2, 2, 2, 3\rangle$ form a complex 1-dimensional family and N is contained in such a quadrangle group. By a theorem of Singerman (1972) every member of this quadrangle group family contains a normal torsion-free genus 3 subgroup with quotient $\cong S_4$, so we have a 1-dimensional Shimura family of Jacobians parametrized by a complex submanifold $\mathbb{H}(L) \cong \mathcal{H}$ of the Siegel upper half space $\mathcal{H}_3$. All these Jacobians are isogenous to a third power of an elliptic curve, and $\mathbb{H}(L)$ intersects the subset of $\mathcal{H}_3$ of period quotients for hyperelliptic curves just in the (transcendental) period quotient for the curve (4). The Fermat curve of exponent 4 gives another (but algebraic) point of this family $\mathbb{H}(L)$.

The eleven non-isomorphic genus 4 curves with many automorphisms need a detailed case-by-case analysis (Wolfart 2001). Eight of them have regular dessins with abelian automorphism group, so Theorem 2 applies. By work of Schindler (1991), their period quotient can be calculated as an algebraic point of $\mathcal{H}_4$, so one can apply Lemma 5 as well. This method works for one further (hyperelliptic) curve $N\backslash\mathcal{H}$, $N \lhd \langle 3, 4, 6\rangle = \Delta$ with automorphism group $\Delta/N \cong SL_2\mathbb{F}_3$ where Theorem 2 fails because it has no regular dessin with abelian automorphism group. The universal covering group of the curve (5) is the kernel of the homomorphism $\Delta = \langle 2, 4, 6\rangle \to G \cong (S_3 \times S_3) \rtimes C_2$. In fact, Manfred Streit calculated the equation (5) using the regular dessin resulting from this homomorphism. Moreover, he found that the curve covers the elliptic curve

$$y^2 = (z - 1)(27z^3 - 27z - 4)$$

which has again non-integral j-invariant, hence no complex multiplication. Finally, there is Bring's curve given by the equations (6) in $\mathbb{P}^4$ and with automorphism group $S_5 \cong \langle 2, 4, 5\rangle /N$. There are several possible proofs (Riera & Rodrigues 1992, Serre 1992, Section 8.3.2) that the Jacobian of Bring's curve is isogenous to a fourth power E^4 of an elliptic curve with – again non-integral – invariant $j(E) = -25/2$.

All these exceptional curves (4), (5), (6) have irreducible canonical representations Φ: see Kuribayashi & Kuribayashi (1990) or Breuer (2000). Since their Jacobians have genus 1 factors, the first case of Theorem 3 applies to prove the last statement of Theorem 5.

Remarks The (finite!) list of all curves with many automorphisms and irreducible representation Φ has been established recently by Breuer (2000). Streit

(2001b) shows that the Schur indicator of Φ provides a sufficient criterion for their Jacobian being of CM type.

Even the existence of automorphism groups of maximal order does not imply that the Jacobian is of CM type: Macbeath's curve X (Macbeath 1965) is uniquely determined by having genus 7 and its automorphism group of order 504, according to Hurwitz the maximal possible order $84(g-1)$ for compact Riemann surfaces of genus $g > 1$. The automorphism group $\operatorname{Aut} X$ is the simple group

$$PSL_2\mathbb{F}_8 = G \cong \Delta/N \quad \text{for} \quad \Delta = \langle 2, 3, 7\rangle ,$$

and by work of Macbeath, his student Jennifer Whitworth and Berry & Tretkoff (1992) it is known that $X = N\backslash\mathcal{H}$ has a Jacobian isogenous to a product of elliptic curves E described by a model

$$y^2 = (x-1)(\zeta x - 1)(\zeta^2 x - 1)(\zeta^4 x - 1) \quad \text{with} \quad \zeta = e^{2\pi i/7}$$

(Theorem 3 applies because the canonical representation Φ is again irreducible – Breuer 2000). As already indicated by Berry & Tretkoff, the question of whether $\operatorname{Jac} X$ has CM type and if its period quotient matrix gives therefore an algebraic point of the Siegel upper half space, reduces to the question 'does E have complex multiplication?' A lengthy calculation gives the Weierstrass model and the invariant $j(E) = 2^8 \cdot 7 = 1792$. This invariant does not belong to the list of 13 rational invariants of elliptic curves with complex multiplication (see e.g. Cremona 1992), whence E has no complex multiplication and the Jacobian of Macbeath's curve X is not of CM type.

References

Baily, W.L. (1962), On the theory of theta functions, the moduli of abelian varieties and the moduli of curves, *Ann. of Math.* **75**, 342–381.

Belyĭ, G. (1980), On Galois extensions of a maximal cyclotomic field, *Math. USSR Izv.* **14** (2), 247–256.

Berry, K. & M. Tretkoff (1992), The period matrix of Macbeath's curve of genus seven. In *Curves, Jacobians, and Abelian Varieties*, Contemporary Mathematics **136**, R. Donagi (ed.), AMS, 31–40.

Breuer, Th. (2000), *Characters and Automorphism Groups of Compact Riemann Surfaces*, London Math. Soc. Lecture Note Series **280** Cambridge University Press.

Cohen, P.B. (1996), Humbert surfaces and transcendence properties of automorphic functions, *Rocky Mountain J. Math.* **26**, 987–1002.

Cohen, P.B., Cl. Itzykson, & J. Wolfart (1994), Fuchsian triangle groups and Grothendieck dessins. Variations on a theme of Belyĭ, *Commun. Math. Phys.* **163**, 605–627.

Cremona, J.E. (1992), *Algorithms for Modular Elliptic Curves*, Cambridge University Press.

Gottschling, E. (1961), Über die Fixpunkte der Siegelschen Modulgruppe, *Math. Ann.* **143**, 111–149.

Grothendieck, A. (1997), Esquisse d'un programme. In *Geometric Galois Actions 1*, L. Schneps & P. Lochak (eds.), London Math. Lecture Note Series **242**, Cambridge University Press, 5–48.

Jones, G. & D. Singerman (1996), Belyǐ Functions, hypermaps and Galois groups, *Bull. London Math. Soc.* **28**, 561–590.

Koblitz, N. & D. Rohrlich (1978), Simple factors in the Jacobian of a Fermat curve, *Can. J. Math.* **30**, 1183–1205.

Kuribayashi, I. & A. Kuribayashi (1990), Automorphism groups of a compact Riemann surface of genera three and four, *J. Pure and Appl. Alg.* **65**, 277–292.

Macbeath, A.M. (1965), On a curve of genus 7, *Proc. Lond. Math. Soc.* **15**, 527–542.

Milne, J.S. (1986), Jacobian varieties. In *Arithmetic Geometry*, G. Cornell & J.H. Silverman (eds.), Springer Verlag, 167–212.

Popp, H. (1972), On a conjecture of H. Rauch on theta constants and Riemann surfaces with many automorphisms, *J. Reine Angew. Math.* **253**, 66–77.

Rauch, H.E. (1970), Theta constants on a Riemann surface with many automorphisms. In *Symposia Mathematica* **III**, Academic Press, 305–322.

Riera, G. & R.E. Rodriguez (1992), The period matrix of Bring's curve, *Pacific J. Math.* **154**, 179–200.

Runge, B. (1999), On algebraic families of polarized Abelian varieties, *Abh. Math. Sem. Hamburg* **69**, 237–258.

Schindler, B. (1991), *Jacobische Varietäten hyperelliptischer Kurven und einiger spezieller Kurven vom Geschlecht 3*. PhD Thesis, Erlangen.

Serre, J.-P. (1992), *Topics in Galois Theory*. Jones & Bartlett.

Shiga, H. & J. Wolfart (1995), Criteria for complex multiplication and transcendence properties of automorphic functions, *J. Reine Angew. Math.* **463**, 1–25.

Shimura, G. (1963), On analytic families of polarized abelian varieties and automorphic functions, *Ann. of Math.* **78**, 149–192.

Shimura, G. & Y. Taniyama (1961), *Complex Multiplication of Abelian Varieties and its Applications to Number Theory*. Publ. Math. Soc. Japan **6**.

Singerman, D. (1972a), Subgroups of Fuchsian groups and finite permutation groups, *Bull. London Math. Soc.* **2**, 29–38.

Singerman, D. (1972b), Finitely maximal Fuchsian groups, *J. London Math. Soc. (2)* **6**, 29–38.

Singerman, D. & R.I. Syddall (1997), Belyĭ Uniformization of elliptic curves, *Bull. London Math. Soc.* **139**, 443–451.

Streit, M. (1996), Homology, Belyĭ functions and canonical curves, *Manuscr. Math.* **90**, 489–509.

Streit. M. (2001a), Symplectic representations of regular hypermaps, `http://www.math.uni-frankfurt.de/~steuding/wolfart.html`

Streit, M. (2001b), Period matrices and representation theory, *Abh. Math. Sem. Hamburg* **71**, 179–290.

Wolfart, J. (1997), The 'obvious' part of Belyĭ's theorem and Riemann surfaces with many automorphisms. In *Geometric Galois Actions 1*, L. Schneps & P. Lochak (eds.), London Math. Lecture Note Series **242**, Cambridge University Press, 97–112.

Wolfart, J. (2001), Triangle groups and Jacobians of CM type, `http://www.math.uni-frankfurt.de/~steuding/wolfart.shtml`

Wüstholz, G. (1986), Algebraic groups, Hodge theory, and transcendence, in *Proc. ICM Berkeley,* **1**, 476–483.

8

Maass Cusp Forms with Integer Coefficients

Peter Sarnak

By a Maass cusp form ϕ we will mean a weight zero cuspidal eigenform of the Laplacian for a congruence subgroup $\Gamma_o(N)$ of $SL(2,\mathbb{Z})$, possibly with a primitive central character χ. We assume that ϕ is a new form and that it is also a Hecke eigenform. Corresponding to ϕ is an automorphic cuspidal representation π of $GL_2(\mathbb{A}_{\mathbb{Q}})$. These ϕ's are quite mysterious, even their existence being a subtle issue (Phillips & Sarnak 1985, Luo 2001). Cusp forms in general are the building blocks of modern automorphic form theory and these Maass forms in particular are especially important in the analytic applications of the theory (Iwaniec 1995, Iwaniec & Sarnak 2000). Unlike their holomorphic weight $k \geq 1$ counterparts, very little is known about the algebraic nature of their coefficients $\lambda_\phi(n)$, $n \geq 1$. Here the coefficients are those in the L-series: $L(s,\phi) = \sum_{n=1}^{\infty} \lambda_\phi(n) n^{-s}$ or equivalently the eigenvalues of the Hecke operators $T(n)$.

Some remarkable Maass forms with integer coefficients are known. They are related to finite subgroups of $GL_2(\mathbb{C})$. The finite subgroups of $PGL_2(\mathbb{C})$ are well known (Klein 1913). Using the classification of these groups and their inverse images in $GL_2^{(m)}(\mathbb{C}) := \{g \in GL_2(\mathbb{C}) | (\det g)^m = 1\}$, $m = 1, 2$, one concludes the following:

Any finite subgroup of $GL_2(\mathbb{C})$ whose elements have integer determinant and trace is conjugate to one of the following maximal such subgroups:

$$
\begin{aligned}
(a)\quad U_2 &= \left\{ \pm\begin{bmatrix} 1 & 0 \\ 0 & 1 \end{bmatrix}, \pm\begin{bmatrix} i & 0 \\ 0 & -i \end{bmatrix}, \pm\begin{bmatrix} 0 & i \\ i & 0 \end{bmatrix}, \pm\begin{bmatrix} 0 & 1 \\ -1 & 0 \end{bmatrix}, \right\} \\
&\subset GL_2^{(1)}(\mathbb{C}) \\
V_2 &= \left\{ \pm\begin{bmatrix} 1 & 0 \\ 0 & 1 \end{bmatrix}, \pm\begin{bmatrix} 1 & 0 \\ 0 & -1 \end{bmatrix}, \pm\begin{bmatrix} 0 & 1 \\ 1 & 0 \end{bmatrix}, \pm\begin{bmatrix} 0 & 1 \\ -1 & 0 \end{bmatrix}, \right\} \\
&\subset GL_2^{(2)}(\mathbb{C})
\end{aligned}
$$

The image of U_2 and of V_2 in $PGL_2(\mathbb{C})$ coincide and is the Klein Four Group, D_2:

$$
\begin{aligned}
(b)\quad U_3 &= \left\{ \pm\begin{bmatrix} 1 & 0 \\ 0 & 1 \end{bmatrix}, \pm\begin{bmatrix} 0 & i \\ i & 0 \end{bmatrix}, \pm\begin{bmatrix} -i & i \\ 0 & i \end{bmatrix}, \right. \\
&\qquad \left. \pm\begin{bmatrix} 0 & 1 \\ -1 & 1 \end{bmatrix}, \pm\begin{bmatrix} 1 & -1 \\ 1 & 0 \end{bmatrix}, \pm\begin{bmatrix} i & 0 \\ i & -i \end{bmatrix}, \right\} \\
V_3 &= \left\{ \pm\begin{bmatrix} 1 & 0 \\ 0 & 1 \end{bmatrix}, \pm\begin{bmatrix} 0 & 1 \\ 1 & 0 \end{bmatrix}, \pm\begin{bmatrix} -1 & 1 \\ 0 & 1 \end{bmatrix}, \pm\begin{bmatrix} 0 & 1 \\ -1 & 1 \end{bmatrix}, \right. \\
&\qquad \left. \pm\begin{bmatrix} 1 & -1 \\ 1 & 0 \end{bmatrix}, \pm\begin{bmatrix} 1 & 0 \\ 1 & -1 \end{bmatrix}, \right\}.
\end{aligned}
$$

Note $U_3 \subset GL_2^{(1)}(\mathbb{C})$, $V_3 \subset GL_2^{(2)}(\mathbb{C})$ and their image in $PGL_2(\mathbb{C})$ is the dihedral group D_3.

$$
(c)\qquad U_4 = \left\{ \frac{1}{2}\begin{bmatrix} x_1 + ix_2 & x_3 + ix_4 \\ -x_3 + ix_4 & x_1 - ix_2 \end{bmatrix} \middle| x_j = \pm 1 \right\} \cup U_2;
$$

$U_4 \subset GL_2^{(1)}(\mathbb{C})$ and its image in $PGL(2, \mathbb{C})$ is the tetrahedral group A_4.

Let G_Q be the Galois group of $\overline{Q}/Q$ and let $\rho : G_Q \longrightarrow GL_2(\mathbb{C})$ be an irreducible 2-dimensional Galois representation. The L-function $L(s, \rho)$ has integer coefficients if and only if the image of ρ can be conjugated to lie in one of the above finite groups. Langlands (1980) has shown that corresponding to any ρ as above (the tetrahedral case being the critical one) is an automorphic cuspidal representation π of $GL_2(\mathbb{A})$ such that $L(s, \pi) = L(s, \rho)$. Moreover it is known, see Casselman (1977), that such a π corresponds to a Maass cusp from ϕ with Laplace eigenvalue equal to $1/4$ if $\epsilon = \det \rho : G_Q \longrightarrow GL_1(\mathbb{C})$ is even (i.e. $\epsilon(\sigma) = 1$, where $\sigma \epsilon G_Q$ is a complex conjugation). If ϵ is odd, then π corresponds to a holomorphic cusp form of weight 1 (see Serre 1977). In particular, if the image of ρ is contained in U_2, U_3 or U_4, then $\det \rho = 1$ and π corresponds to a Maass form. Put another way, if ρ is odd, then its image in $PGL_2(\mathbb{C})$ is either D_2 or D_3 and hence by Deligne & Serre (1994) any weight 1 holomorphic cusp form with integer coefficients must be dihedral.

The above provides us with a rich set of Maass cusp forms with integer coefficients. In what follows we reproduce the content of a letter to H. Kim and F. Shahidi (Sarnak 2001). It shows how their recent works (see Kim & Shahidi 2001a, 2001b; Kim 2000; Henniart 2001) on the 4- and 5-dimensional functorial lifts sym^3 and sym^4 of $GL_2(\mathbb{C}) = {}^LG(G = GL_2)$ allow one to classify Maass cusp forms with integer coefficients.

Theorem *Let ϕ be a Maass cusp form with integer coefficients. Then ϕ corresponds to an even irreducible 2-dimensional Galois representation whose image is contained in one of the groups described in (a), (b), or (c) above. In particular, the Laplace eigenvalue $\lambda_\phi(\infty)$ is $1/4$ and ϕ satisfies the Ramanujan–Selberg Conjectures.*

Proof Let π be the automorphic cuspidal representation of $GL_2(\mathbb{A})$ corresponding to ϕ and let χ be its central character. We examine the cuspidality of the automorphic functorial lifts $\mathrm{sym}^k \pi$ to $GL_{k+1}(\mathbb{A})$, $k = 2, 3, 4$. If $\mathrm{sym}^2 \pi$ is not cuspidal, then according to Gelbart & Jacquet (1978) there is a quadratic character $\eta \neq 1$ of $\mathbb{A}_{Q^\star}/Q^\star$ such that $\pi \otimes \eta \simeq \pi$. In this case η determines a quadratic extension K of Q and π a Hecke character λ of $\mathbb{A}_K^\star/K^\star$ such that $L(s, \lambda) = L(s, \pi)$, see Labesse & Langlands (1979). Since π has integer coefficients it is easy to see that λ must be of finite order. The 2-dimensional Galois representation $\rho = \mathrm{Ind}_{K/Q}\, \epsilon_\lambda$ of G_Q, where ϵ_λ is the character of G_K corresponding to λ via class field theory, satifies $L(s, \rho) = L(s, \lambda) = L(s, \pi)$. This establishes the Theorem in this case. So if $\mathrm{sym}^2 \pi$ is not cuspidal, we are done. Now $\lambda_\phi(n) \in \mathbb{R}$ so that $L(s, \pi \times \tilde{\pi}) = L(s, \mathrm{sym}^2 \pi)L(s, \chi)$ where $\tilde{\pi}$ is the contragredient of π. The Rankin–Selberg L-function $L(s, \pi \times \tilde{\pi})$ has a simple pole at $s = 1$. Hence, if $\chi \neq 1$, $L(s, \mathrm{sym}^2 \pi)$ has a pole at $s = 1$, but then by Gelbart & Jacquet (1978) $\mathrm{sym}^2 \pi$ cannot be cuspidal and we proceed as above. Thus we may assume that $\chi = 1$ and that $\mathrm{sym}^2 \pi$ is cuspidal. Next if $\mathrm{sym}^3 \pi$ is not cuspidal on $GL_4(\mathbb{A})$, then as is shown in Kim & Shahidi (2001a), π corresponds to a representation ρ of the Weil group W_Q of tetrahedral type. However since π has integer coefficients we have that $\det \rho = \pm 1$ (in fact since $\chi = 1$ $\det \rho = 1$). Hence $\rho(W_Q) \subset GL_2^{(2)}(\mathbb{C})$ and its projection in $PGL_2(\mathbb{C})$ is A_4. Thus ρ must be a finite representation of W_Q and hence a representation of G_Q. Thus π corresponds to a 2-dimensional Galois representation ρ having integer trace and determinant and since it is of tetrahedral type, its image must be U_4. To continue we can assume that both $\mathrm{sym}^2 \pi$ and $\mathrm{sym}^3 \pi$ are cuspidal. If $\mathrm{sym}^4\,\pi$ is not cuspidal then as shown in Kim & Shahidi (2001b), π corresponds to a representation of W_Q of octahedral type, that is its image in $PGL(2,\mathbb{C})$ is S_4. However, no lift of this group to $GL(2,\mathbb{C})$ can have integer determinant and trace, and hence this case cannot happen for our π. We are left with a π whose central character χ is trivial and such that $\mathrm{sym}^k\,\pi$, $k = 2, 3, 4$ are all cuspidal. We show that such a π cannot have integer coefficients. Proceeding as done in Kim & Shahidi (2001b) we form the Rankin–Selberg L-functions $L(s, \mathrm{sym}^j\,\pi \times \mathrm{sym}^k\,\pi)$, $2 \le j, k \le 4$, whose analytic properties including their nonvanishing on $\Re(s) = 1$ are known (Shahidi 1990). From the factorizations of these L-functions, see Kim & Shahidi (2001b), we deduce that $L(s, \mathrm{sym}^k\,\pi)$ is analytic and nonvanishing on $\Re(s) = 1$ for $1 \le k \le 8$. Hence by standard analytic methods we have that for any polynomial $T(x)$ of degree at most 8,

$$\lim_{N\to\infty} \frac{1}{N} \sum_{\substack{p \le N \\ p \text{ prime}}} T(\lambda_\pi(p)) \log p = \int_{-2}^{2} T(x)\,d\mu(x)\,, \tag{1}$$

where $d\mu(x) = \sqrt{1 - \frac{x^2}{4}}\,\frac{dx}{\pi}$ is the 'Sato–Tate' measure.

Set

$$T(x) = x^2(x-1)^2(x+1)^2(4-x^2)\,.$$

Note that $T(m) \le 0$ for all $m \in \mathbb{Z}$ while $T(x) \ge 0$ for $x \in [-2, 2]$. From the first we see that if $\lambda(p) \in \mathbb{Z}$ for all p, then the left-hand side of (1) is less than or equal to 0, while from the second we see that the right-hand side of (1) is positive. This contradiction shows that such a π cannot have integral coefficients. □

Remarks

(1) S.D. Miller has pointed out to me that with a little more care one can use a polynomial of degree 6 rather than 8 in the previous argument. Instead of $T(x)$ consider

$$P(x) = (4-x^2)x^2(x^2-1)\,.$$

Again $P(m) \le 0$ for $m \in \mathbb{Z}$. While $P(x)$ is no longer nonnegative on $[-2, 2]$ a calculation shows that $\int_{-2}^{2} P(x)\,d\mu(x) = 1$, which is still positive. So with P replacing T, the argument above shows that for a Maass cusp form π with integer coefficients, $L(s, \mathrm{sym}^k\,\pi)$ must have a pole at $s = 1$ for some $2 \le k \le 6$. This reduces the range in k (which by the way allows one to carry out the analysis above without appealing to the sym^4 lift) to a sharp one. Indeed, if π

corresponds to a tetrahedral Galois representation with image U_4, then it has integer coefficients and $L(s, \mathrm{sym}^k \pi)$ has no pole for $1 \leq k \leq 5$.

(2) If ϕ is a Maass cusp form as above and $\lambda_\phi(\infty) > \frac{1}{4}$, then according to the Theorem, the $\lambda_\phi(n)$'s cannot all be integers. We expect much more, namely that ϕ has some transcendental coefficients. For the special case that $L(s, \phi) = L(s, \chi)$ for χ a Grossencharacter, not of type A_0, Weil (1980), of a real quadratic field K/Q, this is indeed true (these ϕ's correspond to dihedral representations of W_Q with infinite image in $GL_2(\mathbb{C})$). For example, let $K = Q(\sqrt{2})$, and $\epsilon_0 = 1 + \sqrt{2}$ be a fundamental unit in O_K. For m a nonzero integer, define χ_m by

$$\chi_m((\alpha)) = \left|\frac{\alpha}{\alpha'}\right|^{\frac{im\pi}{\log \epsilon_0}}$$

for $0 \neq \alpha \in O_K$ (K has class number 1). Here α' is the Galois conjugate of α. The corresponding Maass form has Laplace eigenvalue

$$\lambda_\phi(\infty) = \frac{1}{4} + \left(\frac{\pi m}{\log \epsilon_0}\right)^2.$$

For p a rational prime which splits in K

$$\lambda_\phi(p) = \left|\frac{\alpha}{\alpha'}\right|^{\frac{im\pi}{\log \epsilon_0}} + \left|\frac{\alpha'}{\alpha}\right|^{\frac{im\pi}{\log \epsilon_0}}.$$

where $\alpha \in O_K$ such that $\alpha\alpha' = p$.

By the Gel'fond–Schneider theorem $\left(\frac{i\pi}{\log \epsilon_0} = \frac{\log(-1)}{\log \epsilon_0} \text{ is transcendental}\right)$, see Waldschmidt (2000), we see that $\lambda_\phi(\infty)$ is transcendental. To see that one of the $\lambda_\phi(p)$'s is transcendental, let $\alpha_1, \alpha_2, \alpha_3 > 0$ be in O_K with $N(\alpha_j) = \alpha_j\alpha_j' = p_j$ being distinct rational primes. Let $x_j = \log(\alpha_j/\alpha_j')$ for $j = 1, 2, 3$ and $y_1 = 1$, $y_2 = \frac{i\pi m}{\log \epsilon_0}$. Clearly y_1, y_2 are linearly independent over Q and one checks that x_1, x_2, x_3 are too. Hence by the six exponential theorem, see Waldschmidt (2000), one of the numbers

$$\frac{\alpha_1}{\alpha_1'}, \frac{\alpha_2}{\alpha_2'}, \frac{\alpha_3}{\alpha_3'}, \left(\frac{\alpha_1}{\alpha_1'}\right)^{\frac{im\pi}{\log \epsilon_0}}, \left(\frac{\alpha_2}{\alpha_2'}\right)^{\frac{im\pi}{\log \epsilon_0}}, \left(\frac{\alpha_3}{\alpha_3'}\right)^{\frac{im\pi}{\log \epsilon_0}}$$

is transcendental. Hence one of $\lambda_\phi(p_1)$, $\lambda_\phi(p_2)$ or $\lambda_\phi(p_3)$ is transcendental.

Acknowledgement I would like to thank S.D. Miller and F. Shahidi for their insightful comments on my letter Sarnak (2001).

Bibliography

Casselman, W. (1977) *Algebraic Number Fields*, A. Frölich (ed.), Academic Press, 663–704.

Deligne, P. and Serre, J-P., (1994) 'Formes modulaires de poids 1' *Ann. Sci. E'colé Norm. Sup.* **4** (7), 507–530.

Gelbart, S. and Jacquet, H. (1978) 'A relation between automorphic forms on $GL(2)$ and $GL(3)$', *Ann. Sci. Ecolé Norm. Sup.* **11**, 471–552.

Henniart, G. (2001) 'Progrès récents en fonctorialité de Langlands' Seminar Bourbaki.

Iwaniec, H. (1995) *Introduction to the Spectral Theory of Automorphic Forms*, Mathematica Iberoamericana.

Iwaniec, H. and Sarnak, P. (2000) 'Perspectives on the analytic theory of L-functions', *GAFA* **11**, 705–741.

Klein, F. (1913) *Lectures on the Icosahedron*, Paul, Trench, Trübner Co.

Kim, H. (2000) 'Functoriality for the exterior square of GL_4 and symmetric fourth power of GL_2', preprint.

Kim, H. and Shahidi, F. (2001a) 'Functorial products for $GL_2 \times GL_3$', *Ann. of Math.*, to appear.

Kim, H. and Shahidi, F. (2001b) 'Cuspidality of symmetric powers with applications', *Duke Math. Jour.*, to appear.

Langlands, R. (1980) *Base Change for $GL(2)$*, Ann. Math. Studies, **96**, Princeton University Press.

Labesse, J-P. and Langlands, R. (1979) 'L–indistinguishability for $SL(2)$', *Can. J. Math.* **31**, 726–785.

Luo, W. (2001) 'Non–vanishing of L–values and the strong Weyl law', *Ann. of Math.*, to appear.

Phillips, R. and Sarnak, P. (1985) 'On cusp forms for cofinite subgroups of $PSL(2, \mathbb{R})$' *Invent. Math.* **80**, 339–364.

Sarnak, P. (2001). Letter to H. Kim and F. Shahidi, May 2001.

Serre, J-P. (1977) 'Modular forms of weight one and Galois representations'. In *Algebraic Number Fields*, A. Frölich (ed.), Academic Press, 193–268.

Shahidi, F. (1990) 'Automorphic L–functions – a survey'. In *Automorphic Forms, Shimura Varieties and L-functions, I*, L. Clozel and J.S. Milne (eds.), Academic Press, 415–437.

Waldschmidt, M. (2000) *Diophantine Approximation on Linear Algebraic Groups*, Springer-Verlag.

Weil, A. (1980) 'On a certain type of characters of the idèle-class group of an algebraic number-field'. *Collected Works* **II**, Springer-Verlag, 255–261.

9

Modular Forms, Elliptic Curves and the ABC-Conjecture

Dorian Goldfeld

1 The ABC-Conjecture

The ABC-conjecture was first formulated by David Masser and Joseph Osterlé (see Osterlé 1998) in 1985. Curiously, although this conjecture could have been formulated in the previous century, its discovery was based on modern research in the theory of function fields and elliptic curves, which suggests that it is a statement about ramification in arithmetic algebraic geometry. The ABC-conjecture seems connected with many diverse and well known problems in number theory and always seems to lie on the boundary of what is known and what is unknown. We hope to elucidate the beautiful connections between elliptic curves, modular forms and the ABC-conjecture.

Conjecture (ABC) *Let A, B, C be non-zero, pairwise relatively prime, rational integers satisfying $A + B + C = 0$. Define*

$$N = \prod_{p|ABC} p$$

to be the square-free part of ABC. Then for every $\epsilon > 0$, there exists $\kappa(\epsilon) > 0$ such that

$$\max(|A|, |B|, |C|) < \kappa(\epsilon) N^{1+\epsilon}.$$

A weaker version of the ABC-conjecture (with the same notation as above) may be given as follows.

Conjecture (ABC (weak)) *For every $\epsilon > 0$, there exists $\kappa(\epsilon) > 0$ such that*

$$|ABC|^{\frac{1}{3}} < \kappa(\epsilon) N^{1+\epsilon}.$$

Oesterlé (1998) showed that if we define

$$\kappa(\epsilon) = \inf_{\substack{A+B+C=0\\(A,B)=1}} \frac{\max(|A|,|B|,|C|)}{N^{1+\epsilon}}$$

then

$$\lim_{\epsilon\to 0} \kappa(\epsilon) = \infty.$$

The best result in this direction, known to date, seems to be in the paper of Stewart & Tijdeman (1986). They prove that for any fixed positive δ there exist infinitely many solutions of

$$A+B+C=0, \quad (A,B)=1, \quad N = \prod_{p|ABC} p > 3$$

with

$$\max(|A|,|B|,|C|) > N \exp\left((4-\delta)\frac{\sqrt{\log N}}{\log\log N}\right).$$

Alan Baker (1996) proposed a more precise version of the ABC-conjecture.

Conjecture (ABC (Baker)) *For every $\epsilon > 0$ there exists a constant $\kappa(\epsilon) > 0$ such that*

$$\max\left(|A|,|B|,|C|\right) < \kappa(\epsilon)\cdot\left(\epsilon^{-\omega}N\right)^{1+\epsilon},$$

where ω denotes the number of distinct prime factors of ABC.

This conjecture would give the best lower bounds one could hope for in the theory of linear forms in logarithms. In the same paper Baker attributes to Granville the following intriguing conjecture.

Conjecture (ABC (Granville)) *Let $\Theta(N)$ denote the number of integers less than or equal to N that are composed only of prime factors of N. Then*

$$\max\left(|A|,|B|,|C|\right) \ll N\Theta(N).$$

At present the best known results in the direction of the ABC-conjecture are exponential in small powers of N and are obtained using machinery from Baker's theory of linear forms in logarithms. The first such result was obtained by Stewart & Tijdeman (1986).

Theorem 1 *Let A, B, C be positive integers satisfying $A+B = C$, $(A, B) = 1$, $C > 2$. Then there exists a constant $\kappa > 0$ (effectively computable) such that $C < e^{\kappa \cdot N^{15}}$.*

This was improved by Stewart & Yu (1991) to:

Theorem 2 *Let A, B, C be positive integers satisfying $A+B = C$, $(A, B) = 1$, $C > 2$. Then there exists a constant $\kappa > 0$ (effectively computable) such that*

$$C < e^{N^{\frac{2}{3}+\frac{\kappa}{\log\log N}}}.$$

Recently, Stewart & Yu (2001), have improved this to

$$z < \exp\left(cN^{\frac{1}{3}} \log(N)^3\right).$$

2 Applications of the ABC-Conjecture

In order to show the profound importance of the ABC-conjecture in number theory, we enumerate some remarkable consequences that would follow if the ABC-conjecture were proven.

Theorem 3 *Assume the ABC-conjecture. Fix $0 < \epsilon < 1$, and fix non–zero integers α, β, γ. Then the diophantine equation*

$$\alpha x^r + \beta y^s + \gamma z^t = 0,$$

has only finitely many solutions in integers x, y, z, r, s, t satisfying

$$xyz \neq 0,\ (x, y) = (x, z) = (y, z) = 1, \qquad r, s, t > 0 \text{ and } \frac{1}{r} + \frac{1}{s} + \frac{1}{t} < 1 - \epsilon.$$

Moreover, the number of such solutions can be effectively computed provided the constant $\kappa(\epsilon)$ in the ABC-conjecture is effective.

Proof Let

$$A = \alpha x^r,\ B = \beta y^s,\ C = \gamma z^t.$$

Without loss of generality, we may assume that $|C|$ is the maximum of $|A|$, $|B$, $|C|$. The ABC-conjecture ($|C| < \kappa(\epsilon)N^{1+\epsilon}$) then implies that

$$|\gamma z^t| < \kappa(\epsilon) \cdot |\alpha\beta\gamma xyz|^{1+\epsilon}. \tag{1}$$

Since $|A|$, $|B| \leq |C|$ it immediately follows that

$$|x| \leq \left|\frac{\gamma}{\alpha}\right|^{\frac{1}{r}} \cdot |z|^{\frac{t}{r}}, \qquad |y| \leq \left|\frac{\gamma}{\beta}\right|^{\frac{1}{s}} \cdot |z|^{\frac{t}{s}}$$

Plugging these bounds into (1) and taking the tth root of both sides, we obtain

$$|z| \ll \kappa(\epsilon) \left| z^{\frac{1}{r}+\frac{1}{s}+\frac{1}{t}} \right|^{1+\epsilon} \ll \kappa(\epsilon)|z|^{1-\epsilon^2}, \qquad (2)$$

where the constants implied by $\ll$ can be effectively computed and depend at most on α, β, γ. The inequality (2) plainly implies that there can be at most finitely many integers z satisfying (2).

Without loss of generality, we may now assume that $|A| \le |B|$. It follows that

$$|x| \le \left| \frac{\beta}{\alpha} \right|^{\frac{1}{r}} \cdot |y|^{\frac{s}{r}}. \qquad (3)$$

Writing the ABC-conjecture in the form

$$|\beta y^s| < \kappa(\epsilon) \cdot |\alpha\beta\gamma xyz|^{1+\epsilon}, \qquad (4)$$

and using the previously proved fact that $|z|$ lies in a finite set, it follows from (3) and (4) that

$$|y| \ll |y|^{(\frac{1}{r}+\frac{1}{s})\cdot(1+\epsilon)} \ll |y|^{1-\epsilon^2}.$$

Thus, y also lies in a finite set. Writing the ABC-conjecture in the form

$$|\beta x^r| < \kappa(\epsilon) \cdot |\alpha\beta\gamma xyz|^{1+\epsilon},$$

and noting that of necessity $r \ge 2$, it immediately follows that

$$|x| \ll x^{\frac{1+\epsilon}{r}},$$

so that x also must lie in a finite set. Finally, we again use the ABC-conjecture to write

$$\max |\alpha x^r|, |\beta y^s|, |\gamma z^t| \ll 1,$$

since x, y, z lie in a finite set. Thus, r, s, t also must lie in a finite set. □

Silverman (1988) proved the following theorem.

Theorem 4 *Assume the ABC-conjecture. Then there exist infinitely many primes p such that*

$$a^{p-1} \not\equiv 1 (\mathrm{mod}\, 3p^2).$$

In 1991 Elkies (see Elkies 1991a,b) proved that the ABC-conjecture implies the Mordell conjecture (first proved by Faltings 1983) which states that every algebraic curve of genus ≥ 2 defined over $\mathbf{Q}$ has only finitely many rational points.

Another interesting application is due to Granville (1998). He proved the following.

Theorem 5 *Let $f(x)$ be a polynomial with integer coefficients which is not divisible by the square of another polynomial. Then there exists a constant $c_f > 0$ such that*

$$\sum_{\substack{n \le x \\ f(n) \text{ is squarefree}}} 1 \sim c_f x \qquad (x \to \infty).$$

The most recent application of ABC is due to Granville & Stark (2000). They show that a very strong uniform ABC-conjecture for number fields implies there are no Siegel zeros for Dirichlet L-functions associated to imaginary quadratic fields $\mathbf{Q}(\sqrt{-\mathbf{d}})$ where $-d < 0$, and d is square-free with $-d \equiv 1(4)$ or $-d \equiv 8, 12(16)$.

Vojta (1987) first showed how to formulate the ABC-conjecture for number fields. Let $K/\mathbf{Q}$ be a number field of degree n with discriminant D_K. For each prime ideal $\mathfrak{p}$ of K define a valuation $|\ |_{\mathfrak{p}}$ normalized so that $|\mathfrak{p}|_{\mathfrak{p}} = \text{Norm}_{K/\mathbf{Q}}(\mathfrak{p})^{-\frac{1}{n}}$. For each embedding $v : K \to \mathbf{C}$ define a valuation $|\ |_v$ by $|\alpha|_v = |\alpha^v|^{\frac{1}{n}}$, for $\alpha \in K$, and where $|\ |$ denotes the ordinary absolute value on $\mathbf{C}$. For $\alpha_1, \alpha_2, \ldots, \alpha_m \in K$ we define the height:

$$H(\alpha_1, \ldots, \alpha_m) = \prod_v \max\Big(|\alpha_1|_v, |\alpha_2|_v, \ldots, |\alpha_m|_v\Big),$$

where the product goes over all places v (prime ideals and embeddings). We also define the conductor:

$$N(\alpha_1, \ldots, \alpha_m) = \prod_{\mathfrak{p} \in I} |\mathfrak{p}|_{\mathfrak{p}}^{-1}$$

where I denotes the set of prime ideals $\mathfrak{p}$ such that $|\alpha_1|_{\mathfrak{p}}, \ldots, |\alpha_m|_{\mathfrak{p}}$ are not all equal. We can now state the

Conjecture 1 (Uniform ABC-Conjecture) *Let $\alpha, \beta, \gamma \in K$ satisfy $\alpha + \beta + \gamma = 0$. Then for every $\epsilon > 0$, there exists $\kappa(\epsilon) > 0$ such that*

$$H(\alpha, \beta, \gamma) \le \kappa(\epsilon)\Big(D_K^{1/n} \cdot N(\alpha, \beta, \gamma)\Big)^{1+\epsilon}.$$

Assuming the uniform ABC-conjecture Stark & Granville obtained the following lower bound for the class number $h(-d)$ of $\mathbf{Q}(\sqrt{-\mathbf{d}})$:

$$h(-d) \geq \left(\frac{\pi}{3} + o(1)\right) \frac{\sqrt{d}}{\log d} \sum_{\substack{(a,b,c)\in\mathbf{Z}^3 \\ -d=b^2-4ac \\ -a<b\leq a<c \quad \text{or} \quad 0\leq b\leq a=c}} \frac{1}{a} \qquad (d \to +\infty).$$

3 Elliptic curves over Q (Global Minimal Models)

An elliptic curve over a field K is a projective non-singular algebraic curve of genus 1 defined over K, furnished with a K-rational point. Every such curve has a generalized Weierstrass equation or model of the form:

$$E : y^2 + a_1xy + a_3y = x^3 + a_2x^2 + a_4x + a_6,$$

where $a_i \in K$, $(i = 1, 2, 3, 4, 6)$ with K-rational point (point at infinity) given in projective coordinates by $(0, 1, 0)$. It was first proved by Mordell (1922), for $K = \mathbf{Q}$, and generalized by Weil (1930) to arbitrary K that the K-rational points on E (denoted $E(K)$) form a finitely generated abelian group (Mordell–Weil group). The rank of the Mordell–Weil group $E(K)$ is defined to be the number of generators of infinite order.

Following Tate's formulaire (Tate 1975), we define

$$\begin{aligned}
b_2 &= a_1^2 + 4a_2 \\
b_4 &= a_1a_3 + 2a_4 \\
b_6 &= a_3^2 + 4a_6 \\
b_8 &= a_1^2a_6 - a_1a_3a_4 + 4a_2a_6 + a_2a_3^2 - a_4^2 \\
\Delta &= -b_2^2b_8 - 8b_4^3 - 27b_6^2 + 9b_2b_4b_6 \\
j &= c_4^3/\Delta,
\end{aligned}$$

where Δ denotes the discriminant of E.

Let

$$E' : y'^2 + a_1'xy + a_3'y = x'^3 + a_2'x'^2 + a_4'x' + a_6'$$

be another elliptic curve defined over K. Then E, E' are isomorphic if and only if there is a coordinate change of the form

$$x = u^2x' + r, \quad y = u^3y' + u^2sx' + t$$

with $r, s, t \in K$ and $u \in K^*$, which transforms E to E'. In this case we have

$$j' = j, \qquad \Delta' = u^{-12}\Delta.$$

For each rational prime number p, consider the local field $\mathbf{Q_p}$. Let v_p denote the p-adic valuation normalized so that $v_p(p) = 1$, $\mathbf{Z_p} = \{\mathbf{x} \in \mathbf{Q_p} \mid \mathbf{v_p(x)} \geq \mathbf{0}\}$, denotes the ring of p-adic integers.

Fix a rational prime p. Among all isomorphic models of a given elliptic curve

$$E : y^2 + a_1xy + a_3y = x^3 + a_2x^2 + a_4x + a_6,$$

defined over $\mathbf{Q_p}$, we can find one where all coefficients $a_i \in \mathbf{Z_p}$, and thus $v_p(\Delta) \geq 0$. This is easily seen by the coordinate change $x \to u^{-2}x$, $y \to u^{-3}y$ which sends each a_i to $u^i a_i$. Choosing u to be a high power of p does what we want. Since v_p is discrete, we can look for an equation with $v_p(\Delta)$ as small as possible.

Definition 1 (Global Minimal Model) Let E be an elliptic curve over $\mathbf{Q}$ with Weierstrass equation given by

$$E : y^2 + a_1xy + a_3y = x^3 + a_2x^2 + a_4x + a_6.$$

Then E is defined to be *minimal* at p if

$a_i \in \mathbf{Z_p} \quad (i = 1, 2, 3, 4, 6)$
$v_p(\Delta)$ is minimal (among all isomorphic models over $\mathbf{Q_p}$).

We define E to be a *global minimal model* if E is minimal at every prime p.

4 Conjectures which are equivalent to ABC

Let E be an elliptic curve defined over $\mathbf{Q}$ (global minimal model) with Weierstrass equation

$$E : y^2 + a_1xy + a_3y = x^3 + a_2x^2 + a_4x + a_6.$$

Then associated to E we have two important invariants:

Discriminant $\Delta = -b_2^2b_8 - 8b_4^3 - 27b_6^2 + 9b_2b_4b_6$,
Conductor $N = \prod_p p^{f_p}$, where

$$f_p = \begin{cases} 0, & \text{if } E(\mathbf{F_p}) \text{ is nonsingular;} \\ 1, & \text{if } E(\mathbf{F_p}) \text{ has a nodal singularity;} \\ 2 + \delta, & \text{if } E(\mathbf{F_p}) \text{ has a cuspidal singularity, with } \delta = 0 \text{ if } p \neq 2, 3. \end{cases}$$

The recipe for the conductor was first shown by Ogg (1967). An algorithm for computing f_p in all cases was proposed by Tate in a letter to Cassels (see Tate 1975). An elliptic curve which never has bad reduction of cuspidal type is said to be semistable, and in this case N is always the square-free part of Δ. This is the bridge between the theory of elliptic curves and the ABC-conjecture.

Conjecture 2 (Szpiro 1981) *Let E be an elliptic curve over $\mathbf{Q}$ which is a global minimal model with discriminant Δ and conductor N. Then for every $\epsilon > 0$, there exists $\kappa(\epsilon) > 0$ such that*

$$\Delta < \kappa(\epsilon) N^{6+\epsilon}.$$

We show that Szpiro's conjecture above is equivalent to the weak ABC-conjecture. Let A, B, C be coprime integers satisfying $A + B + C = 0$ and $ABC \neq 0$. Set $N = \prod_{p|ABC} p$. Consider the Frey–Hellegouarch curve

$$E_{A,B} : y^2 = x(x - A)(x + B).$$

A minimal model for $E_{A,B}$ has discriminant $(ABC)^2 \cdot 2^{-s}$ and conductor $N \cdot 2^{-t}$ for certain absolutely bounded integers s, t, (see Frey 1986). Plugging this data into Szpiro's conjecture immediately shows the equivalence.

Another conjecture equivalent to a version of the ABC-conjecture is the degree conjecture. Let $\Gamma_0(N)$ denote the group of matrices $\begin{pmatrix} a & b \\ c & d \end{pmatrix} \in SL(2, \mathbf{Z})$ with $c \equiv 0 (\bmod 3N)$, and set $X_0(N)$ to be the compactified Riemann surface realized as the quotient of the upper-half plane by $\Gamma_0(N)$. An elliptic curve E defined over $\mathbf{Q}$ is said to be modular if there exists a non-constant covering map

$$\phi : X_0(N) \to E,$$

normalized so that $\phi(i\infty) = 0$, the origin on E. It is now known (by work of Christophe Breuil, Brian Conrad, Fred Diamond, Richard Taylor, and Andrew Wiles) that every elliptic curve over $\mathbf{Q}$ is modular. The degree conjecture concerns the growth in N of the topological degree of the map ϕ as $N \to \infty$.

Conjecture 3 (Degree Conjecture (Frey 1987)) *For every $\epsilon > 0$, there exists $\kappa(\epsilon) > 0$ such that* $\deg(\phi) < \kappa(\epsilon) N^{2+\epsilon}$.

Frey (1987) proved that some bound for the degree implies a weak version of the ABC-conjecture. It was shown by Mai & Murty (1994) that the ABC-conjecture implies the degree conjecture for all Frey–Hellegouarch curves, and by Murty (1996) that the degree conjecture implies the ABC-conjecture. These results use work of Wiles (1995) and Diamond (1996) as well as work of Goldfeld, Hoffstein, Liemann and Lockhart see Hoffstein & Lockhart 1994 on the non–existence of Siegel zeros on $GL(3)$ which are symmetric square lifts from $GL(2)$.

The ABC conjecture is also intimately related to the size of the periods of the Frey–Hellegouarch curve

$$E_{A,B} : y^2 = x(x - A)(x + B).$$

Assume $-B < 0 < A$. This curve has two periods:

$$\Omega_1 = 2 \int_{-B}^{0} \frac{dx}{\sqrt{x(x - A)(x + B)}}$$

and

$$\Omega_2 = 2 \int_{A}^{\infty} \frac{dx}{\sqrt{x(x - A)(x + B)}}.$$

Conjecture 4 (Period Conjecture (Goldfeld 1988)) *Let $E_{A,B} : y^2 = x(x - A)(x + B)$ be the Frey–Hellegouarch curve with $A, B \in \mathbf{Z}$, $(A, B) = 1$, and $-B < 0 < A$. Let N denote the conductor of $E_{A,B}$. Then for every $\epsilon > 0$, there exists $\kappa(\epsilon) > 0$ such that*

$$\min\left(|\Omega_1|, |\Omega_2|\right) > \kappa(\epsilon) N^{-\frac{1}{2}-\epsilon}.$$

It was shown in Goldfeld (1990) that the period conjecture implies the weak ABC-conjecture.

The final conjecture we shall discuss (which is equivalent to ABC) is a conjecture on the size of the Tate–Shafarevich group Ш (see Shafarevich 1957, Tate 1957) of an elliptic curve defined over $\mathbf{Q}$. It was only recently (see Rubin 1987, 1989; Kolyvagin 1988a,b, 1991) that Ш was proved finite for a single elliptic curve and this explains why the ABC conjecture is so intractable. We shall now define Ш from first principles.

Let X be a set. We say a group G acts on X with left set-action $\bullet$ if for all $g \in G$, $x \in X$, the binary operation

$$g \bullet x \in X,$$

and $\bullet$ satisfies (for all $g, g' \in G$, $x \in X$) the identities:

$$e \bullet x = x, \quad (g \cdot g') \bullet x = g \bullet (g' \bullet x),$$

where e is the identity in G and $\cdot$ denotes the group operation in G. If A is an abelian group with internal operation $+$, we say G acts on A with left-group action $\circ$ if $\circ$ is a left set-action which also satisfies

$$g \circ (a + a') = g \circ a + g \circ a'$$

for all $g \in G$ and $a, a' \in A$.

Let A be an abelian group with internal operation $+$ and let G be another group which acts on A with left group-action $\circ$. We define $Z^1(G, A)$ to be the group of all functions (cocycles) $c : G \to A$ which satisfy the cocycle relation

$$c(g \cdot g') = c(g) + g \circ c(g'),$$

where $\cdot$ denotes the group operation in G. The subgroup $B^1(G, A)$ of coboundaries consists of all cocycles of the form $g \circ a - a$ with $a \in A$. We define the first cohomology group $H^1(G, A)$ to be the quotient group $H^1(G, A) = Z^1(G, A)/B^1(G, A)$.

Definition 2 *Fix an abelian group A and another group G acting on A with a left group–action* $\circ$. *A* principal homogeneous action *for* $(G, A, \circ)$ *is a left set-action* $\bullet$ *of G on A which satisfies the identity*

$$g \bullet a - g \bullet a' = g \circ a - g \circ a'$$

for all $g \in G$ *and* $a, a' \in A$.

We now define an equivalence relation on the set of principal homogeneous actions.

Definition 3 *Two principal homogeneous actions* $\bullet, \bullet'$ *for* $(G, A, \circ)$ *are said to be* equivalent *if*

$$g \bullet a - g \bullet' a = g \circ a_0 - a_0$$

for all $g \in G$*, all* $a \in A$*, and some fixed* $a_0 \in A$.

Let $WC(G, A)$ denote the set of equivalence classes of principal homogeneous actions for $(G, A, \circ)$. We will show that $WC(G, A)$ (the Weil–Châtelet group) is in fact a group by demonstrating that there is a bijection (of sets) $\beta : WC(G, A) \to H^1(G, A)$. First, if $\bullet$ is a principal homogeneous action for $(G, A, \circ)$ then for some fixed $a_0 \in A$ we have that $c(g) := g \bullet a_0 - a_0 \in$

$Z^1(G, A)$ because

$$\begin{aligned} c(g \cdot g') &= (g \cdot g') \bullet a_0 - a_0 = g \bullet (g' \bullet a_0) - a_0 \\ &= g \bullet \left(g' \bullet a_0\right) - g \bullet a_0 + g \bullet a_0 - a_0 \\ &= g \circ \left(g' \bullet a_0\right) - g \circ a_0 + c(g) = g \circ c(g') + c(g). \end{aligned}$$

Further, if we replace a_0 by $a_0 + a$ for any $a \in A$ then the cocycle changes to $c(g) + g \circ a - a$ which is equivalent to $c(g)$ mod $B^1(G, A)$. Thus each principal homogeneous action $\bullet$ maps to a unique element of $H^1(G, A)$. One also easily checks that equivalent homogeneous actions map to the same element of $H^1(G, A)$. Finally, to show the surjectivity, let $c(g) \in Z^1(G, A)$. Define a left action $\bullet$ of G on A by $g \bullet a := c(g) + g \circ a$ for all $g \in G$ and $a \in A$. If we change $c(g)$ to the equivalent cocycle $c(g) + g \circ a_0 - a_0$, then this gives rise to a new action $\bullet'$ given by $b \bullet' a = c(g) + g \circ a_0 - a_0 + g \circ a$. Clearly $\bullet$ and $\bullet'$ are equivalent principal homogeneous actions.

Remark 5 *The identity element in the group $WC(G, A)$ is the equivalence class of all actions equivalent to $\circ$. A principal homogeneous action $\bullet$ is equivalent to $\circ$ if and only if G has a fixed point under the left set–action $\bullet$, i.e., if and only if there exists $a_0 \in A$ such that $g \bullet a_0 = a_0$ for all $g \in G$ (clearly true because $g \bullet a_0 - a_0$ is the zero cocycle).*

In order to explicitly realize principal homogeneous actions, it is often convenient to consider a set $X = \phi(A)$, where ϕ is a bijection. The bijection ϕ leads to a transitive right set-action of A on X (denoted X^A) and defined by $x^{a'} = \phi(a + a')$ for all $x = \phi(a) \in X$, and all $a' \in A$. In this situation, the existence of a principal homogeneous action $\bullet$ for $(G, A, \circ)$ gives rise to a left set-action $\bullet'$ of G on X defined by

$$g \bullet' x = \phi(g \bullet a)$$

for all $g \in G$, and $x = \phi(a) \in X$. One checks that $g \bullet' x^{a_1} = \left(g \bullet' x\right)^{g \circ a_1}$ for all $a_1 \in A$. Thus X has the properties of a principal homogeneous space (see Serre 1997), i.e., there is a right set-action of A on X and a left principal homogeneous action $\bullet$ of G on X.

To define the Tate–Shafarevich group Ш for an elliptic curve E defined over $\mathbf{Q}$ we first consider the Weil–Châtelet group $WC(G, E(\overline{\mathbf{Q}}))$ where $G = \mathrm{Gal}(\overline{\mathbf{Q}}/\mathbf{Q})$ which acts on $E(\overline{\mathbf{Q}})$, the group of $\overline{\mathbf{Q}}$-rational points on E. Elements of $WC(G, E(\overline{\mathbf{Q}}))$ can be realized as curves of genus 1, denoted X, defined over $\mathbf{Q}$, which are birationally equivalent to E over $\overline{\mathbf{Q}}$ together with an appropriate action $\bullet$. Note that a curve of genus 1 defined over $\mathbf{Q}$ may not have a point in

Q. Let $\phi : E \to X$ be such a birational equivalence. Then for any $g \in G$ the map

$$(g\phi^{-1})\phi : E \to E$$

is of the type (see Cassels 1981)

$$a \to a + c(g)$$

with $a \in E(\overline{\mathbf{Q}})$, $c(g) \in Z^1(G, E(\overline{\mathbf{Q}}))$, and addition above denoting addition on the elliptic curve E. The right action of $E(\overline{\mathbf{Q}})$ on $X(\overline{\mathbf{Q}})$ is then given by translation (on the elliptic curve E): $x^{a'} = \phi(a + a')$ for $x = \phi(a) \in X(\overline{\mathbf{Q}})$, $a, a' \in E(\overline{\mathbf{Q}})$. The left action $\bullet$ of G on $X(\overline{\mathbf{Q}})$ is given by $g \bullet x = \phi(a + c(g))$ with $x = \phi(a) \in X(\mathbf{Q})$ which is induced from the cocycle $c(g)$ associated to the birational equivalence. The Tate–Shafarevich group Ш for E over **Q** is defined to be the subgroup of $WC(G, E(\overline{\mathbf{Q}}))$ associated to curves X as above which have a point in **R** and in every p-adic field $\mathbf{Q_p}$, or equivalently, the elements of $WC(G, E(\overline{\mathbf{Q}}))$ which have trivial images in $WC(G_p, E(\overline{\mathbf{Q_p}})$ and $WC(G_\infty, E(\mathbf{C})$, where $G_p = \mathrm{Gal}(\overline{\mathbf{Q_p}}/\mathbf{Q_p})$ for all finite primes p, and $G_\infty = \mathrm{Gal}(\mathbf{C}/\mathbf{R})$. If X has a point in **Q** then by the remark above, the action $\bullet$ of G on X is in the identity class of principal homogeneous actions. Thus, Ш measures the obstruction to the Hasse principle (Hasse's principle states that if a curve has points in **R** and in every p-adic field $\mathbf{Q_p}$ then it has a point in **Q**).

Definition 4 Mazur (1993) defined the notion of a *companion* to an elliptic curve E as a curve X of genus 1 which is isomorphic to E over R and over Q_p for all primes p. The Tate–Shafarevich group Ш may then be defined as the set of isomorphism classes over **Q** of companions of E, each endowed (as above) with the structure of a principal homogeneous space.

Conjecture 5 (Bound for Ш) *Let E be an elliptic curve defined over* **Q** *of conductor N with Tate–Shafarevich group* Ш. *Then for every $\epsilon > 0$, there exists $\kappa(\epsilon) > 0$ such that*

$$|Ш| < \kappa(\epsilon) N^{\frac{1}{2}+\epsilon} \qquad (N \to \infty).$$

One of the most remarkable conjectures in number theory is the Birch–Swinnerton-Dyer conjecture (Birch & Swinnerton-Dyer 1963, 1965), henceforth BSD, which relates the rank of the Mordell–Weil group of an elliptic curve E and the Tate–Shafarevich group of E to the special value at $s = 1$ of the Hasse–Weil L-function associated to E (see Silverman 1986 for the definition of the Hasse–Weil L-function). It was shown in Goldfeld & Szpiro (1995) that assuming the BSD (for rank 0 curves only), the above conjectured bound for Ш implies the following version of the ABC-conjecture:

$$|ABC|^{\frac{1}{3}} \ll N^{3+\epsilon}.$$

If one further assumes the generalized Riemann hypothesis (for the Rankin–Selberg zeta function associated to the weight $\frac{3}{2}$ cusp form coming from the Shintani–Shimura lift) then it was also shown in Goldfeld & Szpiro (1995) that the above conjectured bound for Ш (for rank 0 curves only) implies the weak ABC-conjecture:

$$|ABC|^{\frac{1}{3}} \ll N^{1+\epsilon}.$$

Actually, similar implications can be obtained from the following weaker conjecture.

Conjecture 6 (Average Bound for Ш_q) *Let $E : y^2 = x^3 + ax + b$ be an elliptic curve of conductor N with $a, b \in \mathbf{Z}$. For a square-free integer q, define the twisted curve $E_q : y^2 = x^3 + q^2ax + q^3b$ with Mordell–Weil rank r_q and Tate–Shafarevich group Ш_q. Then there exists a constant $c > 0$ and for every $\epsilon > 0$ there exists a constant $\kappa(\epsilon) > 0$ such that*

$$\sum_{\substack{q < N^c \\ r_q = 0}} |\text{Ш}_q| < \kappa(\epsilon) N^{c+\frac{1}{2}+\epsilon} \qquad (N \to \infty).$$

We now sketch the proof that Conjecture 6 plus BSD implies a version of the ABC-conjecture. The BSD conjecture states that the Hasse–Weil L-function $L_E(s)$ of an elliptic curve E defined over $\mathbf{Q}$ has a zero of order r, the rank of the Mordell–Weil group of $E(\mathbf{Q})$, and that the Taylor series of $L_E(s)$ about $s = 1$ is given by

$$L_E(s) = \left(\frac{c_E \Omega_E \cdot |\text{Ш}_E| \cdot \text{vol}(E(\mathbb{Q}))}{|E(\mathbb{Q})_{\text{tor}}|^2} \right) \cdot (s-1)^r + O(s-1)^{r+1}.$$

Here Ω_E is either the real period or twice the real period of E (depending on whether or not $E(\mathbb{R})$ is connected), $|\text{Ш}_E|$ is the order of the Tate–Shafarevich group of $E/\mathbb{Q}$, $\text{vol}(E(\mathbb{Q}))$ is the volume of the Mordell–Weil group for the Néron–Tate bilinear pairing, $|E(\mathbb{Q})_{\text{tors}}|$ is the order of the torsion subgroup of $E/\mathbb{Q}$, and $c_E = \prod_p c_p$ where $c_p = 1$ unless E has bad reduction at E in which case c_p is the order $E(\mathbb{Q}_p)/E_0(\mathbb{Q}_p)$ (Here $E_0(\mathbb{Q}_p)$ is the set of points reducing to non-singular points of $E(\mathbb{Z}/p\mathbb{Z})$.) (see Silverman 1986).

It is known that $c_E \geq 1$,

$$|E(\mathbb{Q})_{\text{tors}}|^2 \leq 256 \qquad \text{(Mazur 1977)},$$

and that $\text{vol}(E(\mathbb{Q})) = 1$ if $r = 0$. So in the rank $r = 0$ situation, a lower bound for $L_E(1)$ together with an upper bound for the order of Ш_E would imply a lower bound for the period Ω_E. If the lower bound for the period were strong enough to give the period conjecture we would get a version of ABC. It is enough to do this for one twisted curve E_q since the period changes by $q^{-\frac{1}{2}}$. Now, by a theorem of Waldspurger (see Waldspurger 1981, Kohnen 1985) one can find enough twists ($q < N^c$ with $c \gg 1$) of E with Mordell–Weil rank 0 where $L_{E_q}(1) \gg 1$, to do what we want. In the case $0 < c \ll 1$ it is necessary to use the generalized Riemann hypothesis.

Conjecture 5 can be proved for CM elliptic curves with $j \neq 0, 1728$ (we actually get better bounds). This was first done in Goldfeld & Lieman (1996) (see Theorem 6 below). For CM elliptic curves E defined over $\mathbf{Q}$ we expect.

Conjecture 7 *Let E be a CM elliptic curve defined over* $\mathbf{Q}$ *with Tate–Shafarevich group* Ш_E. *Then*

$$\begin{aligned} |\text{Ш}_E| &\ll N^{\frac{1}{4}+\epsilon}, && (\text{if } j \neq 0, 1728) \\ |\text{Ш}_E| &\ll N^{\frac{5}{12}+\epsilon}, && (\text{if } j = 0) \\ |\text{Ш}_E| &\ll N^{\frac{3}{8}+\epsilon}, && (\text{if } j = 1728). \end{aligned}$$

The constant implied by $\ll$ depends only on ϵ and is effectively computable.

Theorem 6 (Goldfeld–Lieman) *Let E be a CM elliptic curve defined over* $\mathbf{Q}$ *with Mordell–Weil rank* 0 *and Tate–Shafarevich group* Ш_E. *Then*

$$\begin{aligned} |\text{Ш}_E| &\ll N^{\frac{59}{120}+\epsilon}, && (\text{if } j \neq 0, 1728) \\ |\text{Ш}_E| &\ll N^{\frac{37}{60}+\epsilon}, && (\text{if } j = 0) \\ |\text{Ш}_E| &\ll N^{\frac{79}{120}+\epsilon}, && (\text{if } j = 1728). \end{aligned}$$

The constant implied by $\ll$ depends only on ϵ and is effectively computable.

This result uses the deep work of Rubin (1987), where the BSD conjecture is proved for CM elliptic curves over $\mathbf{Q}$ of Mordell–Weil rank 0, together with the upper bounds for special values of L-functions obtained by Duke *et al.* (1994).

5 Large Tate–Shafarevich groups

Cassels (1964) showed that the Tate–Shafarevich group of an elliptic curve over $\mathbf{Q}$ can be arbitrarily large. Cassels' method actually shows that there exist a fixed constant $c > 0$ and infinitely many integers N for which there exist an elliptic curve of conductor N, defined over $\mathbf{Q}$, with

$$|\text{Ш}| \gg N^{\frac{c}{\log\log N}},$$

This result was obtained by a different method by Kramer (1983). Assuming the Birch–Swinnerton-Dyer conjecture, Mai & Murty (1994) showed that there are infinitely many elliptic curves, defined over $\mathbf{Q}$ for which

$$|\text{Ш}| \gg N^{\frac{1}{4}-\epsilon}.$$

This was improved by de Weger (1996) who showed that

$$|\text{Ш}| \gg N^{\frac{1}{2}-\epsilon}$$

infinitely often, under the assumption of both the generalized Riemann hypothesis and the Birch–Swinnerton-Dyer conjecture.

The connection between the ABC-conjecture and the growth of Ш allows one to construct elliptic curves with large Tate–Shafarevich groups from bad ABC examples. De Weger (1997) has found 11 examples of curves with $|\text{Ш}| > \sqrt{N}$. Cremona (1993), by other methods, had also found several such curves.

The best known example of a Frey–Hellegouarch curve with large Ш is

$$y^2 = x(x - 643641)(x + 2)$$

coming from the ABC example, $A = 3^{10} \cdot 109$, $B = 2$, $C = 23^5$, due to Reyssat with $N = 15042$. In this case

$$\frac{|\text{Ш}|}{\sqrt{N}} = 0.7358\cdots$$

6 Modular symbols

Let $f(z) = \sum_{n=1}^{\infty} a(n)e^{2\pi i n z}$ be a holomorphic Hecke newform of weight 2 for $\Gamma_0(N)$ normalized so that $a(1) = 1$. For $\gamma \in \Gamma_0(N)$ we define the modular symbol

$$\langle \gamma, f \rangle = -2\pi i \int_{\tau}^{\gamma\tau} f(z)\, dz$$

which is independent of $\tau \in \mathfrak{h} \cup \mathbf{Q} \cup \{i\infty\}$. Shimura (1973) showed that the modular symbol is a homomorphism of $\Gamma_0(N)$ into the period lattice associated with $J_0(N)$. More specifically, if the coefficients $a(n)$ all lie in $\mathbf{Q}$ then the homomorphism is into the period lattice of an elliptic curve, i.e.

$$\langle \gamma, f \rangle = m_1 \Omega_q + m_2 \Omega_2,$$

where $m_1, m_2 \in \mathbf{Z}$ and $E = \mathbf{C}/\mathbf{Z}[\Omega_1, \Omega_2]$. For $\gamma = \begin{pmatrix} a & b \\ c & d \end{pmatrix}$, define the height of γ, denoted $H(\gamma)$ to be the maximum of $|a|, |b|, |c|, |d|$.

Conjecture 8 (Modular Symbol Conjecture (Goldfeld 1988)) *Let* $\langle \gamma, f \rangle = m_1 \Omega_1 + m_2 \Omega_2$ *as above. Then* m_1, m_2 *have at most a polynomial growth in* $H(\gamma)$.

It is not hard to show that there exists $\kappa > 0$ such that $\langle \gamma, f \rangle$ is larger than $N^{-\epsilon}$ for some γ with height $H(\gamma) \ll N^{\kappa}$. The above conjecture then implies a lower bound for the periods which can be used (via the period conjecture) to prove a version of the ABC-conjecture. Alternatively, the special value $L_E(1)$ (at the BSD point) can be expressed as a linear combination of modular symbols which also provides a bridge to the growth of Ш.

In order to study the growth properties of modular symbols, we have introduced a new type of Eisenstein series E^* twisted by modular symbols, which is defined as follows:

$$E^*(z, s) = \sum_{\gamma \in \Gamma_\infty \backslash \Gamma_0(N)} \langle \gamma, f \rangle \mathrm{Im}(\gamma z)^s.$$

Now E^* is not an automorphic form, but it satisfies (for all $\gamma \in \Gamma_0(N)$) the following automorphic relation

$$E^*(\gamma z, s) = E^*(z, s) - \langle \gamma, f \rangle E(z, s)$$

where

$$E(z, s) = \sum_{\gamma \in \Gamma_\infty \backslash \Gamma_0(N)} \mathrm{Im}(\gamma z)^s$$

is the classical Eisenstein series. We have shown (Goldfeld 1999a) that $E^*(z, s)$ has a meromorphic continuation to the entire complex s-plane with only one simple pole at $s = 1$ with residue given by

$$\frac{3}{\pi N} \prod_{p|N} \left(1 + \frac{1}{p}\right)^{-1} F(z)$$

where

$$F(z) = 2\pi i \int_z^{i\infty} f(w)\, dw.$$

As a consequence, it follows (see Goldfeld 1999b) that for fixed M, N and $x \to \infty$ that

$$\sum_{\gamma \in \Gamma_\infty \backslash \Gamma_0(N)} \langle \gamma, f \rangle \, e^{-\frac{c^2 M + d^2}{x}} \sim \frac{3}{\pi N} \prod_{p|N} \left(1 + \frac{1}{p}\right)^{-1} \frac{F(iM)}{M} x. \quad (1)$$

This result was recently improved by O'Sullivan (2000) who explicitly evaluated the error term as a function of M, N and found exponential decay in M. An intriguing possibility is to choose M so that $F(iM)$ is precisely the real period of the associated elliptic curve. The problem is that there is a lot of cancellation in the modular symbols so that the asymptotic relation (1) gives no information in the direction of the modular symbols conjecture. It would be of great interest to try to construct other such series which have positive coefficients and have a simple pole at $s = 1$ with residue given by the period of an elliptic curve. If the period were too small, such series would have to have a Siegel zero.

Acknowledgements The author would like to thank Iris Anshel and Shu-Wu Zhang for many helpful conversations.

Supported in part by a grant from the NSF.

References

Baker, A. (1998), Logarithmic forms and the *abc*-conjecture. In *Number Theory. Diophantine, Computational and Algebraic Aspects*, Conference held in Eger, July 29–August 2, 1996, K. Győry, A. Pethő & V.T. Sós (eds.), de Gruyter (1998), 37–44.

Birch, B. & H.P.F. Swinnerton-Dyer (1963), Notes on elliptic curves (I), *J. Reine Angew. Math.* **212**, 7–25.

Birch, B. & H.P.F. Swinnerton-Dyer (1965), Notes on elliptic curves (II), *J. Reine Angew. Math.* **218**, 79–108.

Cassels, J.W.S. (1964), Arithmetic on curves of genus I (VI). The Tate–Safarevic group can be arbitrarily large, *J. Reine. Angew. Math.* **214/215**, 65–70.

Cassels, J.W.S. (1991), *Lectures on Elliptic Curves*, Cambridge University Press.

Cremona, J.E. (1993), The analytic order of Ш for modular elliptic curves, *J. Th. Nombres Bordeaux* **5**, 179–184.

Diamond, F. (1996), On deformation rings and Hecke rings, *Ann. of Math. (2)* **144**, 137–166.

Duke, W., J. Friedlander & H. Iwaniec (1994), Bounds for automorphic L-functions. II, *Invent. Math.* **115**, 219–239.

Elkies, N.D. (1991a), *ABC* implies Mordell, *Int. Math. Res. Notices* **7** (1991), 99–109.

Elkies, N.D. (1991b), *ABC* implies Mordell, *Duke Math. Journ.* **64** (1991).

Faltings, G. (1983), Arakelov's theorem for abelian varieties, *Invent. Math.*, **73**, 337–347.

Frey, G. (1986), Links between stable elliptic curves and certain diophantine equations, *Ann. Univers. Sarav.* **1** (1), (1986), 1–39.

Frey, G. (1987), Links between elliptic curves and solutions of $A - B = C$, *J. Ind. Math. Soc.* **51**, 117–145.

Goldfeld, D. (1990), Modular elliptic curves and Diophantine problems. In *Number Theory*, R. Mollin (ed.), de Gruyter, 157–175.

Goldfeld, D. (1999a), Zeta functions formed with modular symbols. In *Proc. of the Symposia in Pure Math.*, **66**, *1, Automorphic Forms, Automorphic Representations, and Arithmetic*, 111–122.

Goldfeld, D. (1999b), The distribution of modular symbols. In *Number Theory in Progress*, **2**, *Elementary and Analytic Number Theory*, K. Győry, H. Iwaniec & J. Urbanowicz (eds.), de Gruyter, 849–866.

Goldfeld, D. & D. Lieman (1996), Effective bounds on the size of the Tate–Shafarevich group, *Math. Research Letters* **3**, 309–318.

Goldfeld, D. & L. Szpiro (1995), Bounds for the order of the Tate–Shafarevich group, *Compos. Math.* **97**, 71–87.

Granville, A. (1998), ABC means we can count squarefrees, *Int. Math. Res. Notices* **19**, 991–1009.

Granville, A. & H.M. Stark (2000), *ABC* implies no 'Siegel zeros' for *L*-functions of characters with negative discriminant, *Invent. Math.* **139** (3), 509–523.

Hoffstein, J., Lockhart, P. (1994), Coefficients of Maass forms and the Siegel zero, *Ann. of Math.* **140** (2), 161–181. With an appendix by D. Goldfeld, J. Hoffstein & D. Lieman, An effective zero free region.

Kohnen, W. (1985), Fourier coefficients of modular forms of half-integral weight, *Math. Ann.* **271**, 237–268.

Kolyvagin, V.A. (1988a), Finiteness of $E(\mathbb{Q})$ and $Ш(\mathbb{Q})$ for a subclass of Weil curves, *Izv. Akad. Nauk USSR Ser. Mat.* **52**, 522–540; English translation in *Math USSR-Izv.* **32** (1989), 523–543.

Kolyvagin, V.A. (1988b), On the Mordell–Weil group and the Shafarevich–Tate group of elliptic curves, *Izv. Akad. Nauk SSSR Ser. Mat.* **52**, 1154–1180.

Kolyvagin, V.A. (1991), Euler systems. In *The Grothendieck Festschrift: A collection of articles written in honor of the* 60*th birthday of Alexander Grothendieck, volume II*, P. Cartier, L. Illusie, N.M. Katz, G. Laumon, Y. Manin & K.A. Ribet (eds.), Birkhäuser, 435–483.

Kramer, K. (1983), A family of semistable elliptic curves with large Tate–Shafarevich groups, *Proc. Amer. Math. Soc.,* **89**, 379–386.

Mazur, B. (1977), Modular elliptic curves and the Eisenstein ideal, *Pub. IHES Math.* **47**, 33–186.

Mazur, B. (1993), On the passage from local to global in number theory, *Bull. AMS,* **29** (1) , 14–50.

Murty, R. (1999), Bounds for congruence primes. In *Proc. of the Symposia in Pure Math.,* **66**, *1, Automorphic Forms, Automorphic Representations, and Arithmetic*, 177–192.

Mai, L. & R. Murty (1994), A note on quadratic twists of an elliptic curve. In *CRM Proceedings and Lecture Notes* **4**, 121–124.

Mordell, L.J. (1922), On the rational solutions of the indeterminate equations of the third and fourth degrees, *Proc. Camb. Phil. Soc.* **21**, 179–192.

Ogg, A.P. (1967), Elliptic curves and wild ramification, *Amer. J. Math.* **89**, 1–21.

Osterlé, J. (1988), Nouvelles approches du Théorème de Fermat. In *Sem. Bourbaki,* **694** (1987–88), 694-01–694-21.

O'Sullivan, C. (2000), Properties of Eisenstein series formed with modular symbols, *J. Reine Angew. Math.* **518**, 163–186..

Rubin, K. (1987), Tate–Shafarevich groups and $L-$functions of elliptic curves with complex multiplication, *Invent. Math.* **89**, 527–559.

Rubin, K. (1989), The work of Kolyvagin on the arithmetic of elliptic curves. In *Arithmetic of Complex Manifolds*, W.P. Barth & H. Lange (eds.), Lecture Notes in Math. **1399**, Springer-Verlag,, 128–136.

Serre, J.P. (1987), *Galois Cohomology,* Springer-Verlag.

Shafarevich, I.R. (1957), On birational equivalence of elliptic curves, *Dokl. Akad. Nauk SSSR* **114** (2), 267–270. Reprinted in *Collected Mathematical Papers*, Springer-Verlag, (1989), 192–196.

Shimura, G. (1973), On the factors of the jacobian variety of a modular function field, *J. Math. Soc. Japan* **25**, 523–544.

Silverman, J.H. (1986), *The Arithmetic of Elliptic Curves,* Springer-Verlag.

Silverman, J.H. (1988), Wieferich's criterion and the *abc*-conjecture, *J. Numb. Theor.* **30**, 226–237.

Stewart, C.L. & R. Tijdeman (1986) On the Oesterlé–Masser conjecture, *Monatsh. Math.* **102**, 251–257.

Stewart, C.L. & K.R. Yu (1991), On the *abc*-conjecture, *Math. Ann.* **291** (2), 225–230.

Stewart, C.L. & K.R. Yu (2001), On the *abc*-conjecture II, *Duke Math. J.* **108** (1), 169–182.

Tate, J. (1957), WC-groups over $\mathfrak{p}$-adic fields, *Séminaire Bourbaki, Exp.* **156**.

Tate, J. (1975), Algorithm for determining the type of a singular fiber in an elliptic pencil. In *Modular Functions of One Variable IV*, B.J. Birch & W. Kuyk (eds.), Lecture Notes in Math. **476**, 33–52.

Vojta, P. (1987), *Diophantine Approximations and Value Distribution Theory*, Lecture Notes in Math. **1239**, Springer-Verlag.

Waldspurger, J-L. (1981), Sur les coefficients de Fourier des formes modulaires de poides demi-entier, *J. Math. Pures Appl.* **60**, 375–484.

Weil, A. (1930), Sur un théorème de Mordell, *Bull. Sci. Math.,* **54**, 182–191. Reprinted in *Oeuvres Scientifiques, Collected Papers* **1**, Springer-Verlag, (1980), 11–45.

de Weger, B. (1996), $A+B+C$ and big Sha's, Math. Inst. University of Leiden, The Netherlands, Report no. W96–11.

Wiles, A. (1995), Modular elliptic curves and Fermat's last theorem, *Ann. of Math.* **141**, 443–551.

10

On the Algebraic Independence of Numbers

Yu.V. Nesterenko

Abstract

The purpose of this article is to describe results about the algebraic independence of values of analytic functions proved in transcendence theory. In particular we discuss the case of modular and theta functions $\theta(z, \tau)$, $z, \tau \in \mathbb{C}, \Im\tau > 0$, an essential progress has been made in the last five years.

We say that the complex numbers $\omega_1, \ldots, \omega_m$, $m \geq 1$, are algebraically dependent over the field of rational numbers $\mathbb{Q}$ if there exists a nontrivial polynomial $P \in \mathbb{Q}[x_1, \ldots, x_m]$ such that $P(\omega_1, \ldots, \omega_m) = 0$. If such polynomial does not exist we say that numbers $\omega_1, \ldots, \omega_m$ are *algebraically independent*. In the case $m = 1$, the terminology *algebraic* or *transcendental* number is used. For example the numbers sin 1 and cos 1 are algebraically dependent, since $\sin^2 1 + \cos^2 1 - 1 = 0$, but each of them is transcendental. If $\omega_1, \ldots, \omega_m$ are algebraically independent numbers, then for every polynomial $P \in \mathbb{Q}[x_1, \ldots, x_m]$, $P \neq 0$, the number $P(\omega_1, \ldots, \omega_m)$ is transcendental.

1 E-functions

The first result in this area was announced by F. Lindemann (1882) and proved by K. Weierstrass (1885).

Theorem 1 *If* $\alpha_1, \ldots, \alpha_m$ *are algebraic numbers that are linearly independent over* $\mathbb{Q}$*, then the numbers*

$$e^{\alpha_1}, \ldots, e^{\alpha_m}$$

are algebraically independent over the field $\mathbb{Q}$.

As corollaries in the case $m = 1$ we have the transcendence of e (Hermite 1873) and π (Lindemann 1882) and, for any nonzero algebraic α, the transcendence of e^α, $\log\alpha \neq 0$ and $\sin\alpha$ (Lindemann 1882).

Siegel (1929) introduced a class of entire functions which, in his opinion, comprised possible candidates for a generalization of the Lindemann–Weierstrass theorem. He called these functions 'E-functions'. A typical example is

$$f(z) = \sum_{n=0}^{\infty} \frac{(a_1)_n \cdots (a_p)_n}{(b_1)_n \cdots (b_q)_n} z^{n(q-p)}, \qquad q > p \geq 0,$$
$$a_j, b_j \in \mathbb{Q}, \qquad b_j \neq 0, -1, -2, \ldots.$$

where the symbol $(\alpha)_n$ is defined by setting

$$(\alpha)_0 = 1, \qquad (\alpha)_n = \alpha(\alpha+1)\cdots(\alpha+n-1), \qquad n = 1, 2, \ldots,$$

with the numerators in the sum equal to 1 if $p = 0$. It is evident that e^z is an E-function.

Siegel proposed a method of proving the transcendence and algebraic independence of the values of such functions, but he succeeded only in the case of E-functions satisfying second-order linear differential equations over $\mathbb{C}(z)$. The final result was proved in 1955 by A.B. Shidlovskii.

Theorem 2 *Suppose that the E-functions*

$$f_1(z), \ldots, f_m(z), \qquad m \geq 1, \tag{1}$$

form a solution of the system of linear differential equations

$$y_k' = Q_{k0}(z) + \sum_{j=1}^{m} Q_{kj}(z) y_j, \qquad k = 1, \ldots, m,$$

where $Q_{kj}(z) \in \mathbb{C}(z)$. Let α be any algebraic number not equal to 0 *or a pole of one of the functions $Q_{kj}(z)$. Then the numbers*

$$f_1(\alpha), \ldots, f_m(\alpha)$$

are algebraically independent over $\mathbb{Q}$ if and only if the functions (1) *are algebraically independent over $\mathbb{C}(z)$.*

In subsequent years the Siegel–Schidlovskii method was developed and generalized much further. Results on the algebraic independence of the values of E-functions in cases when there are algebraic relations among them, on quantitative results, on methods of proving algebraic independence of the solutions of linear differential equations, and on applications of the general theorems to concrete E-functions were proved.

The Siegel–Shidlovskii method was applied to the values of G-functions. This class of functions with finite radius of convergence contains generalized hypergeometric functions ($p = q$) and was introduced by Siegel in 1929. The study of the values of G-functions is much more complicated then the situation with E-functions because of the much slower convergence of the corresponding series. Typical results concern the linear independence of values of G-functions at rational points close to the origin. Nevertheless in special cases one can prove the transcendence and even algebraic independence of the values. We will discuss these results in Section 5.

A survey of results proved by the Siegel–Shidlovskii method can be found in Shidlovskii (1989) or Feldman & Nesterenko (1998).

2 Mahler functions

Let $d \geq 2$ be an integer and $f_1(z), \ldots, f_m(z)$ be functions single-valued in the neighbourhood of the origin and having the property that for every k, $1 \leq k \leq m$, the function $f_k(z^d)$ is algebraic over the field $\mathbb{C}(f_1(z), \ldots, f_m(z))$. For example the set of functions

$$f_k(z) = \sum_{\nu=0}^{\infty} z^{kd^\nu}, \qquad 1 \leq k < d, \tag{2}$$

with $f_d(z) = z$, satisfies this property because of the relations

$$f_k(z^d) = f_k(z) - z^k, \qquad 1 \leq k < d.$$

One can consider functions of many variables $z_1, \ldots, z_n$ and the transformation

$$z_j \longrightarrow z_1^{d_{j1}} \cdots z_n^{d_{jn}}, \qquad 1 \leq j \leq n,$$

with positive integers d_{ji}.

In 1929 K. Mahler studied transcendence and algebraic independence properties of values of such functions. Under some additional conditions he proved general results of this type and, among other concrete corollaries, the following theorem.

Theorem 3 *If $d \geq 2$ then for any algebraic number α, $0 < |\alpha| < 1$, the values of functions* (2) *at the point α are algebraically independent over $\mathbb{Q}$.*

Another of Mahler's examplez is the transcendence of the value $f(\omega, \alpha)$, where

$$f(\omega, z) = \sum_{\nu=0}^{\infty} [\nu\omega] z^{\nu},$$

for a positive quadratic irrational ω and algebraic α, $0 < |\alpha| < 1$. This result requires the study of functions in two variables.

The subsequent development of Mahler's method and recent results in this area proved by K.K. Kubota, J.H. Loxton, A. van der Poorten, D. Masser, Yu. Nesterenko, Ku. Nishioka, M. Amou, P.G. Becker, T. Töpfer and T. Tanaka can be found in Nishioka (1996). General theorems have rather complicated technical conditions, which is why we note here only the following consequence of a general theorem of Ku. Nishioka

Theorem 4 *Let ω be a positive quadratic number. If α is an algebraic numbers satisfying $0 < |\alpha| < 1$; then the numbers $f^{(k)}(\omega, \alpha)$, $k \geq 0$, are algebraically independent.*

One more example is connected with the modular function $j(\tau)$. It is well known that for every integer $d \geq 2$ the function

$$J(z) = j\left(\frac{\log z}{2\pi i}\right) = z^{-1} + \sum_{\nu=0}^{\infty} c_n z^{\nu}$$

and the functions $J(z^d)$ are algebraically dependent over the field $\mathbb{C}$ (the modular equation). In 1969 Mahler conjectured that for any algebraic α, $0 < |\alpha| < 1$ the number $J(\alpha)$ is transcendental. This fact was proved by Barré-Sirieix *et al.* (1996). Of course the proof uses the modular equation, but surprisingly it has more points in common with the method of Gelfond and Schneider (Section 3) than with Mahler's.

The following function in two complex variables

$$\Theta(z_1, z_2) = 1 + \sum_{n=1} (z_2^n + z_2^{-n}) z_1^{n^2}$$

satisfies the functional equation

$$z_1 z_2 \Theta(z_1, z_1^2 z_2) = \Theta(z_1, z_2)$$

and is connected to the theta function $\theta(z, \tau)$ through the identity

$$\Theta\left(e^{\pi i \tau}, e^{2\pi i z}\right) = \theta(\tau, z).$$

The transcendence properties of these functions are discussed in Section 6.

3 Theorems about functions with addition properties

Hilbert's seventh problem asks about the transcendence of the value e^z at the transcendental point $z = \beta \log \alpha$, for algebraic $\alpha \neq 0, 1$, and algebraic irrational β. It was solved in 1934 by A.O. Gelfond and Th. Schneider. Gelfond was the first to study the algebraic independence of the values of the exponential function at points that are not necessarily algebraic. In this connection he made the following conjectures.

Conjecture 1

1. *Let α be an algebraic number not equal to* 0 *or* 1, *and let $\beta_1, \dots, \beta_m$ be algebraic numbers such that* $1, \beta_1, \dots, \beta_m$ *are linearly independent over $\mathbb{Q}$. Then the numbers*

$$\alpha^{\beta_1}, \dots, \alpha^{\beta_m}$$

are algebraically independent over $\mathbb{Q}$.

2. *Suppose that $\alpha_1, \dots, \alpha_m$ are nonzero algebraic numbers, and that*

$$\log \alpha_1, \dots, \log \alpha_m \tag{3}$$

are branches of their logarithms that are linearly indepedent over $\mathbb{Q}$. Then the numbers (3) *are algebraically independent over $\mathbb{Q}$.*

The first part, for $m = 1$, coincides with Hilbert's seventh problem and is a natural analogue of the Lindemann–Weierstrass theorem. The second part generalizes the Hermite–Lindemann theorem. Both still have not been proved for any $m \geq 2$.

In 1949 Gelfond proved the first conjecture in the case $m = 2$, $\beta_1 = \beta$, $\beta_2 = \beta^2$ and cubic irrational β. The best result in this direction belongs to Diaz (1989).

Theorem 5 *Let α be an algebraic number not equal to* 0 *or* 1, *and let β be an algebraic number with* $\deg \beta = d \geq 2$. *Then*

$$\operatorname{tr}\deg \mathbb{Q}\left(\alpha^{\beta}, \alpha^{\beta^2}, \dots, \alpha^{\beta^{d-1}}\right) \geq \left[\frac{d+1}{2}\right].$$

Here $[x]$ denotes the largest integer less than or equal to x.

This theorem crowns the long line of results following Gelfond and proved by A. Shmelev, R. Tijdeman, D. Brownawell, M. Waldschmidt, S. Lang, G. Chudnovsky, E. Reyssat, Zhu Yao Chen, R. Endell, Yu. Nesterenko and P. Philippon.

Concerning the second of Gelfond's conjectures, we note that all that is currently known is that the numbers (3) and 1 are linearly independent over the field of algebraic numbers. This was proved by A. Baker in 1967 in connection with bounds for linear forms in logarithms of algebraic numbers.

The Weierstrass elliptic function $\wp(z)$ satisfies an algebraic differential equation with constant coefficients, and has an addition law. These properties make $\wp(z)$ similar to the exponential function and enable one to prove theorems about its values that are analogous to those in the exponential case. Schneider in 1937 proved elliptic analogues of the Hermite–Lindemann theorem and Hilbert's seventh problem.

The proof of the following analogue of the Lindemann–Weierstrass theorem was published by Wüstholz (1983) and Philippon (1983).

Theorem 6 *Suppose that $n \geq 1$, that the Weierstrass elliptic function $\wp(z)$ has algebraic invariants g_2, g_3 and complex multiplication over a field* **k**, *and that the algebraic numbers $\alpha_1, \ldots, \alpha_n$ are linearly independent over* **k**. *Then the numbers*

$$\wp(\alpha_1), \ldots, \wp(\alpha_n)$$

are algebraically independent over $\mathbb{Q}$.

There are results which are specific to elliptic functions. The next theorem appeared in Chudnovsky (1984).

Theorem 7 *Let $\wp(z)$ be the Weierstrass elliptic function with invariants g_2, g_3, let ω be a nonzero period of $\wp(z)$ and η the corresponding quasi-period. Then at least two of the numbers*

$$g_2,\ g_3,\ \frac{\omega}{\pi},\ \frac{\eta}{\pi}$$

are algebraically independent over $\mathbb{Q}$.

In particular this theorem implies that in the case of algebraic invariants g_2, g_3 the numbers ω/π, η/π are algebraically independent, and, in the case of complex multiplication, π and ω are algebraically independent.

Let γ be a closed path on the Riemann surface of the elliptic curve

$$y^2 = 4x^3 - g_2 x - g_3, \qquad \Delta = g_2^3 - 27g_3^2 \neq 0 \tag{4}$$

$g_2, g_3 \in \overline{\mathbb{Q}}$. Then the numbers

$$\omega = \int_\gamma \frac{dx}{\sqrt{4x^3 - g_2 x - g_3}}, \qquad \eta = \int_\gamma \frac{x\,dx}{\sqrt{4x^3 - g_2 x - g_3}} \tag{5}$$

are the period and quasi-period of the corresponding elliptic function $\wp(z)$. In particular this implies in the case $y^2 = 4x^3 - 4x$ that the numbers

$$\frac{1}{\pi}\int_0^1 \frac{dx}{\sqrt{x - x^3}} = \frac{1}{2\pi}B(1/4, 1/2) = \frac{\Gamma(1/4)^2}{(2\pi)^{3/2}}$$

and

$$\frac{1}{\pi}\int_0^1 \frac{x\,dx}{\sqrt{x - x^3}} = \frac{1}{2\pi}B(3/4, 1/2) = \frac{2^{3/2}\pi^{1/2}}{\Gamma(1/4)^2}$$

are algebraically independent, or that the numbers π, $\Gamma(1/4)$ are algebraically independent. In particular, $\Gamma(1/4)$ is a transcendental number.

In the same way, applying Theorem 7 to the curve $y^2 = 4x^3 - 4$, one can derive that π and $\Gamma(1/3)$ are algebraically independent and $\Gamma(1/3)$ is a transcendental number.

4 Modular forms

The Eisenstein series of weight $2k$, $k \geq 1$ is defined by

$$E_{2k}(\tau) = \frac{1}{2\zeta(2k)} \sum_{m\in\mathbb{Z}} \sum_{\substack{n \in \mathbb{Z} \\ (m,n) \neq (0,0)}} \frac{1}{(m\tau + n)^{2k}}, \quad k \geq 1, \quad \tau \in \mathbb{C},\ \Im\tau > 0,$$

If $k \geq 2$ this function is a modular form of weight $2k$ with respect of the group $SL(2, \mathbb{Z})$, see Lang (1976). It is well known that any modular form with respect of this group is a polynomial in $E_4(\tau)$, $E_6(\tau)$ over $\mathbb{C}$. The function $E_2(\tau)$ is not a modular form and functions $E_2(\tau)$, $E_4(\tau)$, $E_6(\tau)$ are algebraically independent over $\mathbb{C}$.

Let $\wp(z)$ be the Weierstrass elliptic function with invariants g_2, g_3, periods ω_1, ω_2, corresponding quasi-periods η_1, η_2 and $\tau = \omega_2/\omega_1$, $\Im\tau > 0$. We have, see Lang (1973), Chapter 4,

$$E_2(\tau) = 3\frac{\omega_1}{\pi}\cdot\frac{\eta_1}{\pi}, \qquad E_4(\tau) = \frac{3}{4}\left(\frac{\omega_1}{\pi}\right)^4 \cdot g_2, \qquad E_6(\tau) = \frac{27}{8}\left(\frac{\omega_1}{\pi}\right)^6 \cdot g_3.$$

These formulae imply that Theorem 7 can be reformulated in the form

Let be $\tau \in \mathbb{C}$, $\Im\tau > 0$. Then at least two of the numbers

$$E_2(\tau),\ E_4(\tau),\ E_6(\tau)$$

are algebraically independent over $\mathbb{Q}$.

The following more general result was proved in Nesterenko (1996).

Theorem 8 *Let be $\tau \in \mathbb{C}$, $\Im\tau > 0$. Then at least three of the numbers*

$$e^{\pi i\tau},\ E_2(\tau),\ E_4(\tau),\ E_6(\tau)$$

are algebraically independent over $\mathbb{Q}$.

We distinguish some corollaries of this theorem. The first one is connected to elliptic functions.

Let $\wp(z)$ be the Weierstrass elliptic function with algebraic invariants g_2, g_3 and complex multiplication over the field **k**. *If ω is any period of $\wp(z)$, if η is the corresponding quasi-period, and if $\tau \in$* **k**, *$\Im\tau \neq 0$, then each of sets*

$$\{\pi,\ \omega,\ e^{2\pi i\tau}\}, \qquad \{\omega,\ \eta,\ e^{2\pi i\tau}\}$$

is algebraically independent over $\mathbb{Q}$.

These results in particular imply the algebraic independence of the numbers

$$\left\{\pi, e^{\pi\sqrt{3}}, \Gamma\left(\frac{1}{3}\right)\right\} \quad \text{and} \quad \left\{\pi, e^{\pi}, \Gamma\left(\frac{1}{4}\right)\right\}.$$

For any natural number d there exists a Weierstrass $\wp$-function with algebraic invariants and with complex multiplication field $\mathbb{Q}(\sqrt{-d})$. Thus we obtain the following assertion.

For any natural number d, the numbers

$$\pi,\ e^{\pi\sqrt{d}},$$

are algebraically independent over $\mathbb{Q}$.

Another corollary concerns the values of the modular function

$$j(\tau) = 1728\frac{E_4(\tau)^3}{E_4(\tau)^3 - E_6(\tau)^2}.$$

Let D denote $\frac{1}{2\pi i}\frac{\partial}{\partial\tau}$.

For any $\tau \in \mathbb{C}$, $\Im\tau > 0$, τ not congruent to i and $e^{2\pi i/3}$ with respect to $SL(2,\mathbb{Z})$, at least three of numbers

$$e^{\pi i\tau},\ j(\tau),\ Dj(\tau),\ D^2j(\tau)$$

are algebraically independent over $\mathbb{Q}$.

This is an improvement of result about $e^{\pi i\tau}$ and $j(\tau)$ (Mahler's conjecture) proved in Barré-Sirieix *et al.* (1996); see Section 2.

Let Γ be a congruence subgroup of $SL_2(\mathbb{Z})$, see Lang (1973). A modular form of weight $2k$, $k \in \mathbb{Z}$, $k \geq 0$, relative to Γ will here be a *meromorphic* function $f(\tau)$ on the upper half-plane $\Im\tau > 0$ such that $f(\gamma\cdot\tau) = (c\tau + d)^{2k} f(\tau)$ for any $\gamma = \begin{pmatrix} a & b \\ c & d \end{pmatrix}$ in Γ, and which admits a meromorphic continuation at the cusps of Γ; a modular form of weight 0 is called a *modular function*. We say that $f(\tau)$ is defined over the field $\overline{\mathbb{Q}}$ of algebraic numbers if its Fourier expansion

$$f(\tau) = \sum_{n\geq\nu} a_n e^{2\pi i n\tau/h}, \quad h\in\mathbb{Z},\ h>0,$$

at $i\infty$ has algebraic coefficients a_n.

The following corollary of the Theorem 8 was proved by D. Bertrand (see Nesterenko & Philippon 2001), Chapter 1.

Corollary 1 *Let $f(\tau)$ be a non-constant meromorphic modular form, defined over $\overline{\mathbb{Q}}$. For all $\alpha\in\mathbb{C}$, $\Im\alpha > 0$, distinct from poles of $f(\tau)$, such that $e^{2\pi i\alpha}\in\overline{\mathbb{Q}}$, the numbers $f(\alpha)$, $Df(\alpha)$ and $D^2 f(\alpha)$ are algebraically independent over $\mathbb{Q}$.*

Special cases of it and concrete examples had been stated earlier by D. Bertrand himself and by D. Duverney, Ke. Nishioka, Ku. Nishioka and I. Shiokawa.

Since $\Delta(\tau) = 1728^{-1}\left(E_4(\tau)^3 - E_6(\tau)^2\right)$ is a modular form of weight 12 with respect of $SL(2,\mathbb{Z})$ the corollary is valid for $\Delta(\tau)$ and for the Dedekind eta function

$$\eta(\tau) = \Delta(\tau)^{1/24} = e^{\pi i\tau/12}\prod_{n=1}^{\infty}\left(1 - e^{2\pi i n\tau}\right).$$

Since values of the Rogers–Ramanujan continued fraction

$$RR(\alpha) = 1 + \cfrac{\alpha}{1 + \cfrac{\alpha^2}{1 + \cfrac{\alpha^3}{1 + \ddots}}}$$

can be expressed in terms of the eta function, this implies the transcendence of the number $RR(\alpha)$ for algebraic α, $0 < |\alpha| < 1$.

Another concrete example is connected to the theta function

$$\theta_3 = 1 + 2\sum_{n=0}^{\infty} e^{\pi i n^2\tau}.$$

The function $f(\tau) = \theta_3^2$ is a modular form of weight 1 with respect to the congruence subgroup $\Gamma(4)$. Hence the following assertion holds.

For any algebraic number α, $0 < |\alpha| < 1$, the numbers

$$\sum_{n\geq 0} \alpha^{n^2}, \quad \sum_{n\geq 1} n^2\alpha^{n^2}, \quad \sum_{n\geq 1} n^4\alpha^{n^2}$$

are algebraically independent. In particular we can conclude that the number

$$\sum_{n\geq 0} \alpha^{n^2} \tag{6}$$

is transcendental.

As far back as 1851, in the process of constructing the first examples of transcendental numbers, Liouville presented the series (6) for $\alpha = l^{-1}, l \in \mathbb{Z}, l > 1$ as an example for which his method allowed one to prove only the irrationality.

A leading role in the proof of Theorem 8 is played by Fourier expansions of Eisenstein series

$$E_2(\tau) = P(e^{2\pi i\tau}), \quad E_4(\tau) = Q(e^{2\pi i\tau}), \quad E_6(\tau) = R(e^{2\pi i\tau}), \quad \Im\tau > 0.$$

where

$$P(z) = 1 - 24\sum_{n=1}^{\infty} \sigma_1(n)z^n, \quad Q(z) = 1 + 240\sum_{n=1}^{\infty} \sigma_3(n)z^n,$$

$$R(z) = 1 - 504\sum_{n=1}^{\infty} \sigma_5(n)z^n,$$

and $\sigma_k(n) = \sum_{d|n} d^k$. The coefficients of these series are integer numbers that grow not that rapidly. The functions P, Q, R were specifically considered by Ramanujan in 1916, who stated in particular, the system of differential equations

$$z\frac{\partial P}{\partial z} = \frac{1}{12}(P^2 - Q), \quad z\frac{\partial Q}{\partial z} = \frac{1}{3}(PQ - R), \quad z\frac{\partial R}{\partial z} = \frac{1}{2}(PR - Q^2), \tag{7}$$

The arithmetical properties of the coefficients, the existence of the system (7) and the algebraic independence of $P(z)$, $Q(z)$, $R(z)$ over $\mathbb{C}(z)$ make the situation similar to the E-function case (Section 1). Of course it is more complicated since the system is not linear and the radii of convergence of the series is finite.

5 Hypergeometric functions

There are only a few cases when we can prove the algebraic independence of values of the Gaussian hypergeometric function $F(a, b, c; z)$ defined in the unit circle by the series

$$F(a, b, c; z) = \sum_{n=0}^{\infty} \frac{(a)_n (b)_n}{(c)_n n!} z^n \tag{8}$$

and its derivative. Note that for rational parameters a, b, c this function belongs to the class of Siegel's G-functions.

If we apply Theorem 7 to the elliptic curve

$$y^2 = 4x(1-x)(\lambda^{-1} - x), \quad \lambda \in \overline{\mathbb{Q}},\ \lambda \neq 0,\ 1, \tag{9}$$

we find that the numbers

$$\begin{aligned} &\frac{1}{\pi} \int_{\gamma} x^{-1/2}(1-x)^{-1/2}(1-\lambda x)^{-1/2} dx, \\ &\frac{1}{\pi} \int_{\gamma} x^{1/2}(1-x)^{-1/2}(1-\lambda x)^{-1/2} dx, \end{aligned} \tag{10}$$

are algebraically independent over $\mathbb{Q}$. Here γ is any closed path on the Riemann surface of the elliptic curve (9) that is not homologous to 0.

It is well known (see Kratzer & Franz 1960, 4.2.2) that one can choose paths of integration γ_1, γ_2 in such a way that

$$\frac{1}{\pi} \int_{\gamma_1} x^{-1/2}(1-x)^{-1/2}(1-\lambda x)^{-1/2} dx = F\left(\frac{1}{2}, \frac{1}{2}, 1; \lambda\right) = F_1(\lambda),$$

and

$$\frac{1}{\pi} \int_{\gamma_2} x^{-1/2}(1-x)^{-1/2}(1-\lambda x)^{-1/2} dx = iF\left(\frac{1}{2}, \frac{1}{2}, 1; 1-\lambda\right) = F_2(\lambda),$$

where $|\arg \lambda| < \pi$, $|\arg(1-\lambda)| < \pi$. Since every branch of $F_1(\lambda)$ is a linear combination of $F_1(\lambda)$, $F_2(\lambda)$ with integer coefficients (see Bateman & Erdélyi 1953, 2.7.1) then every branch of $F_1(\lambda)$ can be derived from the first integral (10) by a proper choice of γ.

If $\gamma = \gamma_1$ and $|\arg \lambda| < \pi$, the second integral in (10) coincides with

$$F\left(\frac{3}{2}, \frac{1}{2}, 2; \lambda\right). \tag{11}$$

By the relation

$$F\left(\frac{3}{2}, \frac{1}{2}, 2; \lambda\right) = 2F_1(\lambda) + 4(\lambda - 1)F_1'(\lambda)$$

(see Bateman & Erdélyi 1953, 2.8(25)) one can see that for every $\alpha \in \overline{\mathbb{Q}}$, $\alpha \neq 0, 1$ and every branch of the hypergeometric function $F_1(z)$, the numbers

$$F\left(\frac{1}{2}, \frac{1}{2}, 1; \alpha\right) = F_1(\alpha), \qquad F'\left(\frac{1}{2}, \frac{1}{2}, 1; \alpha\right) = F_1'(\alpha)$$

are algebraically independent.

There exist other Gaussian functions with rational parameters having this property. For example, by the identities

$$\begin{aligned} F\left(\frac{1}{4}, \frac{1}{4}, 1; 4z(1-z)\right) &= F\left(\frac{1}{2}, \frac{1}{2}, 1; z\right), \\ F\left(\frac{1}{4}, \frac{1}{4}, 1; z\right)(1-4z)^{1/4} &= F\left(\frac{1}{12}, \frac{5}{12}, 1; \frac{27z}{(4z-1)^3}\right), \end{aligned}$$

see Bateman & Erdélyi (1953), 2.1.5(27), 2.11(42), one can deduce that *for every algebraic number* $\alpha \neq 0, 1$ *and* $(a, b, c) = (1/4, 1/4, 1)$ *or* $(1/12, 5/12, 1)$ *the values*

$$F(a, b, c; \alpha), \qquad F'(a, b, c; \alpha) \tag{12}$$

are algebraically independent.

In this way, using classical relations between hypergeometric functions, one can deduce the same assertion for functions $F(a, b, c; z)$ with parameters

$$\left(\frac{1}{8}, \frac{3}{8}, 1\right), \left(\frac{1}{6}, \frac{1}{3}, 1\right), \left(\frac{1}{6}, \frac{1}{2}, 1\right), \left(\frac{1}{3}, \frac{1}{3}, 1\right), \left(\frac{1}{4}, \frac{1}{2}, 1\right), \left(\frac{1}{3}, \frac{1}{2}, 1\right).$$

Note that the nine listed cases correspond to nine arithmetic triangle groups of non-compact type, see Takeuchi (1977).

The function $F\left(\frac{1}{12}, \frac{5}{12}, 1; z\right)$ satisfies the differential equation

$$z(1-z)y'' + \left(1 - \frac{3}{2}z\right)y' - \frac{5}{144}y = 0. \tag{13}$$

Another solution of this equation is $F\left(\frac{1}{12}, \frac{5}{12}, \frac{1}{2}; 1-z\right)$. The following proposition describes the uniformization of these functions by Eisenstein series.

Proposition 1 *For any* $\tau \in \mathbb{C}$, $\Im\tau > 0$ *the following identities hold*

$$F\left(\frac{1}{12}, \frac{5}{12}, 1; \frac{1728}{j(\tau)}\right) = E_4(\tau)^{1/4}, \tag{14}$$

$$F\left(\frac{1}{12}, \frac{5}{12}, \frac{1}{2}; \frac{E_6^2}{E_4^3}(\tau)\right) = \frac{\tau + i}{2i}\left(\frac{E_4(\tau)}{E_4(i)}\right)^{1/4}. \tag{15}$$

We use the branch of $E_4(\tau)^{1/4}$ that is real and positive on the imaginary axis $\tau = it, t > 0$. For these values of τ, the numbers $1728/j(\tau) = \Delta(\tau)/E_4(\tau)^3$ and $E_6^2/E_4^3 = 1 - 1728/j(\tau)$ are real, positive and less then 1. For identities (14), (15) we should choose the branches of Gaussian functions given by (8) when $\tau = it, t > 0$.

To prove these identities let us take any solution $y(z)$ of the differential equation (13) and put $f(\tau) = y(z)$ with $z = 1728/j(\tau) = \Delta(\tau)/E_4(\tau)^3$. It is not difficult to check using equation (13) and the system (7) that the function $f(\tau)$ satisfies the differential equation

$$D^2 f - \frac{1}{2}\frac{DE_4}{E_4} Df - \frac{5}{144}\frac{\Delta}{E_4^2} f = 0. \tag{16}$$

On the other hand, using the system (7) one can prove that functions

$$E_4(\tau)^{1/4} \quad \text{and} \quad \tau E_4(\tau)^{1/4}.$$

satisfy the same differential equation (16). Hence the function $f(\tau)$ can be represented in the form $f(\tau) = (c_1 + c_2\tau)E_4(\tau)^{1/4}$.

Since $F(1/12, 5/12, 1; 1728/j(\tau))$ has 1 as a period in a neighbourhood of $i\infty$ and takes value 1 at $i\infty$, as does $E_4(\tau)$, we derive (14). To prove (15) note that $E_6(i) = 0$ and in a neighbourhood of the point $\tau = i$ the function $g(\tau) = F(1/12, 5/12, 1/2; E_6(\tau)^2/E_4(\tau)^3))$ has the property $g(-1/\tau) = g(\tau)$.

The identity (14) and Theorem 8 give another proof of the algebraic independence of values of $F(1/12, 5/12, 1; z)$ and its derivative at any algebraic point α distinct from 0 and 1.

The identity (15) explains another phenomenon, see Beukers & Wolfart (1988),

There exist infinitely many points $\alpha \in \overline{\mathbb{Q}}$ such that the value

$$F(1/12, 5/12, 1/2; \alpha)$$

is an algebraic number.

The reason is that, for every natural n, the numbers $E_4(ni)/E_4(i)$ and $j(ni)$ are algebraic, see Lang (1973). For example, if $\tau = 2i$ we derive from (15) equality

$$F\left(\frac{1}{12}, \frac{5}{12}, \frac{1}{2}; \frac{1323}{1331}\right) = \frac{3}{4}\sqrt[4]{11}.$$

It is possible to uniformize hypergeometric functions by modular forms in all nine cases listed above, see Harnad & McKay (1998). For example the following classical identities hold:

$$F\left(\frac{1}{2},\frac{1}{2},1;\frac{\theta_2^4}{\theta_3^4}(\tau)\right)=\theta_3(\tau)^2,\qquad F\left(\frac{1}{2},\frac{1}{2},1;\frac{\theta_4^4}{\theta_3^4}(\tau)\right)=i\tau\theta_3(\tau)^2.$$

The definition of theta-constants θ_2, θ_3, θ_4 will be given in Section 6. Note that the functions θ_k^2 are modular forms of weight 1 with respect to the group $\Gamma(4)\subset SL(2,\mathbb{Z})$.

More generally, the uniformization of hypergeometric functions is connected to functions that are automorphic under the action of the monodromy group of the corresponding differential equation, see Ford (1929). In this connection it seems very interesting for the future to study transcendence and algebraic independence properties of values of automorphic forms.

6 Values of theta functions

The theta function $\theta(z,\tau)$ is a function of two complex variables z, τ, $\Im\tau>0$, defined by the series

$$\theta(z,\tau)=\sum_{n\in\mathbb{Z}}\exp(\pi i\tau n^2+2\pi inz).$$

For any pair of $a,b\in\mathbb{Q}$ one can define the theta function

$$\theta_{a,b}(z,\tau) = \sum_{n\in\mathbb{Z}}\exp(\pi i\tau(n+a)^2+2\pi i(n+a)(z+b)),$$
$$z,\tau\in\mathbb{C},\ \Im\tau>0$$

with characteristics a, b. The standard notation

$$\theta_1(z,\tau)=\theta_{\frac{1}{2},\frac{1}{2}}(z,\tau),\quad \theta_2(z,\tau)=\theta_{\frac{1}{2},0}(z,\tau),$$
$$\theta_3(z,\tau)=\theta_{0,0}(z,\tau),\quad \theta_4(z,\tau)=\theta_{0,\frac{1}{2}}(z,\tau),$$

are used. In those cases when all the functions involved in the formulas have the same variable τ, we shall use the notation $\theta_k(z)=\theta_k(z,\tau)$, $\theta_k=\theta_k(0,\tau)$. The functions $\theta_2(z)$, $\theta_3(z)$, $\theta_4(z)$ are even, while $\theta_1(z)$ is an odd function. Fourier expansions hold

$$\theta_1=0,\qquad \theta_2=2q^{1/4}\sum_{n=0}^{\infty}q^{n(n+1)},\quad q=e^{\pi i\tau},$$

$$\theta_3=1+2\textstyle\sum_{n=1}^{\infty}q^{n^2},\quad \theta_4=1+2\displaystyle\sum_{n=1}^{\infty}(-1)^n q^{n^2}.$$

Introduce two differential operators

$$\partial = \frac{1}{2\pi i}\frac{\partial}{\partial z}, \qquad D = \frac{1}{\pi i}\frac{\partial}{\partial \tau}.$$

All theta functions with characteristics satisfy the heat equation

$$\partial^2\theta_{a,b}(z,\tau) = D\theta_{a,b}(z,\tau).$$

There are many algebraic and differential relations connecting theta functions. For example,

$$\theta_3^4 = \theta_2^4 + \theta_4^4, \qquad \partial\theta_1 = \partial\theta_1(0,\tau) = \frac{i}{2}\theta_2\theta_3\theta_4.$$

One can check that

$$\theta_{a,b}(z+1,\tau) = \theta_{a,b}(z,\tau)e^{2\pi i a}, \quad \theta_{a,b}(z+\tau,\tau) = \theta_{z,\tau}e^{-\pi i\tau - 2\pi i(z+b)}.$$

These identities imply that

$$\left(\frac{\theta_k(z)}{\theta_4(z)}\right)^2, \quad \partial\left(\frac{\partial\theta_k(z,\tau)}{\theta_k(z,\tau)}\right), \quad k = 1,2,3,4,$$

are elliptic functions in z with period lattice $\Lambda = \mathbb{Z} + \tau\mathbb{Z}$. The Jacobian elliptic functions are connected to these ratio as follows:

$$\operatorname{sn} u = -\frac{\theta_3}{\theta_2}\cdot\frac{\theta_1(z)}{\theta_4(z)}, \quad \operatorname{cn} u = \frac{\theta_4}{\theta_2}\cdot\frac{\theta_2(z)}{\theta_4(z)}, \quad \operatorname{dn} u = \frac{\theta_4}{\theta_3}\cdot\frac{\theta_3(z)}{\theta_4(z)},$$

where $z = u/\pi\theta_3^2$. The periods each of these functions constitute a sublattice in $2K\mathbb{Z} + 2iK'\mathbb{Z}$, where

$$K = \frac{\pi}{2}\theta_3^2, \qquad iK' = \tau K = \tau\frac{\pi}{2}\theta_3^2.$$

Another example is the Weierstrass elliptic function $\mathfrak{P}(v)$ with periods ω_1, $\omega_2 = \tau\omega_1$. It can be expressed in the form

$$\mathfrak{P}(v) = \left(\frac{2\pi i}{\omega_1}\right)^2\left[\left(\frac{\partial\theta_1}{\theta_3}\cdot\frac{\theta_3(z)}{\theta_1(z)}\right)^2 + \frac{1}{12}(\theta_4^4 - \theta_2^4)\right], \quad z = \frac{v}{\omega_1}. \tag{17}$$

Invariants of this function are:

$$g_2 = \frac{2}{3}\left(\frac{\pi}{\omega_1}\right)^4(\theta_2^8 + \theta_3^8 + \theta_4^8), \tag{18}$$

$$g_3 = \frac{4}{27}\left(\frac{\pi}{\omega_1}\right)^6(\theta_4^4 - \theta_2^4)(\theta_2^4 + \theta_3^4)(\theta_3^4 + \theta_4^4). \tag{19}$$

The corresponding zeta function is

$$\zeta(v) = \eta_1 z + \frac{2\pi i}{\omega_1}\cdot\frac{\partial\theta_1(z)}{\theta_1(z)}, \quad z = \frac{v}{\omega_1}.$$

with quasi-periods

$$\eta_1 = \frac{4}{3} \cdot \frac{\pi^2}{\omega_1} \left(\frac{D\theta_2}{\theta_2} + \frac{D\theta_3}{\theta_3} + \frac{D\theta_4}{\theta_4} \right), \qquad \eta_2 = \tau\eta_1 - \frac{2\pi i}{\omega_1},$$

where

$$D = \frac{1}{\pi i} \frac{\partial}{\partial \tau}.$$

Any modular form with respect to $SL(2, \mathbb{Z})$ is a polynomial with constant coefficients in $E_4(\tau)$, $E_6(\tau)$ with

$$\begin{aligned} E_2(\tau) &= 4\left(\frac{D\theta_2}{\theta_2} + \frac{D\theta_3}{\theta_3} + \frac{D\theta_4}{\theta_4} \right), \\ E_4(\tau) &= \frac{1}{2}(\theta_2^8 + \theta_3^8 + \theta_4^8), \\ E_6(\tau) &= \frac{1}{2}(\theta_4^4 - \theta_2^4)(\theta_2^4 + \theta_3^4)(\theta_3^4 + \theta_4^4). \end{aligned} \tag{20}$$

The identities collected above allow us to translate theorems about the transcendence of values of elliptic functions into the language of theta functions. Schneider (1934, 1957) proved several results about the transcendence of numbers connected to elliptic functions. We can present them here as follows.

Theorem 9 *For any $\tau \in \mathbb{C}$, $\Im\tau > 0$:*

(1) *at least one of numbers $\theta_2, \theta_3, \theta_4$ is transcendental;*
(2) *at least one of numbers $\pi\theta_2^2$, $\pi\theta_3^2$, $\pi\theta_4^2$ is transcendental;*
(3) *if $\tau \in \overline{\mathbb{Q}}$, $\Im\tau > 0$ and τ is not an imaginary quadratic then $j(\tau)$ is transcendental.*

The following result of D. Bertrand is a corollary of Theorem 7.

Theorem 10 *For any $k = 2, 3, 4$ and any $\tau \in \mathbb{C}$, $\Im\tau > 0$, at least two of the numbers*

$$\theta_k, D\theta_k, D^2\theta_k \tag{21}$$

are algebraically independent over $\mathbb{Q}$.

According to Theorem 8 one can claim that if $e^{\pi i \tau}$ is an algebraic number then the three numbers (21) are algebraically independent.

The following result is an equivalent form of Theorem 6.

Theorem 11 *Suppose that $n \geq 1$, τ is an imaginary quadratic number and that the algebraic numbers $\alpha_1, \ldots, \alpha_n$ are linearly independent over $\mathbb{Q}(\tau)$; then the numbers*

$$\frac{\theta_2}{\theta_3}(\alpha_1, \tau), \ldots, \frac{\theta_2}{\theta_3}(\alpha_n, \tau)$$

are algebraically independent over $\mathbb{Q}$.

As corollaries of Theorem 8 one can derive

Theorem 12 *For any $\tau \in \mathbb{C}$, $\Im\tau > 0$, the set*

$$e^{\pi i \tau}, \frac{D\theta_2}{\theta_2}, \frac{D\theta_3}{\theta_3}, \frac{D\theta_4}{\theta_4}$$

contains at least three algebraically independent numbers.

We now discuss some conjectures concerning transcendence properties of values of the modular and theta functions. The first combines the assertions of Theorem 8 and Theorem 9(3).

Conjecture 2 *Let be $\tau \in \mathbb{C}$, $\Im\tau > 0$ and assume that the set*

$$\tau,\ e^{\pi i \tau},\ E_2(\tau),\ E_4(\tau),\ E_6(\tau)$$

contains at most three algebraically independent numbers over $\mathbb{Q}$. Then τ is imaginary quadratic and the numbers

$$e^{\pi i \tau},\ E_2(\tau),\ E_4(\tau)$$

are algebraically independent over $\mathbb{Q}$.

Note that in the case of imaginary quadratic τ the last assertion follows from Theorem 8.

The proof of Theorem 9 uses elliptic functions (variable z in the language of theta functions). Schneider himself asked about a proof based on the properties of modular functions (Schneider's second problem). On the other hand the proof of Theorem 8 uses Fourier expansions of the Eisenstein series $E_{2k}(\tau)$ in $q = e^{2\pi i \tau}$. To prove the conjecture one should combine these two different approaches.

Let us denote by $\mathcal{F}_0$ the field generated over $\mathbb{Q}$ by the functions

$$e^{\pi i \tau},\ \theta(0, \tau),\ D\theta(0, \tau),\ D^2\theta(0, \tau), \tag{22}$$

$$e^{2\pi i z},\ \theta(z, \tau),\ \partial\theta(z, \tau),\ \partial^2\theta(z, \tau), \tag{23}$$

and let $\mathcal{F}$ be the field consisting of all functions algebraic over the field $\mathcal{F}_0$. One can prove that the field $\mathcal{F}$ is closed with respect to the differential operators D, ∂, that it contains functions $\theta_{a,b}(cz+d\tau, r\tau)$, for every set of rational numbers a, b, c, d, r, and that the transcendence degree of the field $\mathcal{F}$ over $\mathbb{C}$ equals 8.

Conjecture 3 (D. Bertrand.) *Let be $\tau, u \in \mathbb{C}$, $\Im\tau > 0$, $u \notin \mathbb{Q} + \mathbb{Q}\tau$. Then at least six among the eight values of functions* (22), (23) *at the point* (u, τ) *are algebraically independent.*

Denote by $\mathcal{F}_1$ th subfield of $\mathcal{F}$ consisting of functions algebraic over the field generated by (22). This field contains the functions $E_2(\tau)$, $E_4(\tau)$, $E_6(\tau)$ and, according to Theorem 8, at least three of the values (22) are algebraically independent over $\mathbb{Q}$.

The functions (22) form a transcendence basis of the field $\mathcal{F}_1$. One can choose another convenient transcendence basis consisting of the logarithmic derivatives of theta constants. In 1881 M. Halphen proved that the functions

$$\psi_2 = \frac{D\theta_2}{\theta_2},\ \psi_3 = \frac{D\theta_3}{\theta_3},\ \psi_4 = \frac{D\theta_4}{\theta_4}, \tag{24}$$

satisfy the following system of differential equations.

$$\begin{aligned} D\psi_2 &= 2(\psi_2\psi_3 + \psi_2\psi_4 - \psi_3\psi_4), \\ D\psi_3 &= 2(\psi_3\psi_2 + \psi_3\psi_4 - \psi_2\psi_4), \\ D\psi_4 &= 2(\psi_4\psi_2 + \psi_4\psi_3 - \psi_2\psi_3). \end{aligned} \tag{25}$$

Using the identities (24) one can easily deduce Ramanujan's system (7) from (25) and vice versa. It is possible to prove Theorem 12, as we did Theorem 8, by directly applying the same ideas to the set of functions ψ_2, ψ_3, ψ_4 and using the system (25).

If $u = a + b\tau$, $a, b \in \mathbb{Q}$ then

$$\theta(a + b\tau, \tau) = e^{-\pi i a^2\tau - 2\pi i ab} \cdot \theta_{a,b}(0, \tau),$$

hence the fields generated by the values of the functions (22) and (23) at the point (u, τ) have the same algebraic closure. Since $\theta_{a,b}(0, \tau)^2$ is a modular form for some congruence subgroup of $SL(2; \mathbb{Z})$, Theorem 8 implies that the transcendence degree of the field from Conjecture 3 is at least 3, and it equals 3 if τ is imaginary quadratic.

The two conjectures above are really about $\theta(z, \tau)$ as a function of two variables. A very interesting problem for the future from the transcendence point of view is the study of theta functions of many variables. In this regard we point out that recently Ohyama (1996) stated the system of algebraic differential equations for theta-functions $\theta(z, \tau)$, $z \in \mathbb{C}^2$, $\tau \in \mathbb{C}^3$ analogous to Halphen's one (25). The transcendence degree of the corresponding functional field was computed in a joint article of Bertrand & Zudilin (2000).

References

Bateman, H. & A. Erdélyi (1953), *Higher Transcendental Functions*, volume 1, McGraw-Hill.

Barré-Sirieix, K., G. Diaz, F. Gramain, F. & G. Philibert (1996), Une preuve de la conjecture de Mahler–Manin, *Invent. Math.* **124**, 1–9.

Beukers, F. & J. Wolfart (1988), Algebraic values of hypergeometric functions. In *New Advances in Transcendence Theory*, A. Baker (ed.), Cambridge University Press, 68–81

Bertrand, D. & W. Zudilin (2000), On the transcendence degree of the differential field generated by Siegel modular forms. Prépublication de l'Institut de Mathematiques de Jussieu, no. 248, Mars 2000, 20 pp., `http://xxx.lanl.gov/abs/math.NT/0006176`.

Chudnovsky, G. (1984) *Contributions to the Theory of Transcendental Numbers*, American Mathematical Society.

Diaz, G. (1989), Grands degrés de transcendance pour des familles d'exponentielles, *J. Number Theory* **31**, 1–23.

Feldman, N.I. & Yu.V. Nesterenko (1998), *Transcendental Numbers*, Springer.

Ford, L.R. (1929), *Automorphic Functions*, McGraw-Hill.

Harnad, J. & J. McKay (1998), Modular solutions to equations of generalized Halphen type. Preprint `http://xxx.lanl.gov/abs/solv-int/9804006`.

Kratzer, A. & W. Franz (1960), *Transzendente Functionen*, Teubner.

Lang, S. (1973), *Elliptic Functions*, Addison Wesley.

Lang, S. (1976), *Introduction to Modular Forms*, Springer-Verlag.

Lindemann, F. (1882), Uber die Zahl π, *Math. Ann.* **20**, 213–225.

Nesterenko, Yu.V. (1996), Modular functions and transcendence questions, *Matemat. Sbornik* **187**(9), 65–96 (Russian); English translation in *Sbornik: Mathematics* **187**(9), 1319–1348.

Nesterenko, Yu.V. & P. Philippon (eds.) (2001), *Introduction to Algebraic Independence Theory*, Springer.

Weierstrass, K. (1885), Zu Lindemann's Abhandlung: Uber die Ludolph'sche Zahl, *Sitzungsber. Preuss. Akad. Wiss.*, 1067–1085.

Nishioka, K. (1996), *Mahler Functions and Transcendence*, Lecture Notes in Math., **1631**, Springer.

Ohyama, Y. (1996), Differential equations of theta constants of genus two. In *Algebraic Analysis of Singular Perturbations, Kyoto, 1996*, Sūrikaisekikenkyūsho Kōkyūroku, Kyoto University, 96–103, (Japanese); English translation, Preprint Osaka, Osaka University, (1996).

Philippon, P. (1983), Variétés abéliennes et indépendance algebrique, *Invent. Math.* **72**, 389–405.

Schneider, Th. (1934), Transzendenzuntersuchungen periodischer Functionen, *J. Reine Angew. Math.* **172**, 70–74.

Schneider, Th. (1957), *Einfürung in die Transcendenten Zahlen*, Springer.

Shidlovskii, A.B. (1989), *Transcendental Numbers*, de Gruyter.

Takeuchi, K. (1977), Arithmetic triangle groups, *J. Math. Soc. Japan* **29** (1), 91–106.

Wüstholz, G. (1983), Über das Abelsche Analogon des Lindemannschen Satzes, *Invent. Math.*, **72**, 363–388.

11

Ideal Lattices

Eva Bayer-Fluckiger

Introduction

An *ideal lattice* is a pair (I, b), where I is an ideal of a number field, and b is a lattice, satisfying an invariance relation (see §1 for the precise definition). Ideal lattices naturally occur in many parts of number theory, but also in other areas. They have been studied in special cases, but, as yet, not much in general. In the special case of integral ideal lattices, the survey paper Bayer-Fluckiger (1999) collects and slightly extends the known results.

The first part of the paper (see §2) concerns integral ideal lattices, and states some classification problems. In §3, a more general notion of ideal lattices is introduced, as well as some examples in which this notion occurs.

The aim of §4 is to define twisted embeddings, generalising the canonical embedding of a number field. This section, as well as the subsequent one, is devoted to positive definite ideal lattices with respect to the canonical involution of the real étale algebra generated by the number field. These are also called Arakelov divisors of the number field. The aim of §5 is to study invariants of ideals and also of the number field derived from Hermite type invariants of the sphere packings associated to ideal lattices. This again gives rise to several open questions.

1 Definitions, notation and basic facts

A *lattice* is a pair (L, b), where L is a free $\mathbf{Z}$-module of finite rank, and $b : L \times L \to \mathbf{R}$ is a non-degenerate symmetric bilinear form. We say that (L, b) is an *integral lattice* if $b(x, y) \in \mathbf{Z}$ for all $x, y \in L$. An integral lattice (L, b) is said to be *even* if $b(x, x) \equiv 0 \pmod 2$ for all $x \in L$.

Let K be an algebraic number field. Let us denote by $\mathcal{O}$ its ring of integers, and by D_K its discriminant. Let n be the degree of K.

Set $K_{\mathbf{R}} = K \otimes_{\mathbf{Q}} \mathbf{R}$. Then $K_{\mathbf{R}}$ is an étale $\mathbf{R}$-algebra (i.e. a finite product of copies of $\mathbf{R}$ and $\mathbf{C}$). Let us denote by $\mathrm{N} = \mathrm{N}_{K_{\mathbf{R}}/\mathbf{R}}$ the norm, and by $\mathrm{Tr} = \mathrm{Tr}_{K_{\mathbf{R}}/\mathbf{R}}$ the trace, of this étale algebra. Let $\bar{}\, : K_{\mathbf{R}} \to K_{\mathbf{R}}$ be an $\mathbf{R}$-linear involution.

Definition 1 An *ideal lattice* is a lattice (I, b), where I is a (fractional) $\mathcal{O}$-ideal and $b : I \times I \to \mathbf{R}$ is such that

$$b(\lambda x, y) = b(x, \overline{\lambda} y)$$

for all $x, y \in I$ and for all $\lambda \in O$.

Proposition 1 *Let I be an $\mathcal{O}$-ideal and let $b : I \times I \to \mathbf{R}$ be a lattice. Then the following are equivalent:*

(i) *(I, b) is an ideal lattice;*
(ii) *there exists an invertible element $\alpha \in K_{\mathbf{R}}$ with $\overline{\alpha} = \alpha$ such that*

$$b(x, y) = \mathrm{Tr}(\alpha x \overline{y}).$$

Proof This follows from the fact that $\mathrm{Tr} : K_{\mathbf{R}} \times K_{\mathbf{R}} \to \mathbf{R}$ is non-degenerate. □

The *rank* of an ideal lattice is the degree n of the number field K. As we shall see, the other basic invariants – determinant, signature – are also easy to determine. Let $b : I \times I \to \mathbf{R}$, $b(x, y) = \mathrm{Tr}(\alpha x \overline{y})$, be an ideal lattice.

Proposition 2 *We have*

$$|\det(b)| = \mathrm{N}(I)^2 \mathrm{N}(\alpha) D_K.$$

Proof Straightforward computation. □

In order to determine the signature of ideal lattices, we need the notion of *canonical involution* (or complex conjugation) of an étale $\mathbf{R}$-algebra.

Definition 2 Let E be an étale $\mathbf{R}$-algebra. We have $E = E_1 \times \cdots \times E_r \times F_1 \times \cdots \times F_s$, where $E_i \simeq \mathbf{R}$ and $F_i \simeq \mathbf{C}$. We say that $x = (x_1, \ldots, x_r)$ is *positive*, denoted by $x > 0$, if $x_i \in \mathbf{R}$ and $x_i > 0$ for all $i = 1, \ldots, r$.

Proposition 3 *Let E be an étale $\mathbf{R}$-algebra. and let $* : E \to E$ be an involution. The following properties are equivalent:*

(i) *$x x^* > 0$ for all non-zero $x \in E$;*

(ii) *the the restriction of* $*$ *to* E_i *is the identity for all* $i = 1, \ldots, r$, *and it is complex conjugation on* F_j *for all* $j = 1, \ldots, s$.

Proof This is immediate. □

In particular, this implies that for any étale **R**-algebra E there is exactly one involution such that xx^* is positive for all non-zero $x \in E$.

Definition 3 Let E be an étale **R**-algebra, and let $* : E \to E$ be an involution. We say that $*$ is the *canonical involution* of E if and only if $xx^* > 0$ for all non-zero $x \in E$.

Let $C \subset K_{\mathbf{R}}$ be the maximal étale **R**-subalgebra such that the restriction of $\bar{}$ to C is the canonical involution of C. Set $c = \mathrm{rank}(C)$.

We are now ready to determine the signature of an ideal lattice $b : I \times I \to \mathbf{R}$, $b(x, y) = \mathrm{Tr}(\alpha x \overline{y})$. Let $A \subset C$ be the maximal étale **R**-subalgebra such that all the components of α in A are negative. Let $a = \mathrm{rank}(A)$.

Proposition 4 *The signature of* (I, b) *is* $c - 2a$.

Proof This follows from the definitions. □

Corollary 1 *We have*

$$\det(b) = (-1)^{\frac{n-c+2a}{2}} \mathrm{N}(I)^2 \mathrm{N}(\alpha) D_K.$$

Proof This follows from Propositions 2 and 4. □

We now define some equivalence relations on the set of ideal lattices.

Definition 4 Let (I, b) and (I', b') be two ideal lattices.

(i) We say that (I, b) and (I', b') are *isomorphic*, denoted by $(I, b) \simeq (I', b')$, if there exists $a \in K^*$ such that $I' = aI$ and that $b'(ax, ay) = b(x, y)$ for all $x, y \in I$.
(ii) We say that (I, b) and (I', b') are *equivalent*, denoted by $(I, b) \equiv (I', b')$ (or simply $b \equiv b'$), if there exists an isomorphism of **Z**-modules $f : I \to I'$ such that $b'(f(x), f(y)) = b(x, y)$ for all $x, y \in I$.

Recall that two ideals I and I' are said to be equivalent, denoted by $I \equiv I'$, if there exists $a \in K^*$ such that $I' = aI$.

Proposition 5 *Let (I, b) and (I', b') be two ideal lattices. Suppose that $(I, b) \simeq (I', b')$. Then $I \equiv I'$ and $b \equiv b'$.*

Proof This is clear from the definitions. □

2 Integral ideal lattices

We keep the notation of §1. In particular, K is an algebraic number field, and $\mathcal{O}$ the ring of integers of K. Let $\mathcal{D}_K$ be the different of K, and let D_K be its discriminant.

In this section we suppose that the chosen involution $^-$ preserves K. Let F be the fixed field of this involution. Then either $K = F$ or K is a quadratic extension of F.

The aim of this section is to study *integral* ideal lattices. Recall that a lattice (L, b) is *integral* if $b(x, y) \in \mathbf{Z}$ for all $x, y \in L$; it is said to be *even* if $b(x, x) \in 2\mathbf{Z}$ for all $x \in L$.

Proposition 6 *Let $b : I \times I \to \mathbf{R}$, $b(x, y) = \mathrm{Tr}(\alpha x \overline{y})$, be an ideal lattice. Then (I, b) is integral if and only if*

$$\alpha I \overline{I} \subset \mathcal{D}_K^{-1}.$$

Proof This follows immediately from the definition. □

For any non-zero integer d, let us denote by $\mathcal{L}_d$ the set of integral ideal lattices of determinant d. Set $\mathcal{C}_d(K,^-) = \mathcal{L}_d/\simeq$, and $\mathcal{C}_d(^-) = \mathcal{L}_d/\equiv$. As usual, we denote by $\mathcal{C}(K)$ the ideal class group of K.

We have two projection maps

$$p_1 : \mathcal{C}_d(K,^-) \to \mathcal{C}(K),$$

$$p_2 : \mathcal{C}_d(K,^-) \to \mathcal{C}_d(^-).$$

Several natural questions concerning ideal lattices can be formulated in terms of the sets $\mathcal{C}_d(K,^-)$, $\mathcal{C}_d(^-)$, and of the the projection maps p_1 and p_2. In particular, it is interesting to determine the images and the fibres of these maps. As we will see below, the results are far from complete, especially concerning the map p_2.

Note that if an ideal lattice (I, b) given by $b(x, y) = \mathrm{Tr}(\alpha x \overline{y})$ belongs to $\mathcal{L}_d$, then $\mathrm{N}(\alpha I \overline{I} \mathcal{D}_K) = |d|$. Rather than fixing $|D|$, it turns out that it is more convenient to fix the $\mathcal{O}$-ideal $\alpha I \overline{I}$. The $\mathcal{O}$-ideal $\alpha I \overline{I}$ will be called the *norm* of the ideal lattice (I, b).

For any ideal $\mathcal{A}$, let $\mathcal{L}_{\mathcal{A}}$ be the set of ideal lattices of norm $\mathcal{A}$. Set $\mathcal{C}_{\mathcal{A}}(K,^-) = \mathcal{L}_{\mathcal{A}}/\simeq$, and $\mathcal{C}_{\mathcal{A}}(^-) = \mathcal{L}_{\mathcal{A}}/\equiv$. Again, we have the projection maps

$$p_1 : \mathcal{C}_{\mathcal{A}}(K,^-) \to \mathcal{C}(K),$$

$$p_2 : \mathcal{C}_{\mathcal{A}}(K,^-) \to \mathcal{C}_{\mathcal{A}}(^-).$$

Let us denote by $\mathcal{C}_{\mathcal{A}}(K)$ the image of p_1.

The case where $\mathcal{A}$ is the ring of integers $\mathcal{O}$ of K is especially interesting. If (I, b) and (I', b') are two ideal lattices given by $b(x, y) = \mathrm{Tr}(\alpha x\overline{y})$ and $b'(x, y) = \mathrm{Tr}(\alpha' x\overline{y})$, then we define their product $(I, b)(I', b') = (II', bb')$ by setting II' to be the product of the ideals I and I', and $(bb')(x, y) = \mathrm{Tr}(\alpha.\alpha' x\overline{y})$. If (I, b) and (I', b') have norm $\mathcal{O}$, then their product is again an ideal lattice of norm $\mathcal{O}$, hence we obtain a product on $\mathcal{C}_{\mathcal{O}}(K,^-)$.

Proposition 7 *$\mathcal{C}_{\mathcal{O}}(K,^-)$ is a group with respect to the above product.*

Proof This is clear. □

Proposition 8 *For any ideal $\mathcal{A}$, the set $\mathcal{C}_{\mathcal{A}}(K,^-)$ is either empty, or a principal homogeneous space over the group $\mathcal{C}_{\mathcal{O}}(K,^-)$.*

Proof If (I, b) has norm $\mathcal{A}$ and (I', b') norm $\mathcal{O}$, then the product (II', bb') has norm $\mathcal{A}$. Hence we obtain a structure of homogeneous space of $\mathcal{C}_{\mathcal{A}}(K,^-)$ over $\mathcal{C}_{\mathcal{O}}(K,^-)$. Let us check that it is a principal homogeneous space. This follows from the fact that if $\alpha I\overline{I} = \beta J\overline{J}$, then $\alpha\beta^{-1}(IJ^{-1})\overline{(IJ^{-1})} = O$. □

We denote by U_K be the group of units of K, by U_F the group of units of F, and by $\mathrm{N}_{K/F}$ the norm from K to F.

Proposition 9 (i) *If the involution is trivial, then we have the exact sequence of groups*

$$1 \to U_K/U_K^2 \to \mathcal{C}_{\mathcal{O}}(K,^-) \xrightarrow{p_1} \mathcal{C}_{\mathcal{O}}(K) \to 0.$$

(ii) *If the involution is non-trivial, then we have the exact sequence of groups*

$$1 \to U_F/\mathrm{N}_{K/F}(U_K) \to \mathcal{C}_{\mathcal{O}}(K,^-) \xrightarrow{p_1} \mathcal{C}_{\mathcal{O}}(K) \to 0.$$

Some results about the order of $U_F/\mathrm{N}_{K/F}(U_K)$ are given in Bayer (1982), §2.

Note that if $K = F$, then $\mathcal{C}_{\mathcal{O}}(K)$ is the set of elements of order at most 2 in $\mathcal{C}(K)$. If $K \neq F$, then $\mathcal{C}_{\mathcal{O}}(K)$ is the *relative class group* $\mathcal{C}(K/F)$, that is the kernel of the norm map $\mathrm{N}_{K/F} : \mathcal{C}(K) \to \mathcal{C}(F)$. It is well-known that $\mathrm{N}_{K/F}$ is onto if K/F is ramified, and has cokernel of order 2 if K/F is unramified (see for instance Bayer 1982, propostion 1.2).

Quadratic fields

Suppose that K is a quadratic field, $K = \mathbf{Q}(\sqrt{d})$ where d is a square-free integer, and that the involution $^- : K \to K$ is given by $\overline{\sqrt{d}} = -\sqrt{d}$.

Let $b : I \times I \to \mathbf{Z}$, $b(x, y) = \mathrm{Tr}(\alpha x \overline{y})$, be an integral, even ideal lattice. Then $\alpha \in (1/\mathrm{N}(I))\mathbf{Z}$. In other words, b is an integral multiple of $b_I : I \times I \to \mathbf{Z}$, $b_I(x, y) = \mathrm{Tr}((1/\mathrm{N}(I))x\overline{y})$. Note that the quadratic form associated to b_I is $q_I : I \to \mathbf{Z}$, $q_I(x) = \mathrm{N}(x)/\mathrm{N}(I)$. We have $\det(b_I) = -D_K$.

Set $D = -D_K$. Gauss defined a correspondence between ideal classes of K and binary quadratic forms of determinant D, which sends an ideal I to the quadratic form q_I. The precise statement will be given below, as well as a way of deriving it using the notion of ideal lattice.

Let us first note that $\mathcal{C}_{\mathcal{O}}(K,^-) = \mathcal{C}_D(K,^-)$. Indeed, if (I, b) is an ideal lattice of norm $\mathcal{O}$, then $b = \pm b_I$, hence it has determinant D. Conversely, an ideal lattice of determinant D has norm $\mathcal{O}$.

We can apply the results of the first part of this section to $\mathcal{C}_D(K,^-)$. In particular, by Proposition 7, it is a group. Any ideal I satisfies $(1/\mathrm{N}(I))I\overline{I} = O$, hence $\mathcal{C}_O(K) = \mathcal{C}(K)$. The involution is non-trivial, so we can apply Proposition 9(ii), and obtain the exact sequence

$$1 \to \{\pm\}/\mathrm{N}(U_K) \to \mathcal{C}_D(K,^-) \xrightarrow{p_1} \mathcal{C}(K) \to 0. \qquad (*)$$

We now need information concerning the set $\mathcal{C}_D(^-)$ and the map

$$p_2 : \mathcal{C}_d(K,^-) \to \mathcal{C}_d(^-).$$

Proposition 10 *Let b be an even, binary lattice with determinant D. Then*

(i) *There exists an ideal I in $K = \mathbf{Q}(\sqrt{d})$ such that (I, b) is an ideal lattice.*

(ii) *If I' is another ideal such that (I', b) is an ideal lattice, then $I' \equiv I$ or $I' \equiv \overline{I}$.*

(iii) *K is the only quadratic field over which b is an ideal lattice.*

Proof Let (L, b) be an even binary lattice with determinant D. Let R be the $\mathbf{Z}$-algebra associated to (L, b), that is,

$$R = \{(e, f) \in \mathrm{End}(L) \times \mathrm{End}(L) \mid b(ex, y) = b(x, fy)\}$$

(cf. Bayer-Fluckiger 1987). Recall that the product of R is given by (e, f) $(e'.f') = (ee', f'f)$, and that R is endowed with the involution $(e, f) \mapsto$

(f, e). If (L, b) is an ideal lattice over a quadratic field $K' = \mathbf{Q}(\sqrt{\delta})$, then by definition $b(\sqrt{\delta}x, y) = -b(x, \sqrt{\delta}y)$. Hence there exists $e \in \mathrm{End}(L)$ such that $(e, -e) \in R$.

Let us fix a $\mathbf{Z}$-basis of L, and let

$$\begin{pmatrix} 2A & B \\ B & 2C \end{pmatrix}$$

be the matrix of b in this basis. A straightforward computation shows that the matrix of e in this basis is an integral multiple of

$$E = \begin{pmatrix} B & 2C \\ -2A & -B \end{pmatrix}.$$

As the determinant of this matrix is D, the field K' has discriminant $-D$, hence $K' = K$. This proves (iii).

Let $O = \mathbf{Z}[w]$ be the ring of integers of K, with $w = \sqrt{d}$ if $d \equiv 2,\ 3 \pmod 4$, and $w = (1 - \sqrt{d})/2$ if $d \equiv 1 \pmod 4$. Letting w act by multiplication with E if $d \equiv 2,\ 3 \pmod 4$, and by multiplication with $(1 - E)/2$ if $d \equiv 1 \pmod 4$ provides L with a structure of $\mathcal{O}$-module. This proves that b is an ideal lattice over K, hence assertion (i). Moreover, we see that the only other way of making w act on L is by replacing E with $-E$. This proves (ii). □

Remark 6 Note that the proof of Proposition 10 is constructive: given an even, binary lattice b one constructs the ideals I and $\overline{I}$ over which b is an ideal lattice.

The following are immediate consequences of Proposition 10:

Corollary 2 *The set $\mathcal{C}_D(^-)$ is equal to the set of similarity classes of even binary lattices with determinant D.*

Corollary 3 *If (I, b) and (I', b') are two ideal lattices over K such that $b \simeq b'$, then either $I' \equiv I$ or $I' \equiv \overline{I}$.*

In order to recover the usual statement of Gauss' correspondence between classes of binary quadratic forms and lattices, we need slightly different equivalence relations.

Definition 5 (i) We say that two ideal lattices (I, b) and (I', b') are *strictly isomorphic* if there exists $a \in K^*$ with $\mathrm{N}(a) > 0$ such that $I' = aI$ and $b'(ax, ay) = b(x, y)$.

(ii) We say that two lattices (L, b) and (L', b') with $L \otimes_{\mathbf{Z}} \mathbf{Q} = L' \otimes_{\mathbf{Z}} \mathbf{Q}$ are *strictly equivalent* if there exists a $\mathbf{Z}$-linear isomorphism $f : L \to L'$ with $\det(f) > 0$ such that $b'(fx, fy) = b(x, y)$.

(iii) We say that two ideals I and I' are *strictly equivalent* if there exists $a \in K^*$ with $\mathrm{N}(a) > 0$ such that $I' = aI$.

Let us denote by $\mathcal{C}^s_D(K, ^-)$, respectively $\mathcal{C}^s_D(^-)$, the set of strict isomorphism classes, respectively the set of strict equivalence classes, of ideal lattices of determinant D. Let us denote by $\mathcal{C}^s(K)$ the strict (or narrow) ideal class group.

Let us denote by $\mathcal{L}^+_D$ be the set of positive-definite ideal lattices of determinant D, and set $\mathcal{C}^+_D(K, ^-) = \mathcal{L}/\simeq$, $\mathcal{C}^+_D(^-) = \mathcal{L}/\equiv$.

If K is an imaginary quadratic field, then the exact sequence (*) yields the isomorphism

$$\mathcal{C}^+_D(K, ^-) \simeq \mathcal{C}(K).$$

On the other hand, if K is a real quadratic field, then we obtain from (*) the isomorphism

$$\mathcal{C}^s_D(K, ^-) \simeq \mathcal{C}^s(K).$$

Note that if two ideal lattices are strictly isomorphic, then the corresponding ideals are strictly equivalent. This follows from the proof of Proposition 10.

Using this, we obtain the well-known fact that the ideal class group $\mathcal{C}(K)$ is isomorphic to the set of strict equivalence classes of positive-definite, even binary lattices of determinant D if K is imaginary; the strict ideal class group $\mathcal{C}^s(K)$ is isomorphic to the set of strict equivalence classes of even binary lattices of determinant D if K is real.

Cyclotomic fields of prime power conductor

Let p be a prime number, $r \geq 1$ an integer, and let ζ_{p^r} be a primitive p^rth root of unity. Suppose that $K = \mathbf{Q}(\zeta_{p^r})$, the corresponding cyclotomic field, and that the involution is complex conjugation. Recall that $O = \mathbf{Z}[\zeta_{p^r}]$, that there is exactly one ramified ideal P in the extension $K/\mathbf{Q}$, and that $\mathrm{N}(P) = p$. Hence the different $\mathcal{D}_K$ is a power of P. Let $D = |D_K|$.

Proposition 11 *We have* $\mathcal{C}_O(K, ^-) = \mathcal{C}_D(K, ^-)$.

Proof Let (I, b) be an integral ideal lattice with norm $\alpha I\bar{I}$ and determinant D. Then $\mathrm{N}(\alpha I\bar{I}) = 1$, and $\alpha I\bar{I} \subset \mathcal{D}_K^{-1}$. As $\mathcal{D}_K$ is a power of the single prime ideal P, this implies that $\alpha I\bar{I} = O$. This shows that $\mathcal{C}_D(K, ^-) \subset \mathcal{C}_O(K, ^-)$. Conversely, if (I, b) is an ideal lattice with norm $\mathcal{O}$, then the determinant of

(I, b) is $\pm D$. By Corollary 1, the determinant is positive, hence $\det(b) = D$. This proves the other inclusion, hence the proposition is proved. □

The fixed field of the involution is the maximal totally real subfield F of K. Applying Proposition 9(ii) and the remarks following, we have the exact sequence

$$1 \to U_F/\mathrm{N}_{K/F}(U_K) \to \mathcal{C}_D(K,^-) \xrightarrow{p_1} \mathcal{C}(K/F) \to 0.$$

The order of $U_F/\mathrm{N}_{K/F}(U_K)$ is 2^n, where $n = [K : \mathbf{Q}]$, cf. Bayer (1982), proposition 2.3. and example 2.5. The order of $\mathcal{C}(K/F)$, called the *relative class number of* K, is known in many cases, see for instance Washington (1982).

Let $\mathcal{L}_D^+$ be the set of positive-definite ideal lattices of determinant D, and let $\mathcal{C}_D^+(K,^-)$ be the set of isomorphism classes of these lattices, that is $\mathcal{C}_D^+(K,^-) = \mathcal{L}_D^+/\simeq$.

Let us denote by U_F^+ the set of totally positive units of F. Then we have the exact sequence

$$1 \to U_F^+/\mathrm{N}_{K/F}(U_K) \to \mathcal{C}_D^+(K,^-) \xrightarrow{p_1} \mathcal{C}(K/F).$$

If moreover the relative class number of K is odd, then $U_F^+ = \mathrm{N}_{K/F}(U_K)$ (cf. Shimura 1977, proposition A2), and we have the isomorphism $\mathcal{C}_D^+(K,^-) \simeq \mathcal{C}(K/F)$.

Examples over cyclotomic fields

Let m be an integer, ζ_m a primitive mth root of unity, and suppose that $K = \mathbf{Q}(\zeta_m)$ is the corresponding cyclotomic field. It is not known in general which are the ideal lattices over K, but many examples are available. For instance, the root lattices A_{p-1} (where p is a prime) are ideal lattices for $m = p$, the root lattice E_6 is an ideal lattice for $m = 9$ and E_8 for $m = 15, 20, 24$. A complete description of root lattices that are ideal lattices over cyclotomic fields is given in Bayer-Fluckiger & Martinet (1994), A2. Moreover, the Coxeter–Todd lattice is an ideal lattice for $m = 21$, and the Leech lattice for $m = 35, 39, 52, 56, 84$ (cf. Bayer-Fluckiger 1984, 1994 and the survey in Bayer-Fluckiger 1999). Let us also point out the computations in higher rank cases of Bachoc & Batut (1992), of Batut, Quebbemann & Scharlau (1995), as well as the construction by Nebe (1998) of a unimodular rank-48 lattice with minimum 6 that is an ideal lattice for $m = 65$. Finally, in Bayer-Fluckiger (2000) examples of *modular* ideal lattices are given when m is not a power of a prime p with $p \equiv 1 (\mathrm{mod}\, 4)$.

3 Generalization and examples

The aim of this section is to indicate a possible generalization of the notion of integral ideal lattice, and the usefulness of this notion in some parts of algebra and topology.

Let A be a finite dimensional **Q**-algebra with a **Q**-linear involution $^- : A \to A$. Let $\mathcal{O}$ be an order of A which is invariant under the involution. In this context, an *integral ideal lattice* will be a pair (I, b), where I is a (left) $\mathcal{O}$-ideal and $b : I \times I \to \mathbf{Z}$ is a lattice such that

$$b(\lambda x, y) = b(x, \overline{\lambda} y)$$

for all $x, y \in I$ and for all $\lambda \in O$. This is clearly a generalization of the notion of §2, where $A = K$ was a number field and $\mathcal{O}$ the ring of integers of K.

Proposition 12 *Suppose that A is semi-simple. Let (I, b) be an ideal lattice. Then there exists $\alpha \in A$ such that $b(x, y) = \mathrm{Tr}(x\alpha\overline{y})$.*

Proof This follows from the fact that, as A is semi-simple, $\mathrm{Tr} : A \times A \to \mathbf{Q}$ is non-degenerate. □

Integral ideal lattices naturally appear in several parts of mathematics. In the two examples below A is commutative. However, there are also very interesting examples where A is non-commutative, for instance in the study of polarized abelian varieties.

Knot theory

This example concerns odd-dimensional knots and their algebraic invariants. See Kearton (2000) or Kervaire & Weber (1978) for surveys of the relevant definitions and properties.

Let k be a positive integer, $k \equiv 3 \pmod 4$. Let $\Sigma^k \subset S^{k+2}$ be a fibred knot, and let $\Delta \in \mathbf{Z}[X]$ be the Alexander polynomial of Σ^k. Then Δ is monic, we have $\Delta(X) = X^{\deg(\Delta)}\Delta(X^{-1})$ and $\Delta(1) = \pm 1$. Suppose moreover that Δ has no repeated factors.

Let $A = \mathbf{Q}[X]/(\Delta) = \mathbf{Q}(\tau)$. Then A is a finite-dimensional, semi-simple **Q**-algebra. Let $^- : A \to A$ be the **Q**-linear involution induced by $\overline{\tau} = \tau^{-1}$. Set $O = \mathbf{Z}[X]/(\Delta) = \mathbf{Z}[\tau]$. Then $\mathcal{O}$ is an order of A.

Let M^{k+1} be a minimal Seifert surface of Σ^k. Set $r = k + 1/2$, and set

$$I = H_r(M^{k+1}, \mathbf{Z})/(\text{torsion}).$$

Then I is a rank one $\mathcal{O}$-module, hence isomorphic to an $\mathcal{O}$-ideal. The intersection form $b : I \times I \to \mathbf{Z}$ is a symmetric bilinear form of determinant $\Delta(1) = \pm 1$. Moreover, (I, b) is an ideal lattice. Indeed, the monodromy of the fibration induces an isomorphism $t : I \to I$ that preserves the intersection form. In other words, we have $b(tx, ty) = b(x, y)$ for all $x, y \in I$. The Alexander polynomial is also the characteristic polynomial of t, hence t acts as τ on I. We get $b(\tau x, y) = b(x, \tau^{-1} y)$. Noting that $\tau^{-1} = \overline{\tau}$, we see that $b(\lambda x, y) = b(x, \overline{\lambda} y)$ for all $x, y \in I$ and all $\lambda \in O$. Hence (I, b) is an ideal lattice.

We define the sets $\mathcal{C}_{\Delta(1)}(O, ^-)$, $\mathcal{C}_{\Delta(1)}(^-)$ and $\mathcal{C}_{\Delta(1)}(O)$, as well as the projection maps $p_1 : \mathcal{C}_{\Delta(1)}(O, ^-) \to \mathcal{C}_{\Delta(1)}(O)$, $p_2 : \mathcal{C}_{\Delta(1)}(O, ^-) \to \mathcal{C}_{\Delta(1)}(^-)$ as in §2.

These have topological significance. Indeed, the class of (I, b) in $\mathcal{C}_{\Delta(1)}(O, ^-)$ is an invariant of the isotopy class of the knot. Its image by p_1 is the Alexander module of the knot, and its image by p_2 is an invariant of the homeomorphism class of the minimal Seifert surface.

Moreover, if we suppose that the knot is simple, then these invariants are complete. The usefulness of this approach to solve concrete problems in knot theory is illustrated by several examples in Kearton (2000) and Bayer-Fluckiger (1999).

Symmetric, skew-symmetric and orthogonal matrices with a given characteristic polynomial

Let $f \in \mathbf{Z}[X]$ be a monic polynomial, and set $A = \mathbf{Q}[X]/(f)$, $O = \mathbf{Z}[X]/(f)$. Then A is a finite-dimensional $\mathbf{Q}$-algebra, and $\mathcal{O}$ is an order of A. The involution $^- : A \to A$. will be the *identity* (trivial involution).

Let $b_0 : L \times L \to \mathbf{Z}$ be the *unit lattice*. In other words, there exists a basis of L in which the matrix of b_0 is the identity matrix.

The following proposition is (essentially) due to Bender (1968):

Proposition 13 *There exists an integral symmetric matrix with characteristic polynomial f if and only if b_0 is an ideal lattice.*

Proof Let $M \in \mathrm{M}_n(\mathbf{Z})$ such that $M^t = M$ and that the characteristic polynomial of M is f. Let L be a free $\mathbf{Z}$-module of rank n, and let $(e_1, \dots, e_n)$ be a basis of L in which b_0 is the identity matrix. Let $m : L \to L$ be the endomorphism given by the matrix M in this matrix. Let us endow L with the $\mathcal{O}$-module structure induced by m (that is, the action of X is given by m). As

M is symmetric, we have $b_0(mx, y) = b_0(x, my)$ for all $x, y \in L$. This proves that b_0 is an ideal lattice.

Conversely, suppose that $b_0 : I \times I \to \mathbf{Z}$ is an ideal lattice. Let us denote by $m : I \to I$ the endomorphism given by the image of X in $\mathcal{O}$. Then the characteristic polynomial of m is f. As (I, b_0) is an ideal lattice, we have

$$b_0(mx, y) = b_0(x, my) \tag{$**$}$$

for all $x, y \in I$. Let $(e_1, \ldots, e_n)$ be a basis with respect to which the matrix of b_0 is the identity matrix. The relation $(**)$ then shows that $M^t = M$. This concludes the proof of the proposition. □

Similar results can be proved for skew-symmetric and orthogonal matrices with given characteristic polynomial. In these cases, the involution is non-trivial. It is induced by $X \mapsto -X$ in the first case, and by $X \mapsto X^{-1}$ in the second.

4 Real ideal lattices

So far, we have considered lattices up to isomorphism, rather than embedded in an euclidian space. However, it is often important to find suitable embeddings, and this will be the subject matter of this section.

Let K be a number field of degree n, and let $\mathcal{O}$ be its ring of integers. Let $\bar{\ } : K_{\mathbf{R}} \to K_{\mathbf{R}}$ be the canonical involution (cf. Proposition 4). In this section and the next, all lattices will be supposed *positive-definite*.

Suppose that the number field K has r_1 real embeddings, and r_2 pairs of imaginary embeddings. We have $n = r_1 + 2r_2$. Let $\sigma_1, \ldots, \sigma_{r_1}$ be the real embeddings, and let $\sigma_{r_1+1}, \ldots, \sigma_{r_2}$ be non-conjugate imaginary embeddings.

Let $\alpha = (\alpha_1, \ldots, \alpha_n)$ be a positive element of $K_{\mathbf{R}}$, in other words α_i is real and positive for all i. Let $\sigma_\alpha : K \to \mathbf{R}^n$ be the embedding defined by

$$\begin{aligned}\sigma_\alpha(x) \;=\; \Big(&\sqrt{\alpha_1}x_1, \ldots, \sqrt{\alpha_{r_1}}x_{r_1}, \sqrt{2\alpha_{r_1+1}}\Re(x_{r_1+1}), \sqrt{2\alpha_{r_1+1}}\Im(x_{r_1+1}),\\ &\ldots, \sqrt{2\alpha_{r_2}}\Re(x_{r_2}), \sqrt{2\alpha_{r_2}}\Im(x_{r_2})\Big),\end{aligned}$$

where $x_i = \sigma_i(x)$, $\Re$ denotes the real part and $\Im$ the imaginary part. Note that this definition differs slightly from the one in Bayer-Fluckiger 1999, Definition 5.1).

Proposition 14 *For any ideal I of K and any positive $\alpha \in K_{\mathbf{R}}$, the lattice $\sigma_\alpha(I) \subset \mathbf{R}^n$ is an ideal lattice. Conversely, for any ideal lattice (I, b) there exists an $\alpha \in K_{\mathbf{R}}$ such that the ideal lattice $\sigma_\alpha(I)$ is isomorphic to (I, b).*

Proof It is clear that $\sigma_\alpha(I) \subset \mathbf{R}^n$ is a lattice. A straightforward computation shows that $\sigma_\alpha(I)$ is isomorphic to the lattice $b : I \times I \to \mathbf{R}$ given by $b(x, y) = \mathrm{Tr}(\alpha x \overline{y})$. Hence it is an ideal lattice. Conversely, let (I, b) be an ideal lattice given by $b(x, y) = \mathrm{Tr}(\alpha x \overline{y})$, Then $\sigma_\alpha(I)$ is isomorphic to (I, b). □

The above proposition is useful in information theory. Indeed, let us recall that if $x = (x_1, \dots, x_n) \in \mathbf{R}^n$, then the *diversity* of x, denoted by $\mathrm{div}(x)$, is the number of non-zero x_is. Let $L \subset \mathbf{R}^n$ be a lattice. One defines the *diversity* of L, denoted $\mathrm{div}(L)$, by

$$\mathrm{div}(L) = \min\{\mathrm{div}(x) \mid x \in L, x \neq 0\}.$$

Lattices of high diversity tend to perform better than the Rayleigh fading channel (see Boutros *et al.* 1996, Boutros & Viterbo 1998). The following proposition is proved in Bayer-Fluckiger (1999) in some special cases:

Proposition 15 *Any ideal lattice can be embedded in a Euclidean space with diversity* $r_1 + r_2$.

Proof Let I be an ideal and let $\alpha \in K_{\mathbf{R}}$ be totally real and totally positive. It is easy to see that the lattice $\sigma_\alpha(I)$ has diversity $r_1 + r_2$. By Proposition 14, any ideal lattice can be realised under this form, so the proposition is proved. □

5 Arakelov invariants

We keep the notation of §4. In particular, K is a number field of degree n, and $\mathcal{O}$ its ring of integers. The involution $^- : K_{\mathbf{R}} \to K_{\mathbf{R}}$ is again the canonical involution, and all lattices in this section are supposed positive-definite.

A positive-definite ideal lattice with respect to the canonical involution is also called an *Arakelov divisor* of the number field K. Such a lattice defines a sphere packing in $\mathbf{R}^n$, and the density and thickness of this packing are natural invariants of the lattice. We call these here *Arakelov invariants*.

Definition 6 Let (L, b) be a lattice, and set $q(x) = b(x, x)$. Let $V = L \otimes_{\mathbf{Z}} \mathbf{R}$.

(i) The *minimum* of b is defined by $\min(b) = \inf\{q(x) \mid x \in L, x \neq 0\}$.
(ii) The *maximum* of b is by definition

$$\max(b) = \sup\{\lambda \in \mathbf{R} \mid \forall x \in V, \ \exists y \in L \text{ with } q(x - y) \leq \lambda\}.$$

Note that $R = \sqrt{\max(b)}$ is the *covering radius* of b, and $r = \sqrt{\min(b)}/2$ is its *packing radius*. The thickness and the density of the sphere packing associated to b are defined in terms of these quantities (see Conway & Sloane 1984, chapters I and II). We will here use the related notions of *Hermite invariants*.

Definition 7 Let (L, b) be a lattice. The *Hermite invariants* are defined as follows:

(i) $\gamma(b) = \dfrac{\min(b)}{\det(b)^{1/n}}$.

(ii) $\tau(b) = \dfrac{\max(b)}{\det(b)^{1/n}}$.

It is also useful to consider the best Hermite invariants for lattices of a given rank.

Definition 8

(i) $\gamma_n = \sup\{\ \gamma(b) \mid \text{rank}(b) = n\}$.
(ii) $\tau_n = \inf\{\ \tau(b) \mid \text{rank}(b) = n\}$.

These notions provide us with invariants of the ideal classes of K, and of K itself.

Definition 9 Let I be an ideal. Set

(i) $\gamma_{\min}(I) = \inf\{\gamma(b) \mid (I, b) \text{ is an ideal lattice }\}$.
(ii) $\gamma_{\max}(I) = \sup\{\gamma(b) \mid (I, b) \text{ is an ideal lattice }\}$.
(iii) $\tau_{\min}(I) = \inf\{\tau(b) \mid (I, b) \text{ is an ideal lattice }\}$.
(iv) $\tau_{\max}(I) = \sup\{\tau(b) \mid (I, b) \text{ is an ideal lattice }\}$.

As equivalent ideals carry isomorphic ideal lattices, these are actually invariants of the ideal classes. It is natural to also use these notions to define invariants of the field K, $\gamma_{\min}(K)$, $\gamma_{\max}(K)$, $\tau_{\min}(K)$, and $\tau_{\max}(K)$. Recall that D_K is the discriminant of K. If I is an ideal, let us denote by $\min(I)$ the smallest norm of an integral ideal equivalent to I.

Proposition 16 *Let I be an ideal. Then*

$$\gamma_{\min}(I) \geq \frac{n}{|D_K|^{1/n}} \min(I)^{2/n}.$$

Proof Let (I, b) be an ideal lattice. Set $q(x) = b(x, x)$. We have $q(x) = \text{Tr}(\alpha x \overline{y})$ for some positive $\alpha \in K_{\mathbf{R}}$. Recall that $\det(b) = \text{N}(\alpha)\text{N}(I)^2|D_K|$.

By the inequality between the arithmetic and geometric means, we have

$$\text{Tr}(\alpha x \overline{x}) \geq n\ \text{N}(\alpha x \overline{x})^{1/n} = n\ \det(b)^{1/n}|D_K|^{-1/n}\ \text{N}(I)^{-2/n}\ \text{N}(x)^{2/n}.$$

Hence

$$\frac{q(x)}{\det(b)^{1/n}} \geq \frac{n}{|D_K|^{1/n}} \left(\frac{\mathrm{N}(x)}{\mathrm{N}(I)}\right)^{2/n},$$

and this implies that

$$\frac{\min(b)}{\det(b)^{1/n}} \geq \frac{n}{|D_K|^{1/n}} \min(I)^{2/n}.$$

As $\gamma(b) = \min(b)/\det(b)^{1/n}$, the proposition is proved. □

Corollary 4 *We have*

$$\gamma_{\min}(O) = \frac{n}{|D_K|^{1/n}}.$$

Proof By Proposition 16 we have $\gamma_{\min}(O) \geq n/|D_K|^{1/n}$. On the other hand, the ideal lattice $b : O \times O \to \mathbf{Z}$ given by $b(x, y) = \mathrm{Tr}(x\overline{y})$ has minimum n and determinant $|D_K|$. Hence the equality holds. □

Note that this also implies that $\gamma_{\min}(O) = \gamma_{\min}(K)$.

Corollary 5 *For any ideal I we have*

$$\frac{\gamma_{\min}(I)}{\gamma_{\min}(O)} \geq \min(I)^{2/n}.$$

Proof This follows from Definition 9 and Proposition 16. □

The following is an immediate consequence of Corollary 5.

Corollary 6 *Let I be an ideal. If there exists an ideal lattice (I, b) with $\gamma(b) = \gamma_{\min}(O)$, then I is principal.*

Recall that the field K is said to be *Euclidean with respect to the norm* if for every $a, b \in O$, $b \neq 0$, there exist $c, d \in O$ such that $a = bc + d$ and $|\mathrm{N}(d)| < |\mathrm{N}(b)|$.

Proposition 17 *Suppose that $\tau_{\min}(O) < \gamma_{\min}(O)$. Then K is Euclidean with respect to the norm.*

Proof The argument of Bayer-Fluckiger (1999), proposition 4.1 gives the desired result. □

Example 1 Let $K = \mathbf{Q}(\zeta_{15})$. We have $n = 8$, $D_K = 3^4 5^6$. Hence $\gamma_{\min}(O) = 8/3^{1/2}5^{3/4}$. The root lattice E_8 is an ideal lattice over $\mathcal{O}$ (see for instance Bayer-Fluckiger 1999). We have $\det(E_8) = 1$. The covering radius of E_8 is 1 (cf. Conway & Sloane 1988), hence $\max(E_8) = 1$. This implies that $\tau(E_8) = 1$, therefore $\tau_{\min}(O) \leq 1$. This implies that $\tau_{\min}(O) < \gamma_{\min}(O)$, so by Proposition 17, K is Euclidean with respect to the norm. We have $\gamma(E_8) = 2$. It is known that $\gamma(E_8) = \gamma_8$, hence $\gamma_{\max}(O) = 2$. Summarising, we have

$$\tau_{\min}(O) \leq 1 < \gamma_{\min}(O) < \gamma_{\max}(O) = 2.$$

References

Bachoc, C. & C. Batut (1992), Etude algorithmique de réseaux construits avec la forme trace, *J. Exp. Math.* **1**, 184–190.

Batut, C., H.-G. Quebbemann & R. Scharlau (1995), Computations of cyclotomic lattices, *J. Exp. Math.* **4**, 175–179.

Bayer, E. (1982) Unimodular hermitian and skew-hermitian forms, *J. Algebra* **74**, 341–373.

Bayer-Fluckiger, E. (1984), Definite unimodular lattices having an automorphism of given characteristic polynomial, *Comment. Math. Helv.* **59** (1984), 509–538.

Bayer-Fluckiger, E. (1987), Principe de Hasse faible pour les systèmes de formes quadratiques, *J. Reine Angew. Math.* **378**, 53–59.

Bayer-Fluckiger, E. (1989), Réseaux unimodulaires, in *Séminaire de Théorie des Nombres de Bordeaux* **1**, 189–196.

Bayer-Fluckiger, E. (1999), Lattices and number fields, *Contemp. Math.* **241**, 69–84.

Bayer-Fluckiger, E. (2000), Cyclotomic modular lattices, *J. Théorie des Nombres de Bordeaux*, **12**, 273–280.

Bayer-Fluckiger, E. & J. Martinet (1994), Réseaux liés à des algèbres semisimples, *J. Reine Angew. Math.* **415**, 51–69.

Bender, E. (1968), Characteristic polynomials of symmetric matrices, *Pacific J. Math.* **25**, 433–441.

Boutros, J., E. Viterbo, C. Rastello & J.-C. Belfiore (1996), Good lattice constellations for both Rayleigh fading and Gaussian channels, *IEEE Trans. Information Theory*, **42**, 502–518.

Boutros, J. & E. Viterbo (1998), Signal space diversity: a power and bandwidth efficient diversity technique for the Rayleigh fading channel, *IEEE Trans. Information Theory*, **44**, 1453–1467.

Conner, P. & R. Perlis (1984), *Survey of Trace Forms of Algebraic Number Fields*, World Scientific.

Conway, J.H. & N.J.A. Sloane (1998), *Sphere Packings, Lattices and Groups*, Springer-Verlag.

Craig, M. (1978), Extreme forms and cyclotomy, *Mathematika* **25**, 44–56.

Craig, M. (1978), A cyclotomic construction of Leech's lattice, *Mathematika* **25**, 236–241.

Ebeling, W. (1994), *Lattices and Codes*, Vieweg.

Feit, W. (1978), Some lattices over $\mathbf{Q}(\sqrt{-3})$, *J. Algebra* **52**, 248–263.

van der Geer, G. & R. Schoof (1999), Effectivity of Arakelov divisors and the theta divisor of a number field, preprint.

Kearton, C. (2000), Quadratic forms in knot theory, in *Contemp. Math.* **272**, 135–154.

Kervaire, M. & C. Weber (1978), A survey of multidimensional knots, in Lecture Notes in Math. **685**, 61–134, Springer-Verlag.

Martinet, J. (1995), Structures algébriques sur les réseaux, in *Actes du Séminaire de Théorie des Nombres de Paris, 1992–1993*, London Mathematical Society Lecture Notes **215**, Cambridge University Press, 167–186.

Martinet, J. (1996)*Les Réseaux Parfaits des Espaces Euclidiens*, Masson.

Nebe, G. (1998), Some cyclo-quaternionic lattices, *J. Algebra* **199**, 472–498.

Neukirch, J. (1999), *Algebraic Number Theory*, Springer-Verlag.

Quebbemann, H.-G. (1981), Zur Klassifikation unimodularer Gitter mit Isometrie von Primzahlordnung, *J. Reine Angew. Math.* **326**, 158–170.

Serre, J.-P. (1970), *Cours d'Arithmétique*, P.U.F.

Shimura, G. (1977), On abelian varieties with complex multiplication, *Proc. London Math. Soc.* **34**, 65–86.

Washington, L.C. (1982), *Introduction to Cyclotomic Fields*, Springer-Verlag.

12

Integral Points and Mordell–Weil Lattices

Tetsuji Shioda

Abstract

We study the integral points of an elliptic curve over function fields from the viewpoint of Mordell–Weil lattices. On the one hand, it leads to a surprisingly simple determination of all integral points in some favorable situation. On the other hand, it gives a method to produce elliptic curves with 'many' integral points.

1 Introduction

The finiteness of the set of integral points of an elliptic curve, defined by a Weierstrass equation with integral coefficients in a number field, is due to Siegel; an effective bound was given by Baker.

The function field analogue of this fact is known. It is indeed considerably easier to prove, with stronger effectivity results. See Hindry & Silverman (1988), Lang (1990), Mason (1983) for example.

Yet it will require in general some nontrivial effort to determine all the integral points (e.g. with polynomial coordinates) of a given elliptic curve over a function field.

The purpose of this paper is to study this question from the viewpoint of Mordell–Weil lattices. Sometimes it gives a very simple determination of integral points. For example, we can show that the elliptic curve

$$E : y^2 = x^3 + t^5 + 1$$

defined over $K = \mathbf{C}(t)$ has exactly 240 'integral points' $P = (x, y)$ such that x, y are polynomials in t, and they are all of the form

$$x = gt^2 + at + b, \quad y = ht^3 + ct^2 + dt + e.$$

In fact, it has been known for some time that the structure of the Mordell–Weil lattice in question on $E(K)$ is isomorphic to the root lattice E_8 of rank 8 (see e.g. Shioda 1991a) and the rational points corresponding to the 240 roots of E_8 are integral points of the above form (see Lemma 10.5, Lemma 10.9 and Theorem 10.6 in Shioda 1990). Thus our new assertion here is that there are no more integral points other than those 240.

The proof is very simple, as will be given below, and makes use of the *height formula* and the *specialization map*. It should be emphasized that it does not rely on any previously known bounds (see Example 3.1).

The content of this article is as follows. In §2, we formulate the main results. In §3, we give a few examples including the above one. In §4, we consider the other direction in which we may produce the situation with 'many' integral points.

In this article, we mainly give the examples where the Mordell–Weil lattice is the root lattice E_8, but other lattices are also interesting for the theme of integral points. We hope to come back to this subject elsewhere.

2 Main results

To state our main results, let us recall and fix some standard notation in dealing with Mordell–Weil lattices (cf. Shioda 1990):

$K = k(C)$: the function field of C over k;
C: a smooth projective curve over k;
k: an algebraically closed field of characteristic zero;
$f : S \to C$: an elliptic surface with at least one singular fibre;
S: a smooth projective surface over k;
χ: the arithmetic genus of S (a positive integer);
$O : C \to S$: a given section of f ($f \circ O = \mathrm{id}_C$);
$R_f := \{w \in C | f^{-1}(w) \text{is reducible}\}$;
E: the generic fibre of f, an elliptic curve over K;
$E(K)$: the group of sections of f, which is identified with the Mordell–Weil group of K-rational points of E;
(P): the image curve of a section $P : C \to S$;
(PO): the intersection number of (P) and (O);
$\langle P, Q\rangle$: the height pairing ($P, Q \in E(K)$), as defined by Shioda (1990);
$\mathrm{sp}_v(P)$: the unique intersection point of (P) with the fibre $f^{-1}(v)$, $v \in C$;
$\mathrm{sp}_v : E(K) \to f^{-1}(v)^{\#}$: the specialization map at $v \in C$;

$f^{-1}(v)^{\#}$: the smooth part of $f^{-1}(v)$ given with the structure of a commutative algebraic group (Kodaira 1963, Neron 1964, Tate 1975)

Definition 1 Let Σ be a finite set of points of the base curve C. Given $P \in E(K)$, we say P is Σ-*integral* if (P) and (O) intersect only at the points of S lying over Σ. In other words, P is Σ-integral if and only if $(P) \cap (O) \cap f^{-1}(C - \Sigma) = \emptyset$.

As a special case when $\Sigma = \emptyset$, we call P *everywhere integral* if (P) and (O) do not intersect at all, i.e. $(PO) = 0$. (For instance, any torsion point different from O is everywhere integral (in characteristic zero).)

Theorem 1 *Assume that, for every point $v \in \Sigma$, the specialization map*

$$\mathrm{sp}_v : E(K) \longrightarrow f^{-1}(v)^{\#}$$

is injective. Then any Σ-integral point $P \in E(K)$ is everywhere integral, and the height $\langle P, P \rangle$ is bounded by twice the arithmetic genus χ of S, i.e.

$$\langle P, P \rangle \leq 2\chi.$$

Proof By the explicit height formula (Shioda 1990, Theorem 8.6), we have $\langle P, P \rangle = 2\chi + 2(PO) - \sum_{w \in R_f} \mathrm{contr}_w(P)$, where the summation ranges over $w \in C$ with reducible fibres $f^{-1}(w)$ and where $\mathrm{contr}_w(P)$ is a local contribution at w which is a non-negative rational number.

Now, suppose that a Σ-integral point $P \in E(K)$ is not everywhere integral. Then we should have $P \neq O$ and $(PO) > 0$.

Since P is Σ-integral, we have

$$(PO) = \sum_{v \in \Sigma} (PO)_v$$

where $(PO)_v$ denotes the intersection number of (P) and (O) at a point lying over v.

Then we would have $(PO)_v > 0$ for some $v \in \Sigma$, which would imply that $P \in \mathrm{Ker}(\mathrm{sp}_v)$. But then we must have $P = O$ by the injectivity of the map sp_v. This is a contradiction.

Thus any Σ-integral point P is everywhere integral. The height formula then shows that $\langle P, P \rangle \leq 2\chi$. □

In terms of coordinates, the above result can be rephrased as follows.

Suppose that E/K is given by a (generalized) Weierstrass equation:

$$E : y^2 + a_1xy + a_3y = x^3 + a_2x^2 + a_4x + a_6 \quad (a_i \in K)$$

with the origin $O : (x : y : 1) = (0 : 1 : 0)$ being the point at infinity. The elliptic surface $f : S \to C$ is nothing but the Kodaira–Néron model of E/K.

To fix the idea, we take the simplest case where $K = k(t)$, $C = \mathbf{P}^1$, and suppose $a_i \in k[t]$ for all i. In the case $\Sigma = \{\infty\}$, Σ-integral points are exactly those points $P = (x, y) \in E(K)$ such that x, y are polynomials in t. We will call them simply *integral points* (or sometimes $k[t]$-integral points) rather than $\{\infty\}$-integral points.

Further we assume that the above Weierstrass equation is *minimal* in the sense that, for $h \in k[t]$, if h^i divides a_i for all i, then h must be in k. Then the arithmetic genus χ of the elliptic surface S is given by the smallest integer m such that $\deg a_i \leq im$ for all i (cf. Shioda 1991a, 1993).

Theorem 2 *With the above notation, assume that the specialization map*

$$\mathrm{sp}_\infty : E(K) \longrightarrow f^{-1}(\infty)^\#$$

is injective. Then, for any integral point $P = (x, y) \in E(K)$ such that x, y are polynomials in t, the degree of x, y in t is bounded as follows:

$$\deg(x) \leq 2\chi, \quad \deg(y) \leq 3\chi.$$

Proof Take $\Sigma = \{\infty\}$ in Theorem 1. Then any integral point P satisfies $(PO) = 0$, i.e. the sections (P) and (O) do not intersect at all. First, since O is the point at infinity of E, the rational point $P = (x, y) \in E(K)$ must be finite (i.e. with no poles) over the affine t-line $\mathbf{A}^1 \subset \mathbf{P}^1$. This implies that both x, y are polynomials in t.

Next we look at the fibre at the infinity $t = \infty$. The '∞-model' of our E/K is obtained by the change of variables

$$\bar{t} = \frac{1}{t}, \quad \bar{x} = \frac{x}{t^{2\chi}}, \quad \bar{y} = \frac{y}{t^{3\chi}}.$$

Since (P) and (O) do not intersect at $t = \infty$ either, $\bar{x}$, and $\bar{y}$ must be finite at $\bar{t} = 0$, which implies that $\deg(x) \leq 2\chi$, $\deg(y) \leq 3\chi$. □

3 Examples

Example 3.1 First let us consider the example stated in §1:

$$E : y^2 = x^3 + t^5 + 1.$$

In this case, S is a rational elliptic surface with the arithmetic genus $\chi = 1$. There are 6 singular fibres over $t^5 = -1$ and $t = \infty$, but no reducible fibres. [For example, $f^{-1}(\infty)$ is a singular fibre of type II (a cuspidal cubic $\bar{y}^2 =$

$\bar{x}^3$), and we have $f^{-1}(\infty)^{\#} \simeq \mathbf{G}_a$, the additive group.] Then the Mordell–Weil lattice $E(K)$ is isomorphic to the root lattice E_8 (see Shioda 1990, §8, or Shioda 1991b).

In order to apply Theorem 2, let us show that the specialization map $\mathrm{sp}_\infty : E(K) \to k$ is injective.

For any $P \in E(K)$, $\mathrm{sp}_\infty(P)$ is the unique point of intersection of (P) and the singular fibre $f^{-1}(\infty)$. In particular, if P is one of the 240 points:

$$P : x = gt^2 + at + b, \quad y = ht^3 + ct^2 + dt + e,$$

then

$$u = \mathrm{sp}_\infty(P) = g/h \neq 0, \qquad g = \frac{1}{u^2}, \quad h = \frac{1}{u^3}.$$

Thus, together with u, all ζu belong to the image $\Im(\mathrm{sp}_\infty)$ for any 30th roots of unity $\zeta = \zeta_{30}^{\nu}$, because there are automorphisms

$$(x, y, t) \mapsto (\zeta_3 x, \pm y, \zeta_5 t)$$

of S. Here and below ζ_n denotes a primitive nth root of unity.

It follows that

$$\Im(\mathrm{sp}_\infty) \supset \mathbf{Z}[\zeta_{30}] \cdot u$$

for some $u \neq 0$. Since the latter has rank $\varphi(30) = 8$, we conclude that sp_∞ is injective (remember that $E(K)$ has rank 8).

Hence, by Theorem 2, we have

Corollary 1 *There are exactly* 240 $k[t]$*-integral points for the elliptic curve* $E : y^2 = x^3 + t^5 + 1$ *over* $k(t)$.

Remark (1) In order to explicitly write down these integral points P (i.e. the coefficients $a, b, \ldots, g, h$), we need a cyclic extension of degree 30 of the cyclotomic field $\mathbf{Q}(\zeta_{30})$, called the splitting field of $E/\mathbf{Q}(t)$ (Shioda 1998).

(2) We can also show a slightly stronger result that $k[t, (t^5+1)^{-1}]$-integral points of this elliptic curve E are the 240 points above only. For this, apply Theorem 1 with $\Sigma = \{t | t^5 = -1\} \cup \{\infty\}$.

(3) We compare with some previous work. According to Davenport's theorem (Davenport 1965), any integral point $P = (x, y)$ of the elliptic curve of the form

$$y^2 = x^3 + A(t) \quad (A(t) \in k[t], \deg A(t) = m)$$

has $\deg(x) \leq 2(m-1)$. Thus, in our case ($m = 5$), one would have to examine up to $\deg(x) \leq 8$ if one wants to verify our result. This should be possible with a help of computer, but would require some work. In terms of the lattice points, one needs to examine $P \in E_8$ of norm $\langle P, P \rangle \leq 8$. The number of lattice points of norm 2, 4, 6, 8 in E_8 are respectively, 240, 2160, 6720 and 17520 (cf. Conway & Sloan 1988, Table 4.9).

Example 3.2 More generally, let us consider the following elliptic curve

$$E = E_\lambda : y^2 = x^3 + x(p_0 + p_1 t + p_2 t^2 + p_3 t^3) + q_0 + q_1 t + q_2 t^2 + q_3 t^3 + t^5$$

$$\lambda = (p_0, p_1, p_2, p_3, q_0, q_1, q_2, q_3) \in \mathbf{A}^8.$$

We let k be the algebraic closure of $\mathbf{Q}(\lambda) = \mathbf{Q}(p_i, q_j)$. By Shioda (1991b),§8, the structure of the Mordell–Weil lattice on $E(k(t))$ is the root lattice E_8 for general λ, or more precisely for any λ satisfying the condition $\delta_0(\lambda) \neq 0$, where δ_0 is a certain polynomial in p_i and q_j. The singular fibre at $t = \infty$ is the same as in (1) above.

Now assume that λ is generic over $\mathbf{Q}$, i.e. assume that p_i, q_j are algebraically independent over $\mathbf{Q}$. Then the specialization map $\mathrm{sp}_\infty : E(K) \to k$ is injective. In fact, if $\{P_1, \ldots, P_8\}$ is a basis (or the fundamental roots) of $E(K) = E_8$, and $u_i = \mathrm{sp}_\infty(P_i)$, then $u_1, \ldots, u_8$ are algebraically independent over $\mathbf{Q}$ (see Shioda 1991b, §8). Hence, we have

Corollary 2 *The $k[t]$-integral points of E_λ for generic λ are exactly the* 240 *points of norm* 2.

Example 3.3 We continue the discussion of the previous example, but abandon the assumption that λ is generic.

By letting $q_0 = 1$ and all other p_i, q_j equal 0, we are in Example 3.1.

Similarly, we have the following special cases:

$$E : y^2 = x^3 + t^5 + t \quad \text{or} \quad y^2 = x^3 + x + t^5.$$

In both cases, we get the same conclusion about the integral points as before. See Shioda (1998), where the specialzation map sp_∞ (and the splitting field) is studied more closely for these cases.

4 Many integral points

So far we have examined the situation where the integral points are rather limited, i.e. there are no 'new' integral points.

Now let us look at the other side of the coin. If we are interested in finding elliptic curves with 'more' integral points than usual, then we should consider the case where $\mathrm{Ker}(\mathrm{sp}_v)$ is 'large'. Thus we are naturally led to the $\mathbf{Q}$-split case. To keep this article compact, we avoid generality, and explain the idea by taking the examples related E_8 as in §3.

With the notation of Example 3.2, let us limit the variables $u_1, \ldots, u_8$ to some rational numbers $u_1^0, \ldots, u_8^0$ in such a way that the Mordell–Weil lattice does not degenerate. Then the elliptic curve $E = E_\lambda$ is defined over $\mathbf{Q}(t)$, and we have that $E(k(t)) = E(\mathbf{Q}(t))$ has rank 8 (this is what we call the $\mathbf{Q}$-split situation).

In this case, the specialization map

$$\mathrm{sp}_\infty : E(\mathbf{Q}(t)) \longrightarrow \mathbf{Q}$$

is far from being injective, and in fact, $\mathrm{Ker}(\mathrm{sp}_\infty)$ has rank 7.

Now we choose $u_i^0 = 1$ for all i. This is an admissible choice, i.e. the Mordell–Weil lattice does not degenerate.

The elliptic curve $E/\mathbf{Q}(t)$ arising this way is given in Example (E_8), p. 685 in Shioda (1991b), together with explicit generators $\{P_1, \ldots, P_8\}$ of $E(\mathbf{Q}(t))$.

Proposition 1 *In this case, there are* 62 *new integral points of norm* 4, *in addition to the* 240 *integral points of norm* 2. *Thus there exist at least* 302 $\mathbf{Q}[t]$*-integral points for this elliptic curve.*

This is a consequence of the following two lemmas.

Lemma 1 *If* $P \in E(K)$ *belongs to* $\mathrm{Ker}(\mathrm{sp}_\infty)$ *and has norm* 4, *then* P *is an integral point.*

Proof By the height formula, we have

$$4 = \langle P, P \rangle = 2 + 2(PO).$$

Hence $(PO) = 1$. But since $P \in \mathrm{Ker}(\mathrm{sp}_\infty)$, (P) and (O) intersect only at $t = \infty$. This means that P is an integral point (in the sense that both x, y are polynomials in t). □

Lemma 2 *Let* $\{\alpha_1, \ldots, \alpha_8\}$ *be a basis of the root lattice* E_8 *as in the Dynkin diagram, and suppose* $s : E_8 \to \mathbf{Q}$ *is defined by* $s(\sum_i n_i \alpha_i) = \sum_i n_i$. *Then the number of vectors of norm* 4 *belonging to* $\mathrm{Ker}(s)$ *is* 62, *and they are given up to the sign as follows: (i)* 21 *vectors* $\alpha_i - \alpha_j$, $i < j$, *where* α_i *and* α_j *are not joined by a line; (ii)* 7 *vectors like* $\alpha_1 + \alpha_2 - (\alpha_4 + \alpha_5)$*; (iii)* 2 *vectors* $\alpha_1 + \alpha_2 + \alpha_3 - (\alpha_5 + \alpha_6 + \alpha_7)$, $\alpha_8 + \alpha_2 + \alpha_3 - (\alpha_5 + \alpha_6 + \alpha_7)$.

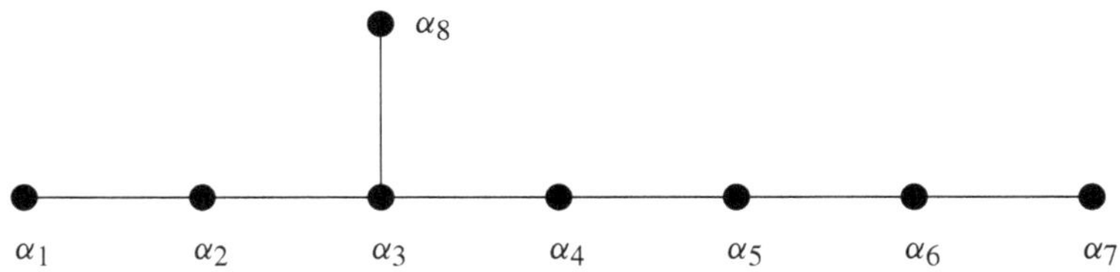

Fig. 1. Dynkin diagram of type E_8

Proof The verification is straightforward (though there are 2160 vectors of norm 4 in the lattice E_8. For this type of question, the construction of Shioda (1995) is useful). □

Remark It is likely that there are no more integral points. We have checked that there are none of norm 6 and 8.

Also we may ask if there exist elliptic curves over $\mathbf{Q}(t)$ (for which the Kodaira–Néron model S is a rational elliptic surface) which have more than 302 integral points.

On the other hand we should note that, even if $\mathrm{Ker}(\mathrm{sp}_\infty)$ is large, it does not guarantee that there are many integral points.

Proposition 2 *There exist elliptic curves $E/\mathbf{Q}(t)$ such that $E(\mathbf{Q}(t)) \simeq E_8$ which have no more integral points other than* 240 *points of norm* 2.

Proof By the finiteness of integral points, there is a positive integer N such that the norm of integral points in $E(\mathbf{Q}(t))$ is bounded by N. Fix such an N; for example, we may take $N = 36$ by Hindry & Silverman (1988), Proposition 8.2.

Let $B(N) = \{P \in E_8 | \langle P, P \rangle \leq N\}$. Choose a hyperplane $H \subset E_8$ which does not contain any points of $B(N) - \{O\}$. Let $s = 0$ be the equation of H; in other words, $s : E_8 \to \mathbf{Q}$ is a linear map with $\mathrm{Ker}(s) = H$. Then, by considering the specialization $u_i^0 = s(P_i)$ in the same way as above, we obtain an elliptic curve $E/\mathbf{Q}(t)$ for which $\mathrm{Ker}(\mathrm{sp}_\infty) \simeq H$. This proves the assertion. □

References

Conway, J. & N. Sloane (1988), *Sphere Packings, Lattices and Groups*, Springer-Verlag; 2nd ed. (1993); 3rd ed. (1999).

Davenport, H. (1965), On $f^3(t) - g^2(t)$, *Norske Vid. Selsk. Forrh.* **38**, 86–87.

Hindry, M. & J.H. Silverman (1988), The canonical height and integral points on elliptic curves, *Invent. Math.* **93**, 419–450.

Kodaira, K. (1963), On compact analytic surfaces II–III, *Ann. of Math.* **77**, 563– 626; **78**, 1–40; *Collected Works, III*, 1269–1372, Iwanami and Princeton University Press (1975).

Lang, S. (1990), Old and new conjectured Diophantine inequalities, *Bull. AMS* **23**, 37–75.

Mason, R.C. (1983), The hyperelliptic equation over function fields, *Math. Proc. Camb. Philos. Soc.* **93**, 219–230.

Néron, A. (1964), Modèles minimaux des variétés abéliennes sur les corps locaux et globaux, *Publ. Math. IHES* **21**.

Shioda, T. (1990), On the Mordell–Weil lattices, *Comment. Math. Univ. St. Pauli* **39**, 211–240.

Shioda, T. (1991a), Mordell–Weil lattices and sphere packings, *Am. J. Math.* **113**, 931–948.

Shioda, T. (1991b), Construction of elliptic curves with high rank via the invariants of the Weyl groups, *J. Math. Soc. Japan* **43**, 673–719.

Shioda, T. (1991c), Theory of Mordell–Weil lattices. In *Proc. ICM, Kyoto, 1990* **I**, 473–489.

Shioda, T. (1993), *Theory of Mordell–Weil Lattices and its Application*, (in Japanese), Lectures in Math. Sci. Univ. Tokyo 1.

Shioda, T. (1995), A uniform construction of the root lattices E_6, E_7, E_8 and their dual lattices, *Proc. Japan Acad.* **71A**, 140–143.

Shioda, T. (1998), Cyclotomic analogue in the theory of algebraic equations of type E_6, E_7, E_8, in *Proc. Seoul Conf. Lattices and Quadratic Forms (June 1998)*, CNM/AMS.

Tate, J. (1975), Algorithm for determining the type of a singular fiber in an elliptic pencil. In *Lecture Notes in Mathematics* **476**, Springer, 33–52.

13

Forty Years of Effective Results in Diophantine Theory

Enrico Bombieri

1 Thue, Siegel and Mahler

When I arrived in Cambridge in the Fall of 1963 to spend a year as a postgraduate with Davenport, I met Alan Baker for the first time. Alan was working very actively on problems of transcendence and diophantine approximation, and it was at that time that he obtained his first results on effective lower bounds for rational approximations to certain algebraic numbers.

The interest of the problem was pointed out by Thue's work in the decade from 1908 to 1918, with his celebrated result (Thue 1909) giving the finiteness of the number of solutions of the equation

$$F(x, y) = m$$

where $F(x, y)$ is a binary form of degree $r \geq 3$, with at least three distinct linear factors $x - \alpha y$ with $\alpha \in \mathbb{C}$, with rational integral coefficients. Here $m \neq 0$ is an integer and the equation is to be solved in rational integers x, y.

Thue showed that if $\alpha \in \overline{\mathbb{Q}}$ is a real algebraic number of degree $r \geq 3$ then for any $\varepsilon > 0$ there are only finitely many rational approximations $p/q \in \mathbb{Q}$ to

$$\left|\alpha - \frac{p}{q}\right| \leq \frac{1}{q^{1+r/2+\varepsilon}}. \tag{1}$$

From this result and a classical argument going back to Liouville one verifies that if $F(x, y) \in \mathbb{Z}[x, y]$ is irreducible over $\mathbb{Q}$ and of degree at least 3, then for every fixed $\varepsilon > 0$ the inequality

$$|F(x, y)| \geq c(F, \varepsilon) \max(|x|, |y|)^{\frac{r}{2}-1-\varepsilon} \tag{2}$$

holds for some positive constant $c(F, \varepsilon)$ and every $x, y \in \mathbb{Z}$. Notwithstanding its importance, the main weakness of this result is that Thue's theorem does not provide a procedure for finding all solution to (1), and as a consequence the constant $c(F, \varepsilon)$ in (2) is *ineffective*.

Why is that? What Thue's argument gives us is that (1) cannot have two solutions p_1/q_1, p_2/q_2 with $q_1 \geq C_1(\alpha, \varepsilon)$ and $\log q_2 \geq C_2(\alpha, \varepsilon) \log q_1$, for two explicitly computable large positive constants $C_1(\alpha, \varepsilon)$ and $C_2(\alpha, \varepsilon)$. The bound $C_2(\alpha, \varepsilon) \log q_1$ we obtain for the solutions depends on the denominator q_1 of the hypothetical solution p_1/q_1, and we know nothing about the size of q_1 beyond $q_1 \geq C_1(\alpha, \varepsilon)$.

Thue arrived at his result following the idea (which originates with Hermite's work on the exponential function) that rational number approximations to algebraic numbers are best understood by specialization of rational function approximations to algebraic functions. At first, he tried to find explicit constructions of best approximations (the so-called Padé approximations), succeeding in the case of algebraic functions of degree 3, such as $\sqrt[3]{1+x}$. In this case, they are expressed by means of hypergeometric polynomials. The case of general algebraic functions did not yield to his efforts, and he took a detour by relaxing what was required from the approximation. The construction of the approximation is indirect, using Dirichlet's pigeonhole principle. This new idea then became the powerful and ubiquitous *Siegel's Lemma*, of which there are today countless versions, as well as geometric interpretations relating it to the deep Riemann–Roch theorem in arithmetic geometry.

The importance and novelty of Thue's work was quickly noticed by Siegel and Mahler, and the subject flourished under them, in the years between 1920 and 1935. Some highlights are:

(1) Siegel's improvement of Thue's exponent $1 + r/2$ in (1) to the exponent $\min(s + r/(s+1))$, for $s = 1, 2, \ldots, r-1$, and its extension to simultaneous approximations;

(2) Siegel's theorem (using (a) above) on the finiteness of integral points on an affine curve C defined over a number field, provided C is of genus $g \geq 1$, or of genus 0 with at least three distinct points at ∞;

(3) Siegel's reduction of certain diophantine equations to the unit equation $A\xi + B\eta = 1$ with ξ, η units in a number field, notably the hyperelliptic equation $y^2 = f(x)$ and the Thue equation;

(4) Mahler's introduction of p-adic diophantine approximation methods, proving the finiteness of the number of solutions of the so-called Thue–Mahler equation

$$F(x, y) = p_1^{a_1} \cdots p_s^{a_s}$$

where F is a form in $\mathbb{Z}[x, y]$ with at least three distinct linear factors $x - \alpha_i y$, $p_1, \ldots, p_s$ are fixed primes, to be solved with $x, y \in \mathbb{Z}$ and $a_1, \ldots, a_s$ positive integers;

(5) the use of the hypergeometric method by Mahler and Siegel to obtain upper bounds for the number of integer solutions of binomial equations $ax^n - by^n = c$.

The ideas involved here are still alive today and they have found a fertile ground in combination with new methods arising from arithmetic algebraic geometry and Arakelov theory. Besides the famous Roth theorem and Schmidt's subspace theorem, which still belong to the 'classical' school, one may mention here, as results of the new school of thought in arithmetic geometry, the deep extension by Faltings & Wüstholz (1994) of Schmidt's subspace theorem, Vojta's new proof (Vojta 1991) of Faltings' theorem (formerly the Mordell conjecture) and Faltings' big theorem (Faltings 1991, 1994), to the effect that the rational points on a subvariety X of an abelian variety A are contained in finitely many translates $x_i B_i \subset X$ of abelian subvarieties $B_i \subset A$.

More recently, Evertse, Schlickewei & Schmidt (2001) have obtained the uniform bound $\exp((6n)^{3n}(r+1))$ for the number of non-degenerate solutions of the equation $x_1 + \cdots + x_n = 1$ in a multiplicative group of rank r and, in a remarkable paper, Rémond (2000) has also proved a uniform bound of a similar nature for the number of translates of abelian subvarieties occurring in Faltings' big theorem.

Unfortunately, except for bounds for the number of solutions, all these results are ineffective in the sense that they don't produce a theoretical algorithm for finding all solutions of the diophantine equations and inequalities considered, for the same reasons as in Thue's original work. The only exception was a paper† by Thue (1918), dealing with integer solutions of certain binomial equations $ax^n - by^n = c$.

2 Baker

The first *effective* results in diophantine approximation to algebraic numbers were obtained by Baker around 1963–64, using the hypergeometric method determining explicit Padé approximation to certain algebraic functions. His results were completely new and quite striking.

First (Baker 1964a), we have a completely explicit lower bound

$$\left| \left(\frac{a}{b}\right)^{1/n} - \frac{p}{q} \right| \geq C(a, b, n, k) q^{-k}$$

† Thue's paper remained unnoticed for a long time, perhaps because his results are stated somewhat awkwardly and because it appeared in an obscure publication.

provided $k > 2$ and a, b are rational integers such that $a > b$ and

$$a > (a-b)^{\rho}(3n)^{2\rho-2}, \qquad \rho = \frac{5}{2}\frac{(2k-1)}{(k-2)};$$

this was the first effective improvement on Liouville's lower bound for a class of algebraic irrationalities of degree greater than 2. It is notable that if a is not much larger than b then the exponent k can be made close to 2, thus approaching Roth's theorem in strength.

A second paper (Baker 1964b) followed, tackling specific irrationalities, with the famous result

$$\left|\sqrt[3]{2} - \frac{p}{q}\right| \geq \frac{10^{-6}}{q^{2.955}}$$

valid for all integers $p, q \geq 1$. In a third paper, Baker (1964c) showed how his effective method applied equally successfully to certain transcendental functions, notably $\log(1+z)$, to yield excellent effective irrationality measures† for a class of numbers $\log(a/b)$ with a/b rather close to 1.

The hypergeometric method, notwithstanding its successes, remained of limited applicability due to lack of control of the height of the Padé approximations, except in a few special cases, and clearly new ideas were needed to tackle the problem of obtaining effective results in these diophantine problems. The main breakthrough came when Baker (1966) removed the major stumbling block in the theory of linear forms in logarithms, nowadays called the theory of logarithmic forms.

In his famous list of problems for the twentieth century, Hilbert proposed, as his seventh, the question of the transcendency of α^{β} for α, β algebraic numbers, $\alpha \neq 0, 1$ and irrational β. The problem is equivalent to proving that if α_1 and α_2 are multiplicatively independent algebraic numbers then $\log \alpha_1$ and $\log \alpha_2$ are linearly independent over the algebraic numbers $\overline{\mathbb{Q}}$. Hilbert's seventh problem was independently settled by Gel'fond and Schneider in 1934, and Gel'fond obtained shortly afterwards an effective good measure of irrationality for α^{β}; his method extends easily to give a good effective lower bound for $L(\beta_1, \beta_2)$ where $L(x_1, x_2) = x_1 \log \alpha_1 + x_2 \log \alpha_2$ and $\beta_1, \beta_2 \in \overline{\mathbb{Q}}$. Here it is understood that $\alpha_1, \ldots, \alpha_n$ are non-zero algebraic numbers and that we fix a determination of the logarithm for $\log \alpha_i$, $i = 1, \ldots, n$.

Unfortunately, Gel'fond's basic construction breaks down if we consider a general logarithmic form $L(x_1, \ldots, x_n)$ in $n > 2$ variables. Until Baker's

† An irrationality measure κ for α is an inequality $|\alpha - p/q| > cq^{-\kappa}$ for some $c > 0$, for $p, q \in \mathbb{N}$, $q \geq 1$. The measure of irrationality is said to be effective if c and κ can be effectively determined.

work, there was a non-trivial lower bound, due to Gel'fond, only in the so-called rational case in which $(x_1, \ldots, x_n) \in \mathbb{Q}^n$, but this result made essential use of Thue's ineffective result on the diophantine equation $F(x, y) = m$, and as a consequence was also ineffective. Moreover, with the work of Gel'fond & Linnik (1948), and Gel'fond & Feldman (1949), it became clear that an effective good lower bound for logarithmic forms in three or more variables had deep consequences in mathematics, far beyond the realm of transcendental number theory.

Baker succeeded by introducing three new ideas. The first was to realize that the auxiliary construction involved not just one function but in fact *several functions* in *several complex variables*, which had to vanish to high order M on a certain set of points S; the second idea consisted in inventing an ingenious and completely new *extrapolation technique* proving that a subset of the original set of auxiliary functions had to vanish on a new set S' much bigger than S, still to an order at least $M/2$; the third, applying the extrapolation step *several times*, obtaining in the end an auxiliary function vanishing on a set so big that the desired conclusion could be reached by using a classical Liouville-type estimate and the non-vanishing of a Vandermonde determinant.

In terms of the quantity

$$\Lambda = \sum_{i=1}^{n} \beta_i \log \alpha_i,$$

assuming $\log \alpha_1, \ldots, \log \alpha_n$ linearly independent over $\mathbb{Q}$, Baker obtained an explicit effective lower bound as follows: *for any fixed $\kappa > n + 1$ there is an effectively computable positive constant $C = C(\alpha_1, \ldots, \alpha_n, n, \kappa, d)$ such that for any set of algebraic numbers β_i, $i = 1, \ldots, n$, not all 0, of degree at most d and height† at most H, $H \geq 2$, we have*

$$\log |\Lambda| \geq -C \cdot (\log H)^\kappa.$$

What matters here is the dependence on H, which is polynomial in $\log H$; the trivial lower bound would be linear in H. Note that any effective lower bound of the type

$$\log |\Lambda| \geq -c(\alpha_1, \ldots, \alpha_n, n, d) f(H)$$

with $f(H) = o(H)$ would suffice for many theoretical applications, but even in this weak form it does not seem to be any easier than Baker's result with $f(H) = (\log H)^\kappa$.

† Here the height of β is the maximum of the coefficients of a minimal equation for β over $\mathbb{Z}$.

In two series of papers, Baker (1966, 1967a, 1967b, 1968a, 1972, 1973, 1975) obtained significant sharpenings of his main result. The so-called rational case, in which $\beta_i \in \mathbb{Z}$, is particularly useful in applications, and the following theorem of Baker & Wüstholz (1993) is the state of the art (except possibly for a better numerical constant). In order to formulate it we need to choose the branch of the logarithms, and it is convenient to introduce a modified absolute logarithmic height†

$$h'(\alpha) = \max\big(h(\alpha), |\log(\alpha)|/\deg(\alpha), 1/\deg(\alpha)\big)$$

on $\overline{\mathbb{Q}}^*$. We have

Theorem 1 *In the rational case $\beta_i \in \mathbb{Z}$, if $\Lambda \neq 0$, we have the lower bound*

$$\log|\Lambda| \geq -c(n,d)\prod_{i=1}^{n} h'(\alpha_i)\log(eB)$$

with $B = \max|\beta_i|$ and $c(n,d) = 2400(15nd)^{2n+4}$.

The main new ingredients in the proof are:

(1) the introduction of division points and Kummer theory;
(2) the use of binomial polynomials $\binom{x}{n}$ as an integral basis for polynomials taking integer values at the integers, and its generalization to the so-called Δ-functions;
(3) sharp zero estimates.

All of these have to be skilfully combined and adapted to the problem at hand, before proving Theorem 1.

Variants of the method (interpolation determinants in place of Siegel's Lemma, Schneider's method in place of Gelfond's) studied by the French school in transcendence (Laurent, Mignotte, Waldschmidt) have been shown to be really useful for explicit numerical applications, especially in the important case $n = 2$. For example, Laurent *et al.* (1995) obtained the much better constant $31d^4$ in place of $c(2,d)$ at the cost of something like $(\log B)^2$ in place of $\log B$, but for many numerical applications the gain in the constant amply compensates for the loss of one logarithm. Generalizations in other directions, involving elliptic and abelian logarithms and studied by several authors (Masser, Chudnovsky, Hirata-Kohno, David) are also of great theoretical interest. The work David (1995), explicit in all constants, has found practical applicability in certain situations where the reduction of the problem to linear

† Here $h(\alpha)$ is the absolute Weil logarithmic height, $h(\alpha) = (1/\deg(\alpha))\log M(\alpha)$, with $M(\alpha)$ the Mahler measure of α.

forms in logarithms turned out to be numerically impractical because of the enormous size of units in the algebraic number fields involved.

The p-adic theory of logarithmic forms is also of great interest. It is easier in certain respects (estimates in a non-archimedean metric lead to clean inequalities), but is much more difficult in other ways (the exponential no longer is an entire function and the extrapolation does not work in the same fashion). The theory was finally put on solid grounds by Kunrui Yu (1999) (earlier attempts to use Kummer theory were flawed by errors), who also obtained for the first time an extension of Theorem 1, of the same strength, to the p-adic case.

3 Applications of Baker's theory

Many mathematicians have contributed to the development of the theory of logarithmic forms. Besides Baker himself, I will mention Stark (introduction of Kummer extensions and division points), Feldman (better arithmetical bases for polynomials), Masser, Wüstholz and Philippon (sharp zero estimates), Waldschmidt and his school (precise numerical results especially for $n = 2$), Van der Poorten, Kunrui Yu (linear forms in p-adic logarithms), Masser, Wüstholz, Hirata-Kohno, David (linear forms in elliptic logarithms).

The applications of Baker's theory are legion. In many cases, a relative weak lower bound suffices, but in other cases one needs very precise estimates, such as the multilinear bound in Theorem 1. I will begin by mentioning two applications in which such a refined bound is essential.

The first application is an improvement of Liouville's classical theorem, stating that a real algebraic number α of degree d has d as an effective measure of irrationality (Roth's celebrated theorem states that $2 + \varepsilon$ is a measure of irrationality for any $\varepsilon > 0$, but this result is ineffective):

$$\left|\alpha - \frac{p}{q}\right| \geq \frac{c(\alpha)}{q^d}$$

for some effective $c(\alpha) > 0$; an explicit elegant lower bound is

$$\left|\alpha - \frac{p}{q}\right| \geq 2^{-d} M(\alpha)^{-1} \frac{1}{\max(|p|, |q|)^d}$$

where $M(\alpha)$ is the Mahler measure of α. Baker's first result provided the first effective general improvement of Liouville's bound, namely

$$\left|\alpha - \frac{p}{q}\right| \geq c'(\alpha)\, \frac{e^{(\log q)^{\gamma(d)}}}{q^d}$$

for some positive constant $\gamma(d)$; this already was enough for an effective solution of the general Thue equation $F(x, y) = m$. However, in order to obtain an effective measure of irrationality for α strictly less than d, it was necessary to obtain a lower bound for $\log|\Lambda|$ which depended linearly on the largest $h'(\alpha_i)$, and this was achieved by Feldman (1971). Theorem 1 yields the Baker–Feldman theorem in a precise form, namely

Theorem 2 *Let α be real algebraic of degree d and let R be the regulator of the field $\mathbb{Q}(\alpha)$. Then α has an effective measure of irrationality at most $d - \eta(d)/R$, where $\eta(d)$ is a positive constant depending only on d.*

The second application is what I regard as one of the most striking applications of Baker's theory of logarithmic forms, namely Tijdeman's theorem (Tijdeman 1976):

Theorem 3 *There is an effectively computable upper bound C for all solutions $x, y, p, q > 1$ in positive integers of the Catalan equation*

$$x^p - y^q = 1.$$

Catalan's conjecture is that $3^2 - 2^3 = 1$ is the only solution. The proof is very clever. One may assume that p, q are prime numbers, and so write the equation in the form $x^p - y^q = \varepsilon$ with $\varepsilon = \pm 1$ and $p \geq q$. Elementary considerations, based on factorizations of $z^n \pm 1$, show that x and y can be written as $x - \varepsilon = p^\gamma r^q$ with $\gamma \in \{0, -1\}$ and $y + \varepsilon = q^\delta s^p$ with $\gamma \in \{0, -1\}$, with positive integers r and s (if $\gamma = -1$ or $\delta = -1$ then r or s must be divisible by p or q). Now the equation $x^p - y^q = \varepsilon$ and elementary inequalities show that

$$|p\gamma \log p - q\delta \log q + pq \log(r/s)| \leq 12p^3 r^{-q}.$$

An application of Theorem 1 with $n = 3$, $B = p^2$, $\alpha_1 = p$, $\alpha_2 = q$, $\alpha_3 = r/s$ shows that†

$$q \ll (\log p)^2 \log\log p. \tag{3}$$

Another inequality proved in a similar fashion is

$$\left| -q\delta \log q + p \log\left(\frac{p^\gamma r^q + \varepsilon}{s^q}\right)\right| \leq 4q^2 s^{-p}.$$

† We use Vinogradov's notation $\ll$ to denote an inequality up to an unspecified constant factor.

This time Theorem 1 with $n = 2$, $B = p, \alpha_1 = q, \alpha_2 = (p^\gamma r^q + \varepsilon)/s^q$ shows that

$$p \ll q \log q \log p. \tag{4}$$

From (3) and (4) it follows that $p \ll (\log p)^3 (\log\log p)^2$, hence p, and *a fortiori* q, are bounded. Then one easily concludes the proof, for example by applying Baker's bounds for the solutions of a superelliptic equation.

Remarkably, no other proof for the finiteness of solutions of Catalan's equation is known.† The method does not extend to the obvious generalization $x^p - y^q = m$ with $m \geq 2$, since the very first step of the proof cannot be carried out, due to the lack of factorizations. One may also ask how far we are from solving completely the Catalan equation by these methods. The current bounds for p, q are of order 10^{31}.

Perhaps the most famous application of Baker's theory was the solution of the famous 'class number 1' problem, namely the determination of all imaginary quadratic fields with class number 1. One finds easily nine imaginary quadratic fields with class number 1, namely those with discriminant -2, -3, -4, -7, -11, -19, -43, -67, -163, and it was conjectured by Gauss that there were no others. Through the work of Deuring, Mordell, Heilbronn, and the calculations of Evelyn and Linfoot, it was determined that there could be at most one more negative discriminant with class number 1, but their argument was non-effective and the existence of a tenth discriminant could not be ruled out.

Heegner published a paper (Heegner 1952) in which, by using methods from the theory of modular functions, he proved Gauss's conjecture. Unfortunately, the validity of Heegner's result was met with skepticism and only in 1966 did a paper by Stark appear, with a clear solution of the problem. Stark's paper ended with the same diophantine equation Heegner had considered, and this led to a reexamination of Heegner's work, which in the end was fully vindicated (see Stark 1969). Baker's solution (Baker 1966) provided a new approach to this problem, and eventually led to an effective bound, by Baker and Stark (Baker 1971, Stark 1971), for the discriminant of imaginary quadratic number fields with class number 2. It is interesting to note that this work on class numbers used logarithmic forms over quadratic fields (rather than the rational case). Finally, the problem of determining effectively, at least theoretically, all imaginary quadratic fields with class number below a given bound was solved by Goldfeld (1976) and Gross & Zagier (1986) using deep techniques from the theory of L-functions and modular forms.

Another important application of the theory is the effective solution of the

† *Note added in proof.* At last a complete solution of Catalan's conjecture has been obtained by Preda Mihailescu using methods from the theory of cyclotomic fields.

unit equation $x+y=1$ (or more generally, a polynomial equation $f(x,y)=0$ without factors of type $x^m Y^n - c$ or $x^m - cy^n$), with x and y S-units in a number field K, for any K, and S any finite set of places of K. This in turn yields the effective solution of the Thue–Mahler equation

$$F(x,y) = m\prod_{i=1}^{s} p_i^{a_i}, \qquad a_i \in \mathbb{N}$$

and hyperelliptic and superelliptic equations

$$y^m = a_0x^n + a_1x^{n-1} + \cdots + a_n,$$

as well as their generalizations to rings of S-integers in number fields. We refer to Győry's article in this volume for a detailed account of the applications of the theory of logarithmic forms to diophantine equations and for a complete bibliography.

There are other applications to completely different areas, such as algebra, algebraic topology, harmonic analysis, dynamical systems and ordinary and partial differential equations, many of them arising from delicate questions in diophantine approximation.

4 The Padé method

The hypergeometric, or Padé, method has been studied in depth by D.V. Chudnovsky and G.V. Chudnovsky, Beukers and several other mathematicians. When successful, it leads to very good bounds, but so far it seems limited to these hypergeometric cases and it seems that there are serious theoretical obstructions to it working in a general setting. We shall comment briefly on this point.

Basically, the exponential growth e^{cn} of the coefficients of the nth Padé approximation to an algebraic function is a *sine qua non* condition for the applicability of the method to diophantine approximation. This occurs in several explicit cases (all arising from hypergeometric functions, as in Baker (1964a,b), but a general attack on the problem yielded only quadratic exponential bounds e^{cn^2}, which were useless for the applications one had in mind.

It turns out that this quadratic exponential behaviour is the norm and the simply exponential behaviour is the exception. The first instance in which this was pointed out explicitly occurs in the work of Chudnovsky & Chudnovsky (1984). Consider an elliptic curve in Weierstrass form $y^2 = 4x^3 - g_2x - g_3$, with algebraic invariants $g_2, g_3 \in \overline{\mathbb{Q}}$, and consider y as an algebraic function of x, in a neighbourhood of an algebraic *non-torsion* point (x_0, y_0) on the elliptic curve. Then we may consider the Padé approximations of order n relative to the Taylor series of $y = y_0 + y_1(x-x_0) + \cdots$ of y, with centre x_0. In this case, they prove that the height of the associated Padé polynomials grows like e^{cn^2}, with $c > 0$.

This appears to be a general phenomenon. Consider an arbitrary algebraic function of one variable given by an equation $f(x, y) = 0$ of degree d in y, and the Padé polynomials $p_i(x)$ of degree n associated to the expansion of $p_0(x) + p_1(x)y + \cdots + p_{d-1}(x)y^{d-1}$ near $x = 0$. Bombieri, Cohen & Zannier (1997) generalizing the previous result, showed that the question of whether the height of these polynomials grows like e^{cn} or e^{cn^2} is strictly related to the geometric question of whether certain divisors of degree 0 on the curve $f(x, y) = 0$ are torsion points on the Jacobian of the curve or not. Moreover, in view of the recent solution of a conjecture of Bogomolov on small points on curves in abelian varieties, one can show that the jump from the e^{cn} behaviour (the torsion case) to the e^{cn^2} behaviour (the non-torsion case) is quite abrupt, ruling out intermediate results.

A characterization of conditions under which the Padé approximations of a single algebraic function y have height e^{cn} remains an open, and interesting, problem.

5 An alternative effective method

An alternative approach to effective results (Bombieri 1993, Bombieri & Cohen 1997), has been obtained through an extension of the original Thue–Siegel method.

Let $g_1, \ldots, g_n$ be multiplicatively independent algebraic numbers in a field K of degree d. We define

$$\Gamma = \langle g_1, \ldots, g_n \rangle \oplus \mathrm{tors}(K),$$

so Γ has rank n. We want to know how close elements Ag of a coset $A\Gamma$ can come to 1, assuming $Ag \neq 1$. An easy lower bound is†

$$|Ag - 1| \geq (2H(Ag))^{-d}$$

and the goal is to improve it to

$$|Ag - 1| \geq (2H(Ag))^{-\kappa d},$$

for any $\kappa > 0$, provided $H(Ag)$ is sufficiently large.

The basic idea is a reduction to the case $n = 1$ (linear forms in only two logarithms!), which is achieved by means of an elementary trick, as follows.

† Here $H(g)$ is the absolute Weil height, $H(g) = e^{h(g)}$.

Write $g = \varepsilon g_1^{m_1} \cdots g_n^{m_n}$. Now let Q, N be positive integers and let $L = \mathrm{lcm}(1, 2, \ldots, Q)$. By Dirichlet's theorem on simultaneous approximation, there are integers p_i and q, $1 \le q \le Q$, such that

$$\left| \frac{m_i}{LN} - \frac{p_i}{q} \right| \le \frac{1}{Q^{1/n} q}, \qquad i = 1, \ldots, n.$$

Now $r = LN/q$ is an integer, therefore we have the following statement: *there exists an integer* $r \equiv 0 \,(\mathrm{mod}\, N)$, *with* $LN/Q \le r \le LN$, *such that*

$$|m_i - rp_i| \le rQ^{-1/n} \quad \text{for} \quad i = 1, \ldots, n.$$

Therefore, if we set $a = \varepsilon A \prod g_i^{m_i - rp_i}$ and $g' = \prod g_i^{p_i}$ we have

$$|a(g')^r - 1| = |Ag - 1|,$$

whence

$$|a^{1/r} g' - 1| \ll H(a^{1/r} g')^{-\kappa d r},$$

if the inequality we want to prove is not satisfied. This means that $(g')^{-1}$ is a remarkably good approximation to $a^{1/r}$.

Since $h(a^{1/r}) \le (1/r)h(A) + nQ^{-1/n} \max h(g_i)$ is also small we have good control on the height of the number to be approximated, and direct methods based on diophantine approximation can be used here, for example an equivariant version of the original Thue–Siegel method for approximating roots of algebraic numbers. Note that, contrary to the non-equivariant case, we do not have a good Padé method for roots in the equivariant setting, and, even if we had the natural conjectural estimates, they would not be useful for our purposes here.

As a consequence, one can prove the following theorem.

Theorem 4 *Let K be a number field of degree d and let v be a place of K, lying over the rational prime p if v is finite.*

Let Γ be a finitely generated subgroup of K^* and let $g_1, \ldots, g_n$ be generators of Γ/tors. Let $g \in \Gamma$, $A \in K^*$ and $\kappa > 0$ be such that

$$0 < |1 - Ag|_v < H(Ag)^{-\kappa}.$$

Define $Q = \prod h'(g_i)$. Then we have

$$h(Ag) \le c(\kappa, n, d, v)\, Q \max(h'(A), Q)$$

where $c(\kappa, n, d, v)$ is an explicit function of κ, n, d and v.

An easy application of Theorem 4 is a new proof of the Baker–Feldman theorem, Theorem 2.

So far, this method seems to be less efficient than Baker's, but it handles uniformly both the archimedean and non-archimedean cases, and it seems to have potential for extensions to treating several places simultaneously.

6 The future: the *abc*-conjecture

A favourite motto of André Weil was *"Nihil est in arithmetico quod non prius fuerit in algebraico"*, which may be translated as: "There is nothing in number theory which did not previously exist in algebra." This has been vindicated in many instances, but there is an area in which the algebraic and geometric origin of number theory remains in the dark, namely *arithmetic ramification.*

We recall the *abc*-conjecture of Masser and Oesterlé.

Conjecture 1 *For every fixed $\varepsilon > 0$ there is a positive constant $C(\varepsilon)$ with the following property.*

If a, b, c are positive coprime integers with $a + b = c$, then

$$c \leq C(\varepsilon) \prod_{p|abc} p^{1+\varepsilon}, \tag{5}$$

where the product runs over the set of distinct prime divisors of abc.

For example, is it always true that

$$c \leq \prod_{p|abc} p^2 \ ? \tag{6}$$

Nothwithstanding the simplicity of this statement, such a result, if correct, has startling implications. Consider the notorious Fermat equation $x^n + y^n = z^n$, and apply the conjecture taking $a = x^n$, $b = y^n$, $c = z^n$. Then

$$\prod_{p|x^n y^n z^n} p = \prod_{p|xyz} p \leq z^3,$$

hence we would get

$$z^n \leq C(\varepsilon) z^{3(1+\varepsilon)}$$

and, taking $\varepsilon = 1$ and assuming $n > 6$, we would get $z^{n-6} \leq C(1)^2$. Since $z \geq 2$, this would give $n \leq 6 + [2 \log C(1)/ \log 2]$. In particular, one would obtain Fermat's Last Theorem for all sufficiently large exponents n, and for $n \geq 7$ assuming (6). It is not much more difficult to apply (5) or (6) to the

Catalan equation $x^m + 1 = y^n$ we have encountered before. For example, (6) immediately implies

$$y^n \leq (xy)^2 < y^{2\frac{n}{m}+2}.$$

This leaves us with the possibilities $m = 2$, or $n = 2$, or $n \leq 5$ and $m \leq 5$, which in fact have been already analyzed in the mathematical literature. The conclusion would be Catalan's conjecture that $1 + 2^3 = 3^2$ is the only solution of Catalan's equation.

The generalized Fermat equation

$$(GFE) \qquad x^p + y^q = z^r$$

with p, q, r positive integers and pairwise coprime x, y and z, falls easily under the scope of the *abc*-conjecture (5), provided

$$\frac{1}{p} + \frac{1}{q} + \frac{1}{r} < 1. \tag{7}$$

A few solutions of (GFE) satisfying (7) are known: Five 'small' solutions

$$1 + 2^3 = 3^2, \quad 2^5 + 7^2 = 3^4, \quad 7^3 + 13^2 = 2^9,$$

$$2^7 + 17^3 = 71^2, \quad 3^5 + 11^4 = 122^2,$$

and five 'large' solutions

$$\begin{aligned} 17^7 + 76271^3 &= 21063928^2, \\ 1414^3 + 2213459^2 &= 65^7, \\ 43^8 + 96222^3 &= 30042907^2, \\ 33^8 + 1549034^2 &= 15613^3, \\ 9262^3 + 15312283^2 &= 113^7, \end{aligned}$$

the first four found by Beukers and the last one by Zagier. These large solutions look quite unusual, and their existence begs some theoretical explanation.

As interesting as it may be, the *abc*-conjecture is too vague when we come to the constant factor $C(\varepsilon)$. Some theoretical basis has been provided by Baker (1998) for the following more precise form:

Conjecture 2 *There is an absolute constant K such that, with a, b, c as before, we have*

$$c \leq K \cdot \prod_{p|abc} (p/\varepsilon)^{1+\varepsilon}$$

for every ε in the range $0 < \varepsilon \leq 1$.

We recall the definition of the Weil height relative to a divisor D of a variety X over a number field K, which we suppose geometrically irreducible. Let D be a Cartier divisor on X, with associated line sheaf $\mathcal{O}(D)$ and rational section σ_D with divisor of zeros and poles $\mathrm{div}(\sigma_D) = D$. There are basepoint-free line sheaves $\mathcal{L}$, $\mathcal{M}$ such that $\mathcal{O}(D) \cong \mathcal{L} \otimes \mathcal{M}^{-1}$. Choose generating sections $\mathbf{s} = \{\mathbf{s_0}, \ldots, \mathbf{s_m}\}$ of $\mathcal{L}$ and $\mathbf{t} = \{t_0, \ldots, t_n\}$ of $\mathcal{M}$, and call the data

$$\mathcal{D} = (\sigma_D; \mathcal{L}, \mathbf{s}; \mathcal{M}, \mathbf{t})$$

a *presentation* of the divisor D. Let L/K be a finite extension of K and $|\ |_v$ be an absolute value on L, normalized so that for $a \in \mathbb{N} - 0$ we have

$$\log |a|_v = \frac{[L_v : \mathbb{Q}_v]}{[L : \mathbb{Q}]} \log \|a\|_p,$$

where $\|\ \|_p$ is the usual p-adic or real absolute value of $\mathbb{Q}$ such that $v|p$. Then for $P \in X(L)$ the *local height* of P relative to the presentation $\mathcal{D}$ and $v \in M_L$ is

$$\lambda_{\mathcal{D}}(P, v) = \max_i \min_j \log \left| \frac{s_j}{t_j \sigma_D}(P) \right|_v .$$

Since $\mathcal{L}$ and $\mathcal{M}$ are basepoint-free generated by the sections $\mathbf{s}$ and $\mathbf{t}$, we see that

$$\lambda_{\mathcal{D}}(P, v) = \begin{cases} = -\infty & \text{if } P \text{ is a pole of } \sigma_D \\ \in \mathbb{R} & \text{if } P \notin \mathrm{supp}(D) \\ = \infty & \text{if } P \text{ is a zero of } \sigma_D. \end{cases}$$

Consider the case in which $X = \mathbb{P}^1$ and D is the divisor $D = [0]+[1]+[\infty]$. Let $(x_0 : x_1)$ be standard homogeneous coordinates on $\mathbb{P}^1$, which we view as global sections of $\mathcal{O}_{\mathbb{P}^1}(1)$. Then D has a presentation

$$\mathcal{D} = (x_0 x_1 (x_0 - x_1);\ \mathcal{O}_{\mathbb{P}^1}(3), x_0^3, x_1^3;\ \mathcal{O}_{\mathbb{P}^1}, 1).$$

Let P be a point of $\mathbb{P}^1$, not $0, 1, \infty$. In affine coordinates, we can write P as $(1 : x)$ and then the local height relative to $\mathcal{D}$ is simply

$$\lambda_{\mathcal{D}}(P, v) = \max \left(\log \left| \frac{1}{x(1-x)} \right|_v, \log \left| \frac{x^2}{(1-x)} \right|_v \right).$$

The corresponding global height of the point $P = (1 : x)$ is

$$\begin{aligned} h_{\mathcal{D}}(P) &= \sum_v \lambda_{\mathcal{D}}(P, v) \\ &= \sum_v \max \left(\log \left| \frac{1}{x(1-x)} \right|_v, \log \left| \frac{x^2}{(1-x)} \right|_v \right) \end{aligned}$$

$$= \sum_v \max\left(0, \log\left|x^3\right|_v\right)$$
$$= 3h(P),$$

as one sees using the product formula

$$\sum_v \log|x(1-x)|_v = 0.$$

Thus with this presentation the global height $h_{\mathcal{D}}(P)$ is exactly $3h(P)$, and for a general presentation it is $3h(P) + O(1)$.

The divisor $K = -2[0]$ is canonical divisor of $\mathbb{P}^1$, and admits a presentation $\mathcal{K}$ such that $h_{\mathcal{K}}(P) = -2h(P)$. Thus we have

$$h(P) = h_{\mathcal{D}}(P) + h_{\mathcal{K}}(P).$$

Now let a, b, c are pairwise coprime positive integers with $a + b = c$. We take $P = (1 : x)$ with $x = a/c$, hence $1 - x = b/c$ and $h(x) = \log c$. Let p be a prime dividing abc to order k. Then we see that

$$\lambda_{\mathcal{D}}(P, p) = \max\left(\log\left|\frac{c^2}{ab}\right|_v, \log\left|\frac{a^2}{bc}\right|_v\right) = k\log p.$$

Thus the abc-conjecture can be stated in the following new form.

Conjecture 3 *Let D and K be the divisors on $\mathbb{P}^1$ given by $D = [0]+[1]+[\infty]$ and $K = -2[0]$ and let $\mathcal{D}$, $\mathcal{K}$ be presentations of these divisors. Then for every fixed $\varepsilon > 0$ there is $C = C(\varepsilon, \mathcal{D}, \mathcal{K}) < \infty$, such that for every $P \in \mathbb{P}^1(\mathbb{Q}) - D$ we have*

$$h_{\mathcal{D}}(P) + h_{\mathcal{K}}(P) \le \sum_{\lambda_{\mathcal{D}}(P,p)>0} \log p + \varepsilon h(P) + C.$$

There is nothing special about $\mathbb{P}^1$ and the divisor $[0] + [1] + [\infty]$. A general formulation is as follows.

Let X be a curve defined over an algebraic number field L and for $P \in X(\overline{L})$ define

$$d(P) = \frac{1}{[L(P) : \mathbb{Q}]} \log D_{L(P)/\mathbb{Q}}$$

where $D_{L(P)/\mathbb{Q}}$ is the absolute discriminant of the extension $L(P)/\mathbb{Q}$. Then one may optimistically pose the $(X/L, D)$-conjecture:

Conjecture 4 *Let X be a projective non-singular curve defined over a number field L, and let D, K, A on X be, respectively, an effective divisor sum of distinct points on X, a canonical divisor, an ample divisor, all defined over*

L. Let us fix presentations $\mathcal{D}$, $\mathcal{K}$, $\mathcal{A}$ of these divisors. Then for every fixed $\varepsilon > 0$ there is a constant $C = C(\varepsilon, X, L, \mathcal{D}, \mathcal{K}, \mathcal{A}) < \infty$ such that for $P \in (X - D)(\overline{L})$ we have

$$h_{\mathcal{D}}(P) + h_{\mathcal{K}}(P) \leq \sum_{\lambda_{\mathcal{D}}(P,v)>0} \log |1/p|_v + d(P) + \varepsilon h_{\mathcal{A}}(P) + C.$$

Here v runs over all finite places of $L(P)$ and p is the rational prime such that $v|p$.

However, it is unclear that this is a good generalization of the conjecture. The problem comes from the the definition of the conductor (the sum over finite places on the right-hand side) and from the introduction of the discriminant. The conductor comes from ramification, and the question arises whether there should be a contribution from ramification at the infinite places, perhaps involving the regulator of the field $L(P)$, and also whether the places of wild ramification for the extension $L(P)/L$ could contribute much more than $\sum \log |1/p|_v + d(P)$ to the right-hand side of the inequality†.

In this context, the *abc*-conjecture is a very special case of the $(X/L, D)$-conjecture for X a curve. Moreover, the *abc*-conjecture is a consequence of earlier conjectures of Vojta (1987), which were originally motivated as an arithmetic analogue of the Second Main Theorem (with ramification term) in Nevanlinna theory. In Vojta's theory, it turns out that both Roth's Theorem in diophantine approximation and Faltings' Theorem (the Mordell Conjecture) are instances of such an analogue, but without the ramification term.

Surprisingly, a beautiful argument by Elkies (1991) shows that the $(\mathbb{P}^1/L, [0]+[1]+[\infty])$-conjecture implies the $(X/L, D)$-conjecture above. The proof is not geometric and depends in an essential way on a characterization, due to Belyĭ [Bel], of the set of algebraic curves defined over $\overline{\mathbb{Q}}$ as the set of coverings of $\mathbb{P}^1$ unramified over $0, 1, \infty$.

Perhaps Belyĭ's theorem is a first sign that understanding arithmetic ramification may require methods which go beyond Weil's dream of finding in algebra and geometry the source of everything in number theory. However it may be, it is likely that future progress in this direction will eventually bring a beautiful harvest.

References

Baker, A., Rational approximations to certain algebraic numbers, *Proc. London Math. Soc.* **4** (1964), 385–398.

† I owe these remarks to some convincing arguments of Wüstholz.

Baker, A. (1964a), Rational approximations to $\sqrt[3]{2}$ and other algebraic numbers, *Quart. J. Math. Oxford* **15**, 375–383.

Baker, A. (1964b), Approximations to the logarithms of certain algebraic numbers, *Acta Arith.* **10**, 315–323.

Baker, A.(1966), Linear forms in the logarithms of algebraic numbers I, *Mathematika* **13**, 204–216.

Baker, A.(1967a), Linear forms in the logarithms of algebraic numbers II, *Mathematika* **14**, 102–107.

Baker, A. (1967b), Linear forms in the logarithms of algebraic numbers III, *Mathematika* **14**, 220–228.

Baker, A. (1968), Linear forms in the logarithms of algebraic numbers IV, *Mathematika* **15**, 204–216.

Baker A. (1971), Imaginary quadratic fields with class number 2, *Annals of Math.* **94**, 139–151.

Baker, A. (1972), A sharpening of the bounds for linear forms in logarithms I, *Acta Arith.* **21**, 117–129.

Baker, A. (1973), A sharpening of the bounds for linear forms in logarithms II, *Acta Arith.* **24** (1973), 33–36.

Baker, A. (1975), A sharpening of the bounds for linear forms in logarithms III, *Acta Arith.* **27**, 247–252.

Baker, A., Logarithmic forms and the *abc*-conjecture, in *Number theory (Eger 1996)*, K. Győry, A. Pethő & V. Sós (eds.), de Gruyter (1998), 37–44.

Baker, A. & G. Wüstholz (1993), Logarithmic forms and group varieties, *J. Reine Angew. Math.* **442**, 19–62.

Belyĭ, G.V. (1979), On the Galois extensions of the maximal cyclotomic field, (in Russian) *Izv. Akad. Nauk SSSR*, 267–276.

Bombieri, E., Effective diophantine approximation on $\mathbb{G}_m$, *Annali Sc. Norm. Sup. Pisa Cl. Sc., S. IV* **XX** (1993), 61–89.

Bombieri, E. & P.B. Cohen (1997), Effective diophantine approximation on $\mathbb{G}_m$ II, *Ann. Sc. Norm. Sup. Pisa Cl. Sc., S. IV* **XXIV**, 205–225.

Bombieri, E. & P.B. Cohen, with an Appendix by U. Zannier (1997), Siegel's Lemma, Padé approximations and Jacobians, *Ann. Sc. Norm. Sup. Pisa Cl. Sc., S. IV* **XXV**, 155–178.

Chudnovsky, D.V. & G.V. Chudnovsky (1984), Padé approximations to solutions of linear differential equations and applications to diophantine analysis. In *Number Theory, New York 1982*, Lect. Notes Math. **1052** Springer-Verlag, 85–167.

David, S. (1995), *Minorations de formes linéaires de logarithmes elliptiques*, Mém. Soc. Math. France (N.S.) **62**, iv+143 pp.

Elkies, N.D. (1991), ABC implies Mordell, *Int. Math. Res. Notices* **1**, 127–132.

Evertse, J.-H., Schlickewei, H.P. & W.M. Schmidt (2001), Linear equations in variables which lie in a multiplicative group, *Annals of Math.*, submitted.

Faltings, G. (1991), Diophantine approximation an Abelian varieties, *Annals of Math.* **133**, 549–576.

Faltings, G. (1994), The general case of Lang's conjecture. In *Barsotti Symposium in Algebraic Geometry*, V. Cristante & W. Messing (eds.), Academic Press, 175–182.

Faltings, G. & G. Wüstholz (1994), Diophantine approximation in projective spaces, *Invent. Math.* **136**, 109–138.

Feldman, N.I. (1971), An effective refinement of the exponent in Liouville's theorem, *Izv. Akad. Nauk* **35**, 973–990 (in Russian). English translation: *Math. USSR Izv.* **5** (1971), 985–1002.

Goldfeld, D.M. (1976), The class number of quadratic fields and the conjectures of Birch and Swinnerton-Dyer, *Ann. Sc. Norm. Sup. Pisa Cl. Sc. IV* **III**, 624–663.

Gross, B.H. & D. Zagier (1986), Heegner points and derivatives of L-series, *Invent. Math.* **84**, 225–320.

Heegner, K. (1952), Diophantische Analysis und Modulfunktionen, *Math. Z.* **56**, 227–253.

Laurent, M., M. Mignotte & Yu. Nesterenko (1995), Formes linéaires en deux logarithmes et déterminants d'interpolation, *J. Number Theory* **55**, 285–321.

Rémond, G. (2000), Décompte dans une conjecture de Lang, *Inventiones Math.*, **142**, 513–545.

Stark, H.M. (1969), On the 'gap' in a theorem of Heegner, *J. Number Th.* **1**, 16–27.

Stark, H.M. (1971), A transcendence theorem for class-number problems, *Annals of Math.* **94**, 153–173.

Thue, A. (1909), Über Annäherungwerte algebraischer Zahlen, *J. Reine Angew. Math.* **135**, 284–305.

Thue, A.(1918), Berechnung aller Lösungen gewisser Gleichungen von der form $ax^r - by^r = f$, *Skr. u. Vidensk. Kristiania*.

Tijdeman, R. (1976), On the equation of Catalan, *Acta Arith.* **29**, 197–209.

Vojta, P. (1987), *Diophantine Approximation and Value Distribution Theory*, Lect. Notes Math. **1239**, Springer-Verlag.

Vojta, P. (1991), Siegel's theorem in the compact case, *Annals of Math.* **133**, 549–576.

Yu, K. (1999), *p*-adic logarithmic forms and group varieties, *Acta Arith.* **89**, 337–378.

14

Points on Subvarieties of Tori

Jan-Hendrik Evertse

1 Introduction

Denote by $\mathbf{G}_{\mathbf{m}}^N$ the N-dimensional torus. Let K be any algebraically closed field of characteristic 0. Further, let Γ be a finitely generated subgroup of $\mathbf{G}_{\mathbf{m}}^N(K) = (K^*)^N$ and $\overline{\Gamma}$ its division group. We survey results about the structure of sets

$$X \cap \overline{\Gamma},$$

where X is an algebraic subvariety of $\mathbf{G}_{\mathbf{m}}^N$ defined over K.

We recall that $\mathbf{G}_{\mathbf{m}}^N$ consists of points $(x_1, \ldots, x_N)$ with $x_1 \cdots x_N \neq 0$. For $\mathbf{x} = (x_1, \ldots, x_N)$, $\mathbf{y} = (y_1, \ldots, y_N) \in \mathbf{G}_{\mathbf{m}}^N$ and $m \in \mathbf{Z}$ we define coordinatewise multiplication $\mathbf{x} * \mathbf{y} = (x_1 y_1, \ldots, x_N y_N)$ and exponentiation $\mathbf{x}^m = (x_1^m, \ldots, x_N^m)$. By a subvariety of $\mathbf{G}_{\mathbf{m}}^N$ defined over a field K we mean an irreducible Zariski-closed subset of $\mathbf{G}_{\mathbf{m}}^N$, that is a set $\{\mathbf{x} \in \mathbf{G}_{\mathbf{m}}^N : f_1(\mathbf{x}) = 0, \ldots, f_M(\mathbf{x}) = 0\}$ where $f_1, \ldots, f_M$ are polynomials in $K[x_1, \ldots, x_N]$ generating a prime ideal. By a *subtorus* of $\mathbf{G}_{\mathbf{m}}^N$ we mean a subvariety which is a subgroup of $\mathbf{G}_{\mathbf{m}}^N$, i.e., which is closed under coordinatewise multiplication. Thus, a subtorus is the set of solutions of a system of equations $X_1^{a_1} \cdots X_N^{a_N} = X_1^{b_1} \cdots X_N^{b_N}$ where the a_i, b_i are non-negative integers, and a subtorus is isomorphic to $\mathbf{G}_{\mathbf{m}}^{N'}$ for some $N' \leq N$. By a *torus coset* over K we mean a translate of a subtorus, i.e. $\mathbf{u} * H = \{\mathbf{u} * \mathbf{x} : \mathbf{x} \in H\}$ where $\mathbf{u} \in \mathbf{G}_{\mathbf{m}}^N(K)$ and where H is a subtorus. For more basic facts about subtori and torus cosets we refer to Schmidt (1996b), Section 2.

As before, let K be an algebraically closed field of characteristic 0, X a subvariety of $\mathbf{G}_{\mathbf{m}}^N$ defined over K, Γ a finitely generated subgroup of $\mathbf{G}_{\mathbf{m}}^N(K) = (K^*)^N$ and $\overline{\Gamma}$ its division group, i.e., the group of $\mathbf{x} \in \mathbf{G}_{\mathbf{m}}^N(K)$ for which there is a positive integer m with $\mathbf{x}^m \in \Gamma$. We define the rank of Γ to be the rank of $\Gamma/\Gamma_{\text{tors}}$. Chabauty (1938) proved the following result about the set $X \cap \Gamma$ (i.e. not with the division group).

Theorem A *Suppose that* $K = \overline{\mathbf{Q}}$ *and that* rank $\Gamma < N - \dim X$. *Then if* $X \cap \Gamma$ *is infinite, there is a torus coset* $\mathbf{u} * H \subset X$ *such that* $(\mathbf{u} * H) \cap \Gamma$ *is infinite.*

In his proof, Chabauty used a method based on p-adic power series which was introduced by Skolem.

Chabauty's work inspired Lang to formulate a general conjecture (cf. Lang 1983, p. 221) the following special case of which was proved by Laurent (1984).

Theorem B $X \cap \overline{\Gamma}$ *is contained in a finite union of torus cosets* $\mathbf{u}_1 * H_1 \cup \cdots \cup \mathbf{u}_t * H_t$ *with* $\mathbf{u}_i * H_i \subset X$ *for* $i = 1, \ldots, t$.

Laurent deduced his theorem from a result on linear equations. Let $a_1, \ldots, a_N \in K^*$ and consider the equation

$$a_1 x_1 + \cdots + a_N x_N = 1 \qquad \text{in } \mathbf{x} = (x_1, \ldots, x_N) \in \overline{\Gamma}. \tag{1}$$

To avoid easy constructions of infinite sets of solutions, we consider only *non-degenerate* solutions of (1), these are solutions with

$$\sum_{i \in I} a_i x_i \neq 0 \quad \text{for each non-empty subset } I \text{ of } \{1, \ldots, N\}. \tag{2}$$

It follows from work of the Evertse, van der Poorten & Schlickewei (see Evertse 1984), and Laurent (1984) that equation (1) has at most finitely many non-degenerate solutions.

The ingredients going into the proof of this result were:

> W.M. Schmidt's Subspace Theorem, see below, (with which one can handle equations (1) with solutions $\mathbf{x} \in \Gamma$ where $\Gamma \subset \mathbf{G}_{\mathbf{m}}^N(\overline{\mathbf{Q}})$);
> a specialization argument (with which one can extend the result to equations with solutions $\mathbf{x} \in \Gamma$ where $\Gamma \subset \mathbf{G}_{\mathbf{m}}^N(K)$ for some arbitrary field K of characteristic 0);
> some Kummer theory (to pass from Γ to $\overline{\Gamma}$).

Laurent proved his Theorem B by taking polynomials $a_1 M_1 + \cdots + a_s M_s$ vanishing identically on X, where the a_i are constants and the M_i are monomials, and applying the result on linear equations to $a_1 M_1 + \cdots + a_s M_s = 0$ where the M_i are considered to be the unknowns.

We now discuss quantitative versions of Theorem B, i.e., explicit upper bounds for the number of torus cosets t. This is joint work of Evertse & Schlickewei. We keep our conventions that K is an algebraically closed field of characteristic 0, Γ a finitely generated subgroup of $\mathbf{G}_{\mathbf{m}}^N(K)$, $\overline{\Gamma}$ its division group, and X a subvariety of $\mathbf{G}_{\mathbf{m}}^N$ defined over K. A linear subvariety of $\mathbf{G}_{\mathbf{m}}^N$ is defined by a set of polynomials of degree 1, which may have constant terms. The degree $\deg X$ of X is the number of points in the intersection of X with a general linear subvariety of $\mathbf{G}_{\mathbf{m}}^N$ of dimension $N - \dim X$. (In other words, if we embed $\mathbf{G}_{\mathbf{m}}^N$ into projective space $\mathbf{P}^N$ by means of the map $\iota : (x_1, \ldots, x_N) \mapsto (1 : x_1 : \cdots : x_N)$ and Y is the Zariski closure of $\iota(X)$ in $\mathbf{P}^N$, then we define $\deg X := \deg Y$, with the usual definition for the latter, cf. Hartshorne 1977, p. 52.)

The main tool is the following result of Evertse, Schlickewei & Schmidt (2002), which gives an explicit upper bound for the number of non-degenerate solutions of the linear equation (1):

Theorem 1 *Suppose Γ has rank $r \geq 0$. Then equation* (1) *has at most $e^{(6N)^{3N}(r+1)}$ non-degenerate solutions.*

For a historical survey of equation (1) we refer to Evertse & Schlickewei (1999).

By making explicit the arguments in Laurent's proof, it is possible to prove the following quantitative version of Theorem B:

Theorem 2 *Suppose* $\operatorname{rank}\Gamma = r \geq 0$, $\dim X = n$, $\deg X = d$. *Then $X \cap \overline{\Gamma}$ is contained in some union of torus cosets $\mathbf{u}_1 * H_1 \cup \cdots \cup \mathbf{u}_t * H_t$ where $\mathbf{u}_i * H_i \subset X$ for $i = 1, \ldots, t$ and where*

$$t \leq c(n,d)^{r+1} \qquad \text{with } c(n,d) = \exp\left(\left(6d\binom{n+d}{d}\right)^{5d\binom{n+d}{d}}\right). \tag{3}$$

The main features of this upper bound are its good dependence on r and its uniform dependence on n and d. It should be noted that the bound depends on $n = \dim X$ and not on N. However, if L is the smallest linear subvariety of $\mathbf{G}_{\mathbf{m}}^N$ containing X and X has codimension δ in L then $d \geq \delta + 1$ (cf. Griffiths & Harris, p. 173); hence the upper bound depends implicitly on δ.

Theorem 2 is the first result giving an explicit upper bound for the number of torus cosets in the most general case, but such explicit bounds have been given before in certain special cases. Let S be a finite set of places in some number field F. Denote by U_S the group of S-units and by U_S^N the N-fold direct product. From Győry (1992), Theorem 9, it follows that if X is defined over F then $X \cap (U_S)^N$ is contained in the union of at most $c_1(N, d, \#S, [F : \mathbf{Q}])$ torus cosets contained in X, with some explicit expression for c_1. From Schmidt (1996b), Theorem 2, it follows that if $\operatorname{rank}\Gamma = 0$, i.e. if $\overline{\Gamma} = U^N$, where U is the group of roots of unity in some algebraically closed field K of characteristic 0, then $X \cap \overline{\Gamma}$ is contained in the union of at most $c_2(N, d)$ torus cosets contained in X, with some explicit expression for c_2.

We deduce some corollaries of Theorem 2, keeping its notation. Let X^{exc} (the exceptional set of X) be the union of all torus cosets $\mathbf{u} * H$ of dimension ≥ 1 which are contained in X and let $X^0 = X \backslash X^{\text{exc}}$. For instance, if X is the variety given by equation (1), then X^0 consists precisely of the non-degenerate points of X, i.e., with (2). Since zero-dimensional torus cosets are simply points, we obtain at once from Theorem 2

Corollary 1 *Let Γ, X be as in Theorem 2. Then $X^0 \cap \overline{\Gamma}$ has cardinality at most $c(n, d)^{r+1}$.*

A special case of this is:

Corollary 2 *Let Γ be as in Theorem 2 and let X be an irreducible curve of degree d in $\mathbf{G}_{\mathbf{m}}^N$ defined over K. Suppose X is not a torus coset. Then $X \cap \overline{\Gamma}$ has cardinality at most $e^{(6d(d+1))^{5d(d+1)}(r+1)}$.*

A qualitative version of this result (giving only the finiteness) follows from work of Lang (1960) and Liardet (1974).

We now consider points that 'lie almost in $\overline{\Gamma}$'. To make this precise we need heights. Therefore we have to restrict ourselves to the case that X is defined over $\overline{\mathbf{Q}}$ and that $\Gamma \subset \mathbf{G}_{\mathbf{m}}^N(\overline{\mathbf{Q}}) = (\overline{\mathbf{Q}}^*)^N$.

Denote by h the usual logarithmic Weil height on $\mathbf{P}^N(\overline{\mathbf{Q}})$ (see below) and for $\mathbf{x} = (x_1, \ldots, x_N) \in \mathbf{G}_{\mathbf{m}}^N(\overline{\mathbf{Q}})$ put $h(\mathbf{x}) := h(1 : x_1 : \cdots : x_N)$. Let Γ be a finitely generated subgroup of $\mathbf{G}_{\mathbf{m}}^N(\overline{\mathbf{Q}})$ and $\overline{\Gamma}$ its division group. For $\varepsilon > 0$, we define the following sets:

$$T(\overline{\Gamma}, \varepsilon) = \{\mathbf{x} \in \mathbf{G}_{\mathbf{m}}^N(\overline{\mathbf{Q}}) : \exists\, \mathbf{y}, \mathbf{z} \text{ with } \mathbf{x} = \mathbf{y} * \mathbf{z}, \quad \mathbf{y} \in \overline{\Gamma},\ \mathbf{z} \in \mathbf{G}_{\mathbf{m}}^N(\overline{\mathbf{Q}}),\ h(\mathbf{z}) \leq \varepsilon\}, \tag{4}$$

$$C(\overline{\Gamma}, \varepsilon) = \{\mathbf{x} \in \mathbf{G}_{\mathbf{m}}^N(\overline{\mathbf{Q}}) : \exists\, \mathbf{y}, \mathbf{z} \text{ with } \mathbf{x} = \mathbf{y} * \mathbf{z}, \quad \mathbf{y} \in \overline{\Gamma},\ \mathbf{z} \in \mathbf{G}_{\mathbf{m}}^N(\overline{\mathbf{Q}}),\ h(\mathbf{z}) \leq \varepsilon \cdot (1 + h(\mathbf{y}))\}. \tag{5}$$

We may view $T(\overline{\Gamma}, \varepsilon)$ as a thickening of $\overline{\Gamma}$ and $C(\overline{\Gamma}, \varepsilon)$ as a truncated cone centered around $\overline{\Gamma}$. It is obvious that $T(\overline{\Gamma}, \varepsilon) \subset C(\overline{\Gamma}, \varepsilon)$. For instance, if $\operatorname{rank} \Gamma = 0$ then $T(\overline{\Gamma}, \varepsilon) = C(\overline{\Gamma}, \varepsilon)$ is the set of points of height $\leq \varepsilon$.

We mention the following result of Evertse, Schlickewei & Schmidt (2002).

Theorem 3 *Let $0 < \varepsilon < N^{-1} e^{-(4N)^{3N}}$. Suppose Γ has rank $r \geq 0$. Then the set of vectors $\mathbf{x} = (x_1, \ldots, x_N)$ satisfying*

$$x_1 + \cdots + x_N = 1, \quad \mathbf{x} \in C(\overline{\Gamma}, \varepsilon) \tag{6}$$

is contained in the union of at most $e^{(5N)^{3N}(r+1)}$ *proper linear subspaces of* $\overline{\mathbf{Q}}^N$.

One may wonder whether it is possible to deduce a quantitative result similar to Theorem 2 for sets

$$X \cap C(\overline{\Gamma}, \varepsilon)$$

if, in the proof of Theorem 2, one uses Theorem 3 instead of Theorem 1. This approach does not work. A problem is that Theorem 3 deals only with equations all of whose coefficients are equal to 1, whereas by going through the proof of Theorem 2 one arrives at equations of the shape

$$a_1x_1 + \cdots + a_Nx_N = 1 \quad \text{in } \mathbf{x} \in C(\overline{\Gamma}, \varepsilon) \tag{7}$$

with coefficients $a_1, \ldots, a_N$ over which one has no control.

One may try to reduce (7) to (6) by working with the tuple of variables $\mathbf{w} = (w_1, \ldots, w_N)$ where $w_1 = a_1x_1, \ldots, w_N = a_Nx_N$, and with the group Γ' generated by Γ and $\mathbf{a} = (a_1, \ldots, a_N)$. Then Γ' has rank $\leq r+1$. We clearly have $w_1 + \cdots + w_N = 1$. But then the problem remains that in general $\mathbf{x} \in C(\overline{\Gamma}, \varepsilon)$ does not imply that $\mathbf{w} \in C(\overline{\Gamma'}, \varepsilon)$. At this point our argument breaks down.

The situation is quite different if we restrict ourselves to points $\mathbf{x}$ belonging to the smaller set $T(\overline{\Gamma}, \varepsilon)$. Notice that for such $\mathbf{x}$ we do have $\mathbf{w} \in T(\overline{\Gamma'}, \varepsilon)$. Thus, by applying Theorem 3 but restricted to solutions in $T(\overline{\Gamma}, \varepsilon)$, it is possible to obtain the following analogue of Theorem 2 for sets $X \cap T(\overline{\Gamma}, \varepsilon)$:

Theorem 4 *Let* Γ *be a finitely generated subgroup of* $\mathbf{G}_{\mathbf{m}}^N(\overline{\mathbf{Q}})$ *of rank* $r \geq 0$. *Further, let* X *be a subvariety of* $\mathbf{G}_{\mathbf{m}}^N$ *defined over* $\overline{\mathbf{Q}}$ *of dimension* n *and degree* d. *Let* $c(n,d)$ *be the quantity from Theorem* 2. *Suppose that* $0 < \varepsilon < c(n,d)^{-1}$.

Then $X \cap T(\overline{\Gamma}, \varepsilon)$ *is contained in a union of torus cosets* $\mathbf{u}_1 * H_1 \cup \cdots \cup \mathbf{u}_t * H_t$ *where* $\mathbf{u}_i * H_i \subset X$ *for* $i = 1, \ldots, t$ *and where* $t \leq c(n,d)^{r+1}$.

Theorem 4 implies that in particular, $X^0 \cap T(\overline{\Gamma}, \varepsilon)$ has cardinality at most $c(n,d)^{r+1}$. Previously, Bombieri & Zannier (1995), Theorem 1, and in a more precise form Schmidt (1996b), Theorem 4, and David & Philippon (1999), Theorem 1.3, obtained a similar result in the special case that $r = 0$, i.e., that $T(\overline{\Gamma}, \varepsilon)$ is just the set of points with small height in $\mathbf{G}_{\mathbf{m}}^N(\overline{\mathbf{Q}})$. The result of Schmidt was one of the ingredients in the proofs of the results mentioned above.

The best one can obtain at present for the set $X \cap C(\overline{\Gamma}, \varepsilon)$ is the following 'semi-effective' result. By $h(X)$ we denote the logarithmic height of X (see below). Given a subtorus H of $\mathbf{G}_{\mathbf{m}}^N$, let X^H denote the union of all torus cosets $\mathbf{u}*H$ contained in X. The set X^H is Zariski-closed in X.

Theorem 5 (i). *Let Γ, X be as in Theorem 4. There are an ineffective constant $\alpha = \alpha(N, d, \Gamma) > 0$, depending only on N, d and Γ, and an effective constant $\beta = \beta(N, d) > 0$ depending only on N and d, such that for every ε with $0 < \varepsilon < 1/(\alpha + \beta h(X))$, the set $X^0 \cap C(\overline{\Gamma}, \varepsilon)$ is finite.*
(ii). *Let H be a positive-dimensional subtorus such that $X^H \neq \emptyset$ and such that $H \cap \Gamma$ is not a torsion group. Then for every $\varepsilon > 0$, $X^H \cap C(\Gamma, \varepsilon)$ is Zariski-dense in X^H.*

The proof of part (ii) is straightforward. Part (i) is a consequence of a 'semi-effective' version of Theorem B proved by Laurent (1974). The dependence on $h(X)$ of the upper bound for ε is necessary. It is an interesting open problem to prove a version of part (i) such that both constants α, β are effective and depend only on N and d.

Theorems 2, 4, 5 are unpublished work of Evertse & Schlickewei. Recently these results have been improved and generalized by Rémond (2001, 2002).

A semi-abelian variety is a commutative group variety A which has a subgroup variety T such that $T \cong \mathbf{G}_{\mathbf{m}}^N$ for some $N \geq 0$ and such that the factor group variety A/T is an abelian variety. Thus, a semi-abelian variety is a common generalization of a torus and an abelian variety. Lang (1983), p. 221, posed the following conjecture, which includes Theorem B as a special case. *If A is a semi-abelian variety defined over an algebraically closed field K of characteristic 0, X is a subvariety of A defined over K, Γ is a finitely generated subgroup of $A(K)$ and $\overline{\Gamma}$ its division group, then $X \cap \overline{\Gamma}$ is contained in the union of finitely many translates of semi-abelian subvarieties of A which are all contained in X.*

As is well-known, Faltings (1983) was the first to give a proof of Mordell's conjecture, which may be viewed as Lang's conjecture in the case that X is a curve of genus ≥ 2 and A is the Jacobian of X. Vojta (1991) gave a very different proof of this, thereby introducing new and powerful techniques from Diophantine approximation. By extending Vojta's ideas, Faltings (1991, 1994) proved Lang's conjecture in the case that A is an abelian variety and with '$X \cap \overline{\Gamma}$' replaced by '$X \cap \Gamma$'. For a more detailed treatment of Faltings' proof, see Edixhoven & Evertse (1993). Vojta (1996) generalized Faltings' result to arbitrary semi-abelian varieties, but still only for sets $X \cap \Gamma$. Finally, McQuillan (1995) extended this to sets $X \cap \overline{\Gamma}$ and thereby completed the proof of Lang's conjecture. McQuillan combined Vojta's result with ideas of Hindry (1988).

A subject for future research is of course to obtain quantitative analogues of Theorems 2–5 for (semi-)abelian varieties. Recently, Rémond (2000a, 2000b) proved the following quantitative analogue of Theorem 2 for abelian varieties. Let A be an abelian variety of dimension N defined over $\overline{\mathbf{Q}}$. Fix a symmetric, ample line bundle $\mathcal{L}$ on A. Let X be a subvariety of A of dimension n. Suppose that the degree of X with respect to $\mathcal{L}$ (i.e., the intersection number $\mathcal{L}^n \cdot X$) is equal to d. Further, let Γ be a finitely generated subgroup of $A(\overline{\mathbf{Q}})$ of rank r and $\overline{\Gamma}$ the division group of Γ. Then $X(\overline{\mathbf{Q}}) \cap \overline{\Gamma}$ is contained in some union $\bigcup_{i=1}^{t}(x_i + B_i)$ where $x_i + B_i$ $(i = 1, \ldots, t)$ is a translate of an abelian subvariety of A with $x_i + B_i \subset X$ and where $t \leq \left(c_{A,\mathcal{L}} d\right)^{N^{5(n+1)^2}(r+1)}$ with $c_{A,\mathcal{L}}$ an effectively computable number depending on A and $\mathcal{L}$. Recently, Rémond (2001) proved a generalization of Theorem 5 for semi-abelian varieties.

In order to give an overview of the main ingredients going into the proofs of the above mentioned results, we will sketch in the next section a proof of a weaker version of Corollary 1. We will deduce this weaker version directly from the basic results from Diophantine approximation, and not follow the route via the linear equation (1).

2 Proof of a weaker version of Corollary 1

We consider the special case that X is a subvariety of $\mathbf{G}_{\mathbf{m}}^N$ defined over $\overline{\mathbf{Q}}$ and that Γ is a finitely generated subgroup of $\mathbf{G}_{\mathbf{m}}^N(\overline{\mathbf{Q}})$. We will sketch a proof of the following result.

Theorem 6 *Suppose* $\deg X = d$, $\operatorname{rank} \Gamma = r$. *Then the set* $X^0 \cap \overline{\Gamma}$ *has cardinality at most* $c_1(N, d)^{r+1}$, *where* $c_1(N, d)$ *is an effectively computable constant depending only on* N *and* d.

We first show that it suffices to prove the result for the set $X^0 \cap \Gamma$ instead of $X^0 \cap \overline{\Gamma}$. Notice that in order to prove Theorem 6 it suffices to prove that every finite subset M of $X^0 \cap \overline{\Gamma}$ has cardinality at most $c_1(N, d)^{r+1}$. Let Γ' be the multiplicative group generated by M. Then Γ' is finitely generated and has rank $\leq r$. Now assuming Theorem 6 to be true for the set $X^0 \cap \Gamma'$ we get the required upper bound for the cardinality of M. Notice that to pass from Γ to $\overline{\Gamma}$, no Kummer theory is needed.

By means of a specialization argument we may extend Theorem 6 to the case that X is defined over any field K of characteristic 0 and that $\Gamma \subset \mathbf{G}_{\mathbf{m}}^N(K)$. We shall not work this out.

Theorem 6 is deduced from the following result.

Theorem 7 *Let X be a subvariety of $\mathbf{G}_\mathbf{m}^N$ defined over $\overline{\mathbf{Q}}$ and let Γ be a finitely generated subgroup of $\mathbf{G}_\mathbf{m}^N(\overline{\mathbf{Q}})$. Suppose $\deg X = d$, $\operatorname{rank}\Gamma = r$. Then $X^0 \cap \Gamma$ is contained in the union of at most $c_2(N,d)^{r+1}$ proper subvarieties of X, each of degree at most $c_3(N,d)$, where $c_2(N,d)$, $c_3(N,d)$ are explicitly computable constants depending only on N and d.*

Noticing that for each subvariety Y of X we have $X^0 \cap Y \subset Y^0$, we easily obtain by induction on $\dim X$ that $X^0 \cap \Gamma$ has cardinality at most $c_1(N,d)^{r+1}$. Together with the reduction argument explained above this gives Theorem 6.

Absolute values and heights

We give some basic facts about absolute values and heights. Let K be an algebraic number field and denote its ring of integers by $\mathcal{O}_K$. Denote by $\mathcal{M}(K)$ the set of places of K. Every archimedean place $v \in \mathcal{M}(K)$ corresponds either to an isomorphic embedding $\sigma : K \hookrightarrow \mathbf{R}$ or to a pair of complex conjugate embeddings $\{\sigma, \overline{\sigma} : K \hookrightarrow \mathbf{C}\}$. The non-archimedean places of K correspond to the prime ideals of $\mathcal{O}_K$. We define normalized absolute values $|\cdot|_v$ ($v \in \mathcal{M}(K)$) on K by

$$\begin{aligned}
|x|_v &= |\sigma(x)|^{1/[K:\mathbf{Q}]} \quad \text{if } v \text{ corresponds to } \sigma : K \hookrightarrow \mathbf{R};\\
|x|_v &= |\sigma(x)|^{2/[K:\mathbf{Q}]} = |\overline{\sigma}(x)|^{2/[K:\mathbf{Q}]} \quad \text{if } v \text{ corresponds to } \{\sigma, \overline{\sigma} : K \hookrightarrow \mathbf{C}\};\\
|x|_v &= (N\wp)^{-w_\wp(x)/[K:\mathbf{Q}]} \quad \text{if } v \text{ corresponds to the prime ideal } \wp,
\end{aligned}$$

where $N\wp = \#\mathcal{O}_K/\wp$ denotes the norm of $\wp$ and $w_\wp(x)$ the exponent of $\wp$ in the prime ideal decomposition of x. These absolute values satisfy the product formula

$$\prod_{v \in \mathcal{M}(K)} |x|_v = 1 \quad \text{for } x \in K^*.$$

For $\mathbf{x} = (x_0, \ldots, x_N) \in K^{N+1}$, $v \in \mathcal{M}(K)$ we define

$$||\mathbf{x}||_v = ||x_0, \ldots, x_N||_v := \max(|x_0|_v, \ldots, |x_N|_v).$$

Finally we define the logarithmic Weil height $h(\mathbf{x}) = h(x_0, \ldots, x_N)$ of $\mathbf{x} \in \overline{\mathbf{Q}}^{N+1}$ by picking a number field K with $\mathbf{x} \in K^{N+1}$ and putting

$$h(\mathbf{x}) := \sum_{v \in \mathcal{M}(K)} \log ||\mathbf{x}||_v .$$

This is independent of the choice of K. Further by the product formula it defines a height on $\mathbf{P}^N(\overline{\mathbf{Q}})$.

For a polynomial P with coefficients in $\overline{\mathbf{Q}}$ we define $h(P) := h(\mathbf{p})$, where $\mathbf{p}$ is the vector consisting of all coefficients of P.

We define the height of $\mathbf{x} = (x_1, \ldots, x_N) \in \mathbf{G}_{\mathbf{m}}^N(\overline{\mathbf{Q}})$ by $h(\mathbf{x}) = h(1 : x_1 : \cdots : x_N)$. We introduce also another height $\hat{h}(\mathbf{x}) := \sum_{i=1}^N h(1 : x_i)$. This latter height has the convenient properties

$$\hat{h}(\mathbf{x}) = 0 \Longleftrightarrow \mathbf{x} \text{ is torsion}, \quad \hat{h}(\mathbf{x}^m) = |m|\hat{h}(\mathbf{x}), \quad \hat{h}(\mathbf{x}{*}\mathbf{y}) \leq \hat{h}(\mathbf{x}) + \hat{h}(\mathbf{y}) \quad (8)$$

for $\mathbf{x}, \mathbf{y} \in \mathbf{G}_{\mathbf{m}}^N(\overline{\mathbf{Q}})$, $m \in \mathbf{Z}$. Further we have

$$h(\mathbf{x}) \leq \hat{h}(\mathbf{x}) \leq N \cdot h(\mathbf{x}) \quad \text{for } \mathbf{x} \in \mathbf{G}_{\mathbf{m}}^N(\overline{\mathbf{Q}}). \tag{9}$$

Let Y be a projective subvariety (i.e., irreducible and Zariski closed) of $\mathbf{P}^N$ defined over $\overline{\mathbf{Q}}$. Let $\dim Y = n$, $\deg Y = d$. Denote by F_Y the Chow form of Y (cf. Shafarevich 1977, pp. 65–69). We define the height of Y by $h(Y) := h(F_Y)$. In particular, if Y is linear then we have

$$h(Y) = h(\mathbf{a}_0 \wedge \cdots \wedge \mathbf{a}_n)\,, \tag{10}$$

where $\mathbf{a}_0, \ldots, \mathbf{a}_n$ is a basis of $Y(\overline{\mathbf{Q}})$ considered as a vector space and where $\mathbf{a}_0 \wedge \cdots \wedge \mathbf{a}_n$ denotes the usual exterior product.

There is a more advanced height h_{F} for varieties, introduced by Faltings (1991), which is defined by means of arithmetic intersection theory. We need only (cf. Bost *et al.* 1994, Theorem 4.3.8) that there is a constant $C_1(N)$ depending only on N such that

$$|h_{\mathrm{F}}(Y) - h(Y)| \leq C_1(N) \deg Y\,. \tag{11}$$

Let ι be the map of $\mathbf{G}_{\mathbf{m}}^N$ into $\mathbf{P}^N$ given by

$$(x_1, \ldots, x_N) \mapsto (1 : x_1 : \ldots : x_N).$$

Let X be a subvariety of $\mathbf{G}_{\mathbf{m}}^N$ of dimension n and degree d defined over $\overline{\mathbf{Q}}$. Let Y denote the Zariski closure of $\iota(X)$ in $\mathbf{P}^N$. We define $h(X) := h(Y)$, $h_{\mathrm{F}}(X) := h_{\mathrm{F}}(Y)$. David & Philippon (1999) introduced another, more natural height $h_{\mathrm{DP}}(X)$, which has the property that $h_{\mathrm{DP}}(X) = 0$ if and only if X is the translate of a subtorus by a torsion point of $\mathbf{G}_{\mathbf{m}}^N$. By (11) and David & Philippon (1999), Proposition 2.1(v), there is a constant $C_2(N)$ depending only on N such that

$$|h_{\mathrm{DP}}(X) - h(X)| \leq C_2(N) \deg X\,. \tag{12}$$

A much more involved result (David & Philippon 1999, Theorem 1.2) states, that if X is not a torus coset, then

$$h_{\mathrm{DP}}(X) \geq \frac{1}{2^{41}(\deg X)^2\{\log(\deg X + 1)\}^2}\,. \tag{13}$$

Points of small height

Let Y be an n-dimensional linear subvariety of $\mathbf{P}^N$ defined over $\overline{\mathbf{Q}}$. Take a basis $\mathbf{a}_0, \ldots, \mathbf{a}_n$ of $Y(\overline{\mathbf{Q}})$ considered as vector space. Then from (10) and elementary height computations it follows that

$$h(Y) \leq c(n) + h(\mathbf{a}_0) + \cdots + h(\mathbf{a}_n)$$

where $c(n)$ is some constant depending only on n. This implies that if $\lambda < 1/(n+1)$ and $h(Y)$ is sufficiently large, then the set of $\mathbf{y} \in Y(\overline{\mathbf{Q}})$ with $h(\mathbf{y}) < \lambda \cdot h(Y)$ is contained in a proper linear subspace of Y.

The following generalization is due to Zhang (1995), Theorem 5.8:

Theorem 8 *Let Y be a projective subvariety of $\mathbf{P}^N$ defined over $\overline{\mathbf{Q}}$ with* $\dim Y = n$, $\deg Y = d$. *Then for every $\lambda < 1/((n+1)d)$ the set of $\mathbf{y} \in Y(\overline{\mathbf{Q}})$ with $h(\mathbf{y}) < \lambda \cdot h_{\mathrm{F}}(Y)$ is not Zariski-dense in Y.*

David & Philippon (1999) Proposition 5.4, proved the following result for subvarieties of $\mathbf{G}_{\mathbf{m}}^N$, which is basically a quantitative version of Theorem 8 for small λ:

Theorem 9 *Let X be a subvariety of $\mathbf{G}_{\mathbf{m}}^N$ defined over $\overline{\mathbf{Q}}$ which is not a torus coset. Suppose* $\dim X = n$, $\deg X = d$. *Put*

$$\begin{aligned} \alpha(n,d) &= 2(4e)^{n+1}d\,, \\ \beta(N,n,d) &= 2^{4N+90}(4e)^{2n+2}(n+1)^2 \cdot d^7 \log(d+1)^4\,. \end{aligned}$$

Then the set of $\mathbf{x} \in X(\overline{\mathbf{Q}})$ with $h(\mathbf{x}) \leq \alpha(n,d)^{-1} h_{\mathrm{DP}}(X)$ is contained in a proper Zariski-closed subset of X, the sum of the degrees of the irreducible components of which is at most $\beta(N,n,d)$.

We apply Theorem 9 to the set $X^0 \cap \Gamma$, where X, Γ are as in Theorem 7, i.e., with $\dim X = n$, $\deg X = d$, $\operatorname{rank} \Gamma = r$. We observe that for any translate $\mathbf{u} * X = \{\mathbf{u} * \mathbf{x} : \mathbf{x} \in X\}$ we have $\deg(\mathbf{u} * X) = \deg X$. This implies that the statement of Theorem 7 does not change if we replace X by a translate $\mathbf{u} * X$ with $\mathbf{u} \in \Gamma$. We replace X by such a translate of minimal height. Thus, we may assume without loss of generality that

$$h_{\mathrm{DP}}(\mathbf{u} * X) \geq h_{\mathrm{DP}}(X) \quad \text{for every } \mathbf{u} \in \Gamma. \tag{14}$$

The following lemma is more or less routine.

Lemma 1 *Assume* (14). *Then for every $C \geq 1$, the set of points $\mathbf{x} \in X^0 \cap \Gamma$ with*

$$h(\mathbf{x}) \leq C \cdot h_{\mathrm{DP}}(X)$$

is contained in the union of at most $c_4(N,d)\big(c_4(N,d)\cdot C\big)^r$ proper subvarieties of X, each of degree at most $c_5(N,d)$, where $c_4(N,d)$ and $c_5(N,d)$ are constants depending only on N and d.

Proof We may assume that X is not a torus coset since otherwise X^0 is empty. It is slightly more convenient to work with the height $\hat{h}(\mathbf{x})$ introduced above. Define the distance function $\delta(\mathbf{u}_1,\mathbf{u}_2) := \hat{h}(\mathbf{u}_1 * \mathbf{u}_2^{-1})$. Let $\alpha(n,d)$, $\beta(N,n,d)$ have the meaning of Theorem 9.

In view of (9), we have to consider the set of points $\mathbf{x} \in X^0 \cap \Gamma$ with $\hat{h}(\mathbf{x}) \le B$ with $B = NC \cdot h_{\mathrm{DP}}(X)$. Let $\mathcal{S}$ be a maximal subset of this set, with the property that any two distinct points $\mathbf{u}_1, \mathbf{u}_2 \in \mathcal{S}$ satisfy $\delta(\mathbf{u}_1,\mathbf{u}_2) \ge \varepsilon$ where $\varepsilon = \alpha(n,d)^{-1} h_{\mathrm{DP}}(X)$. According to, e.g., Lemma 4 of Schmidt (1996a) (which is valid for any function with properties (8) defined on an abelian group of rank r), the set $\mathcal{S}$ has cardinality at most $(1+(2B/\varepsilon))^r \le (3N\alpha(n,d)\cdot C)^r$.

Our choice of $\mathcal{S}$ implies that for every $\mathbf{x} \in X^0 \cap \Gamma$ with $\hat{h}(\mathbf{x}) \le B$, there is a $\mathbf{u} \in \mathcal{S}$ with $\delta(\mathbf{x},\mathbf{u}) < \varepsilon$. Consider the points $\mathbf{x}$ corresponding to a fixed $\mathbf{u} \in \mathcal{S}$. By (9) and (14) we have $h(\mathbf{u}^{-1} * \mathbf{x}) \le \alpha(n,d)^{-1} h_{\mathrm{DP}}(\mathbf{u}^{-1} * X)$. By applying Theorem 9 with $\mathbf{u}^{-1} * X$ and the points $\mathbf{u}^{-1} * \mathbf{x}$ and then passing from $\mathbf{u}^{-1} * \mathbf{x}$ to $\mathbf{x}$ we infer that the set of vectors $\mathbf{x}$ under consideration lies in a finite union of proper subvarieties of X, the sum of the degrees of which is at most $\beta(N,n,d)$. Together with our estimate for the cardinality of $\mathcal{S}$ this implies Lemma 1. □

The Subspace Theorem

Let K be an algebraic number field. Let S be a finite set of places of K. For $v \in S$, let $L_0^{(v)}, \ldots, L_n^{(v)}$ be linearly independent linear forms in $K[x_0,\ldots,x_n]$. The Subspace Theorem, first proved by Schmidt (1972) for archimedean absolute values and later extended by Schlickewei (1977) to arbitrary sets of absolute values, reads as follows:

For every $\kappa > n+1$ the set of points $\mathbf{x} = (x_0 : \cdots : x_n) \in \mathbf{P}^n(K)$ satisfying

$$\log\left(\prod_{i=0}^{n}\prod_{v\in S}\frac{|L_i^{(v)}(\mathbf{x})|_v}{||\mathbf{x}||_v}\right) \le -\kappa h(\mathbf{x}) \tag{15}$$

is contained in the union of finitely many proper linear subspaces of $\mathbf{P}^N$.

Instead of (15) we deal with systems of inequalities

$$\log\left(\frac{|L_i^{(v)}(\mathbf{x})|_v}{||\mathbf{x}||_v}\right) \le -c_{iv} h(\mathbf{x}) \quad (v \in S, i = 0,\ldots,n) \qquad \text{in } \mathbf{x} \in \mathbf{P}^n(K). \tag{16}$$

Clearly, the solutions of (16) lie in only finitely many proper linear subspaces if

$$\sum_{i=0}^{n}\sum_{v\in S} c_{iv} > n+1\,. \tag{17}$$

Let $\{L_0,\ldots,L_N\}$ be the union of the sets of linear forms $\{L_1^{(v)},\ldots,L_n^{(v)}\}$ $(v\in S)$. Define the map $\mathbf{x}\mapsto\mathbf{y}=(y_0:\cdots:y_N)$ by $y_i=L_i(\mathbf{x})$ for $i=0,\ldots,N$ and let Y be the image of $\mathbf{P}^n$ under this map. Then Y is an n-dimensional linear projective subvariety of $\mathbf{P}^N$ defined over K. For $\mathbf{x}\in\mathbf{P}^n$, we have that $\mathbf{y}\in Y(K)$ and that $L_i^{(v)}(\mathbf{x})$ is a coordinate of $\mathbf{y}$. This leads us to consider systems of inequalities

$$\log\left(\frac{|y_i|_v}{||\mathbf{y}||_v}\right)\le -c_{iv}h(\mathbf{y})\quad (i=0,\ldots,N,\ v\in S)\qquad \text{in } \mathbf{y}\in Y(K). \tag{18}$$

Let $\mathcal{I}(Y)$ be the set of $(n+1)$-tuples $\mathbf{i}=\{i_0,\ldots,i_n\}$ such that the variables $y_{i_0},\ldots,y_{i_n}$ are linearly independent on Y, i.e., there is no non-trivial linear combination $\sum_{k=0}^{n} c_k y_{i_k}$ vanishing identically on Y. Notice that condition (17) translates into

$$\frac{1}{n+1}\left(\sum_{v\in S}\max_{\mathbf{i}\in\mathcal{I}(Y)}\sum_{i\in\mathbf{i}} c_{iv}\right)=1+\delta\quad \text{with } \delta>0. \tag{19}$$

Schmidt (1989) was the first to prove a quantitative version of his Subspace Theorem, giving an explicit upper bound of the number of subspaces. For an overview of further history we refer to the survey paper Evertse & Schlickewei (1999). Below we state a consequence of a result of Evertse & Schlickewei (2002), Theorem 2.1.

Theorem 10 *Let Y be a linear projective subvariety of $\mathbf{P}^N$ of dimension n defined over the number field K. Assume* (19). *Then the set of solutions $\mathbf{y}\in Y(K)$ of* (18) *with*

$$h(\mathbf{y}) > (1+\delta^{-1})(N+1)^n\cdot(1+h(Y)) \tag{20}$$

lies in some finite union $T_1\cup\cdots\cup T_t$ of proper linear subspaces of Y where

$$t\le 4^{(n+9)^2}(1+\delta^{-1})^{n+4}\log 4N\log\log 4N\,. \tag{21}$$

We would like to emphasize that for applications it is very crucial that the quantities in (20) and (21) are independent of K and S and that the quantity in (21) is independent of Y.

The method of proof of Theorem 10 is basically Schmidt's (cf. Schmidt 1989), but with some technical innovations. Instead of Roth's lemma (a non-vanishing result for polynomials) used by Schmidt, the proof uses a very special case of an explicit version of Faltings' Product Theorem (Faltings 1991, Theorems 3.1 and 3.3). This led to a considerable improvement upon the upper bound for the number of subspaces given by Schmidt (1989). Further, the basic geometry of numbers used by Schmidt was replaced by the 'geometry of numbers over $\overline{\mathbf{Q}}$' developed independently by Roy & Thunder (1996), Theorem 6.3, and Zhang (1995), Theorem 5.8. This was of crucial importance to remove the dependence on the number field K which was still present in earlier versions of Theorem 10. For further comments we refer to Evertse & Schlickewei (1999).

In their fundamental paper, Faltings & Wüstholz (1994) gave a proof of the Subspace Theorem very different from Schmidt's. Their argument does not use the geometry of numbers, but instead the full power of Faltings' Product Theorem. Moreover, Faltings & Wüstholz treated systems of inequalities (18) where the solutions $\mathbf{y}$ may be taken from an arbitrary projective subvariety Y of $\mathbf{P}^N$, not just a linear subvariety.

Ferretti (1998) obtained a quantitative version of the result of Faltings & Wüstholz. Among others, Ferretti considered systems (18) for arbitrary varieties Y. Under suitable conditions imposed on the exponents c_{iv}, he gave explicit constants C_1, C_2, C_3 such that the set of solutions $\mathbf{y}$ of (18) with $h(\mathbf{y}) \geq C_1$ lies in the union of at most C_2 proper subvarieties of Y, each of degree $\leq C_3$. Unfortunately, Ferretti's constants C_1, C_2 and C_3 depend on K and S which is an obstacle for applications. Very recently Evertse & Ferretti (2001) proved a version of Ferretti's result with constants C_1, C_2, C_3 independent of K and S.

We have to apply Theorem 10 to the set $X^0 \cap \Gamma$. Recall that for a number field K and a finite set of places $S \subset \mathcal{M}(K)$ containing the archimedean places, the group of S-units is given by $U_S = \{x \in K^* : |x|_v = 1 \text{ for } v \notin S\}$. Let X, Γ be as in Theorem 7 but assume that X is linear. Choose the number field K and the set of places $S \subset \mathcal{M}(K)$ such that X is defined over K and $\Gamma \subset U_S^N$. Let Y be the Zariski closure of $\iota(X)$ in $\mathbf{P}^N$ (where as before $\iota : (x_1, \ldots, x_N) \mapsto (1 : x_1 : \cdots : x_N)$) so that Y is also linear. Given $\mathbf{x} = (x_1, \ldots, x_N) \in X^0 \cap \Gamma$ with $h(\mathbf{x}) > 0$ put $y_0 := 1$, $y_i := x_i$ for $i = 1, \ldots, N$ and $\mathbf{y} = (y_0, \ldots, y_N)$. Then by definition, $h(\mathbf{y}) = h(\mathbf{x})$. Define reals c_{iv} by

$$\log\left(\frac{|y_i|_v}{\|\mathbf{y}\|_v}\right) = -c_{iv} h(\mathbf{y}) \quad \text{for } v \in S, i = 0, \ldots, N. \tag{22}$$

The following result is a consequence of Evertse (1995), Lemma 15. Its proof involves only elementary combinatorics.

Lemma 2 *Assume that X is linear and that* $\mathrm{Stab}(X) := \{\mathbf{u} \in \mathbf{G}_{\mathbf{m}}^N : \mathbf{u} * X = X\}$ *is trivial. Then there are constants $c_6(N), c_7(N) \geq 1$ depending only on N such that for every $\mathbf{x} \in X^0 \cap \Gamma$ with $h(\mathbf{x}) \geq c_6(N) \cdot (1 + h(X))$, the reals c_{iv} defined by* (22) *satisfy* (19) *with $\delta \geq c_7(N)^{-1}$.*

Proof of Theorem 7

Let K be the number field and S the finite set of places introduced in the previous subection. For the moment we assume that X is linear, i.e., $d = 1$. Further we may assume that $\mathrm{Stab}(X)$ is trivial (and in particular that X is not a torus coset) since otherwise $X^0 = \emptyset$. Lastly, we assume (14) which is no loss of generality. Let $c_8(N), c_9(N), \ldots$ denote explicitly computable constants, depending only on N.

In view of (12) and (13) there is a constant $c_8(N)$ such that $c_8(N)h_{\mathrm{DP}}(X)$ exceeds the lower bounds for $h(\mathbf{x})$ required in Theorem 10 and Lemma 2. Take $\mathbf{x} = (x_1, \ldots, x_N) \in X^0 \cap \Gamma$ with $h(\mathbf{x}) \geq c_8(N)h_{\mathrm{DP}}(X)$. Let $y_0 = 1$, $y_i = x_i$ for $i = 1, \ldots, N$, $\mathbf{y} = (y_0, \ldots, y_N)$. By Lemma 2 the reals c_{iv} ($v \in S, i = 0, \ldots, N$) defined by (22) satisfy (19) with $\delta \geq c_7(N)^{-1}$. A problem is that the c_{iv} depend on $\mathbf{x}$. But we may approximate the c_{iv} by reals c'_{iv} from a finite set independent of $\mathbf{x}$ which still satisfy (19) with a slightly smaller lower bound for δ. By means of an elementary combinatorial argument, which we do not work out, one can show that there is a set $\mathcal{C} \subset \mathbf{R}^{(N+1)\#S}$ of cardinality at most $c_9(N)^r$ independent of $\mathbf{x}$ with the following property: there is a tuple $(c'_{iv} : v \in S, i = 0, \ldots, N) \in \mathcal{C}$ such that $c'_{iv} \leq c_{iv}$ for $v \in S$, $i = 0, \ldots, N$, and which satisfies (19) with $\delta \geq c_{10}(N)^{-1}$. (One has to use the fact that the tuple $(c_{iv} : v \in S, i = 0, \ldots, N)$ belongs to a translate of an r-dimensional linear subspace of $\mathbf{R}^{(N+1)\#S}$.) This means that for every $\mathbf{x} \in X^0 \cap \Gamma$ with $h(\mathbf{x}) \geq c_8(N)h_{\mathrm{DP}}(X)$ the corresponding vector $\mathbf{y}$ satisfies one of at most $c_9(N)^r$ systems of inequalities (18), with $\delta \geq c_{10}(N)^{-1}$.

By applying Theorem 10 to the systems just mentioned, we obtain that the set of $\mathbf{x} \in X^0 \cap \Gamma$ with $h(\mathbf{x}) \geq c_8(N) \cdot h_{\mathrm{DP}}(X)$ lies in the union of at most $c_{11}(N) \cdot c_9(N)^r$ proper linear subvarieties of X. Further, Lemma 1 implies that the set of $\mathbf{x} \in X^0 \cap \Gamma$ with $h(\mathbf{x}) < c_8(N) \cdot h_{\mathrm{DP}}(X)$ lies in the union of at most $c_{12}(N)^{r+1}$ proper subvarieties of X of degree at most $c_{13}(N)$. By combining these two estimates we get Theorem 7 in the case that X is linear.

Now assume that X has degree $d > 1$. For example, by Faltings (1991), Proposition 2.1, X is the set of zeros of a set of polynomials in $K[x_1, \ldots, x_N]$

of degree at most d. Let φ_d be the Veronese embedding from $\mathbf{G}_{\mathbf{m}}^N$ into $\mathbf{G}_{\mathbf{m}}^{N'}$ with $N' = \binom{N+d}{d}$, mapping $\mathbf{x}$ to the vector consisting of all monomials of degree $\leq d$. Then $\varphi_d(X^0 \cap \Gamma) \subset \tilde{X}^0 \cap \tilde{\Gamma}$, where $\tilde{X}$ is a linear subvariety defined over K of $\mathbf{G}_{\mathbf{m}}^{N'}$ and where $\tilde{\Gamma}$ is a finitely generated subgroup of $\mathbf{G}_{\mathbf{m}}^{N'}(K)$ of rank r. We know already that Theorem 7 holds for the set $\tilde{X}^0 \cap \tilde{\Gamma}$. By applying φ_d^{-1} we get Theorem 7 for $X^0 \cap \Gamma$.

References

Bombieri, E. & U. Zannier (1995), Algebraic Points on Subvarieties of $\mathbf{G}_m^n$, *Intern. Math. Res. Not.* **7**, 333–347.

Bost, J.B., H. Gillet & C. Soulé (1994), Heights of projective varieties and positive Green forms, *J. Amer. Math. Soc.* **7**, 903–1027.

Chabauty, C. (1938), Sur les équations diophantiennes liées aux unités d'un corps de nombres algébriques fini, *Annali di Math.* **17**, 127–168.

David, S. & P. Philippon (1999), Minorations des hauteurs normalisées des sous-variétés des tores, *Ann. Scuola Norm. Sup. Pisa Cl. Sci. (4)* **28**, 489–543. *Erratum*, ibid. **29**, 729–731.

Edixhoven, B. & J.-H. Evertse (eds.) (1993), *Diophantine Approximation and Abelian Varieties, Introductory Lectures*, Lecture Notes in Mathematics **1566**, Springer Verlag.

Evertse, J.-H. (1984), On sums of S-units and linear recurrences, *Compos. Math.* **53**, 225–244.

Evertse, J.-H. (1995), The number of solutions of decomposable form equations, *Invent. Math.* **122**, 559–601.

Evertse, J.-H. & R.G. Ferretti (2001), Diophantine inequalities on projective varieties, preprint, University of Leiden. Submitted for publication. `http://www.math.leidenuniv.nl/~evertse/publicaties.shtml`

Evertse, J.-H. & H.P. Schlickewei (1999), The Absolute Subspace Theorem and linear equations with unknowns from a multiplicative group. In *Number Theory in Progress,* **I**, *Proc. of a Conference in Zakopane in Honour of A. Schinzel, June 30-July 9, 1997*, K. Győry, H. Iwaniec & J. Urbanowicz (eds.), de Gruyter, 121–142.

Evertse, J.-H. & H.P. Schlickewei (2002), A quantitative version of the Absolute Subspace Theorem, *J. Reine Angew. Math.*, to appear.

Evertse, J.-H., H.P. Schlickewei & W.M. Schmidt (2002), Linear equations in variables which lie in a finitely generated group, *Ann. of Math.*, to appear.

Faltings, G. (1983), Endlichkeitssätze für abelsche Varietäten über Zahlkörpern, *Invent. Math.* **73**, 349–366.

Faltings, G. (1991), Diophantine approximation on abelian varieties, *Ann. Math.* **133**, 549–576.

Faltings, G. (1994), The general case of S. Lang's conjecture. In *Barsotti Symposium in Algebraic Geometry*, V. Christante & W. Messing (eds.), Academic Press, 175–182.

Faltings, G. & G. Wüstholz (1994), Diophantine approximations on projective spaces, *Invent. Math.* **116**, 109–138.

Ferretti, R.G. (1998) Quantitative Diophantine approximations on projective spaces, Preprint, ETH Zürich.

Griffiths, P. & J. Harris (1978), *Principles of Algebraic Geometry*, Wiley-Interscience.

Győry, K. (1992), Some recent applications of S-unit equations, *Astérisque* **209**, 17–38.

Hartshorne, R. (1977), *Algebraic Geometry*, Springer Verlag.

Hindry, M. (1988), Autour d'une Conjecture de Serge Lang, *Invent. Math.* **94**, 575–603.

Lang, S. (1990), Integral points on curves, *Pub. Math. IHES.*

Lang, S. (1983), *Fundamentals of Diophantine Geometry*, Springer Verlag.

Laurent, M. (1984), Équations diophantiennes exponentielles, *Invent. Math.* **78** (1984), 299–327.

Liardet, P. (1974), Sur une conjecture de Serge Lang, *C.R. Acad. Sci. Paris* **279**, 435–437.

McQuillan, M. (1995), Division points on semi-abelian varieties, *Invent. Math.* **120**, 143–159.

van der Poorten, A.J. & H.P. Schlickewei (1991), Additive relations in fields, *J. Austral. Math. Soc. Ser. A* **51**, 154–170.

Rémond, G. (2000a), Inégalité de Vojta en dimension supérieure, *Ann. Scuola Norm. Sup. Pisa Cl. Sci. (4)* **29**, 101–151.

Rémond, G. (2000b), Décompte dans une conjecture de Lang, *Invent. Math.* **142**, 513–545.

Rémond, G. (2001), Approximation diophantienne sur les variétés semi-abéliennes, preprint Institut Fourier.

Rémond, G. (2002), Sur les sous-variétés des tores, *Composito Math.*, to appear.

Roy, D. & J.L. Thunder (1996), An absolute Siegel's Lemma, *J. Reine Angew. Math.* **476**, 1–26.

Schlickewei, H.P. (1977), The $\wp$-adic Thue–Siegel–Roth–Schmidt theorem, *Arch. Math. (Basel)* **29**, 267–270.

Schmidt, W.M. (1972), Norm form equations, *Ann. of Math.* **96**, 526–551.

Schmidt, W.M. (1980), *Diophantine Approximation*, Lecture Notes in Mathematics **785**, Springer Verlag.

Schmidt, W.M. (1989), The Subspace Theorem in Diophantine approximations, *Compos. Math.* **69**, 121–173.

Schmidt, W.M. (1996a), Heights of algebraic points lying on curves or hypersurfaces, *Proc. Amer. Math. Soc.* **124**, 3003–3013.

Schmidt, W.M. (1996b), Heights of points on subvarieties of $\mathbf{G}_m^n$, in *Number Theory, Papers from the Séminaire de Théorie des Nombres de Paris, 1993–94*, S. David (ed.), Cambridge University Press, 157–187.

Shafarevich, I.R. (1977), *Basic Algebraic Geometry*, Springer Verlag.

Vojta, P. (1991), Siegel's theorem in the compact case, *Ann. Math.* **133**, 509–548.

Vojta, P. (1996), Integral points on subvarieties of semiabelian varieties, I, *Invent. Math.* **126**, 133–181.

Zhang, S. (1995), Positive line bundles on arithmetic varieties, *J. Amer. Math. Soc.* **8**, 187–221.

15

A New Application of Diophantine Approximations

G. Faltings

The method of diophantine approximation has yielded many finiteness results, as the theorems of Thue and Siegel or the theory of rational points on subvarieties of abelian schemes. Its main drawback is non-effectiveness. In the present overview I first recall some progress made in the last decade, and the remaining problems. After that I explain how to extend the known methods to some new cases, proving finiteness of integral points on certain affine schemes.

1 Known results

Before stating them we have to introduce some terminology. Recall that for a rational point $x \in \mathbb{P}^n(\mathbb{Q})$ in projective n-space we define its height as follows:

> Represent $x = (x_0 : \ldots : x_n)$ as a vector with integers x_i such that their greatest common divisor is 1. Then the (big) height $H(x)$ is the length of this vector, and the (little) height $h(x)$ its logarithm.

This definition can be made more sophisticated using Arakelov theory, and extends to number fields. The height measures the arithmetic complexity of the point x, and for a given bound c the number of points x with $H(x) \leq c$ is finite.

By restriction we get a height function on the rational points of any subvariety $X \subset \mathbb{P}^n$. Up to bounded functions it only depends on the ample line bundle $\mathcal{L} = \mathcal{O}(1)$ on X. Also one can define heights for subvarieties $Z \subset X$, or for effective algebraic cycles (see Faltings 1991). For example one can use the Chow variety, but there is a more direct definition using Arakelov theory. Namely, one adds to the cycle an infinite component (Green's current) to obtain an Arakelov cycle, and uses the intersection number with a power of the ample hermitian bundle.

Having introduced this machinery, the method of diophantine approximation can be described as follows:

> Suppose we are given a projective algebraic scheme X over a number field K, and we want to show that X has only finitely many K-rational points, or that there are only finitely many K-rational points with a given property (for example integral points on an open subset). If this is not the case the logarithmic heights $h(x)$ of these points tend to infinity. One chooses r such points $\{x_1, \ldots, x_r\}$ with the property that the height $h(x_1)$ as well as all the ratios $h(x_{i+1})/h(x_i)$ are very big. Here the number r as well as the meaning of 'very big' can be made precise depending on the initial problem. This depends on certain choices, starting with a model for X over the integers of K, and so on.
>
> Then consider r-tuples of integers $(d_1, \ldots, d_r)$ such that d_r is big and the ratio d_i/d_{i+1} is approximately equal to $h(x_{i+1})/h(x_i)$. Thus d_1 is much bigger than d_2, which in turn is much bigger than d_3, etc.. After that construct (using Siegel's lemma) an integral section
>
> $$F \in \Gamma(\underbrace{X \times \cdots \times X}_{r\,\text{factors}}, \mathcal{O}(d_1, \ldots, d_r))$$
>
> of the line bundle $\mathcal{O}(d_1, \ldots, d_r)$ on $X \times \cdots \times X$ whose norm is suitable bounded. By local estimates it follows that F has high index $i(F, x)$ in
>
> $$x = (x_1, \ldots, x_r).$$
>
> Here the index is the order of vanishing at $(x_1, \ldots, x_r)$, with weight $1/d_i$ for ith coordinate. Again the precise meaning of 'bounded norm' and 'high index' can be infered from the initial data.
>
> Next consider the subvarieties
>
> $$Z(\sigma) \subset X \times \cdots \times X,$$
>
> where F has index $\geq \sigma$. Then for some small $\epsilon > 0$ and an integer n bounded by $r \cdot \dim(X)$ the schemes $Z(i(F, x) - n \cdot \epsilon)$ and $Z(i(F, x) - (n+1) \cdot \epsilon)$ must have a common irreducible component Z. By the product theorem Z is a product of irreducible subvarieties
>
> $$Z = X_1 \times \cdots \times X_r,$$
>
> provided ϵ is bigger than a certain multiple of the ratios d_{i+1}/d_i (which are supposed to be small). Furthermore one can bound the degrees of the X_i by a constant depending only on the initial data, not on the choice of points x_i or degrees d_i. Similarly one bounds their heights by a multiple of d_1/d_i.

Now one applies induction (the dimensions decrease) to the product $X_1 \times \cdots \times X_r$, constructing a new section of $\Gamma(X_1 \times \cdots \times X_r, \mathcal{O}(d_1, \ldots, d_r))$, and ultimately new $X_i{}' \subset X_i$. After finitely many steps the X_i become empty, which contradicts the fact that they contain x_i. Of course one needs to control degrees and arithmetic constants (that is heights), which needs some machinery (see Faltings 1991).

What kind of results have been shown that way? First of all we can reprove the classical results of Roth, and W. Schmidt's generalisation to higher dimensions. Second, for subvarieties X of abelian varieties A, all rational points lie on finitely many translates $b + B \subset X$, where B is an abelian subscheme. This was a conjecture of S. Lang, and has been extended to semiabelian varieties by P. Vojta (1996,1999). Finally affine open subschemes in an abelian variety A have only finitely many integral points.

As a technical point one might note that for abelian varieties one has to consider more sophisticated line bundles on $A \times \cdots \times A$ than just products. This technique has been discovered by Vojta (for curves, see Vojta 1991) and allows us to make effective use of the Mordell–Weil theorem.

For general subvarieties of projective space we know no reasonable improvements to Schmidt's results (which hold for the full projective space), that is we do not know how to make effective use of the fact that rational points lie on a proper subvariety. Similarly for affine varieties (and integral points), say complements of hypersurfaces $D \subset \mathbb{P}^n = P$. The only exception is if D is not geometrically irreducible: namely suppose that over $\overline{K}$ D decomposes into irreducibles $\{D_1, \ldots, D_s\}$. Then in the process above one considers Fs which vanish to a high order (better index) on the union of the subvarieties $D_i^r \subset P^r$, instead of all on D^r as required by the general procedure. This is easier to achieve, thus one obtains better constants and non-trivial results. The content of the present paper is to give some examples with irreducible divisors $D \subset P$ for which the complement $P - D$ has still only finitely many integral points. The idea is to pass to a covering of P ramified only along D, and such that the preimage of D becomes highly reducible. This cannot happen if D is smooth, because then the fundamental group of $P - D$ is cyclic abelian and the preimage of D in the universal cover is isomorphic to it (Fulton 1980). In fact for us D will be a plane curve with cusps and simple double points. Thus consider a smooth projective algebraic surface X and a generic projection $X \to P = \mathbb{P}^2$. The ramification divisor of this projection will be smooth, but its image D in P has the two types of singularities mentioned above. The cusps arise because the projection (restricted to the ramification divisor) may not be an immersion, and the double points because two points on the ramification divisor may have the same image in P. The number of cusps and double points can be computed by a Pluecker formula. Now if $Y' \to P$ is the Galois hull of $X \to P$ one shows that the Galois group $\mathrm{Gal}(X/P)$ is the full symmetric group S_n, with n the degree of X/P. The irreducible components of the preimage of D correspond to pairs $\{i, j\}$ of integers between 1 and n. Thus there are $n(n-1)/2$ such components, and we can hope to apply our machinery.

In the coming sections I shall first consider the geometric theory of such coverings. After that I treat the arithmetic machinery of diophantine approximation. Finally we shall compute some numerical invariants, which are used to show that in some cases our main result cannot be reduced to questions about semiabelian schemes. Our main example will be the case $X = \mathbb{P}^1 \times \mathbb{P}^1$ and $\mathcal{L} = \mathcal{O}(a, b)$. On the way we will encounter various numerical restrictions on the data, which we work out for this case. In particular they all hold if $a, b \geq 5$.

2 Geometry of projections

Suppose that X is a projective smooth geometrically irreducible algebraic surface over an algebraically closed field K of characteristic 0, and that $\mathcal{L}$ is an ample line bundle on X. We assume that for any closed point $x \in X$ the global sections $\Gamma(C, \mathcal{L})$ generate the fibre

$$\mathcal{L}_x / m_x^4 \cdot \mathcal{L}_x,$$

that for any pair $\{x, y\}$ of different point the global sections generate the direct sum

$$\mathcal{L}_x / m_x^3 \cdot \mathcal{L}_x \oplus \mathcal{L}_y / m_y^3 \cdot \mathcal{L}_y,$$

and finally that for three different points $\{x, y, z\}$ they generate

$$\mathcal{L}_x / m_x^2 \cdot \mathcal{L}_x \oplus \mathcal{L}_y / {m_y}^2 \cdot \mathcal{L}_y \oplus \mathcal{L}_2 / {m_2}^2 \cdot \mathcal{L}_2.$$

For example this holds if $\mathcal{L}$ is the tensor product of five ample line bundles. We also assume that $\mathcal{K}_X \otimes \mathcal{L}^{\otimes 3}$ is ample. Consider three-dimensional subspaces $E \subset \Gamma(X, \mathcal{L})$. These are parametrised by a Grassmannian G. Over a suitable open subset, E generates $\mathcal{L}$, and thus a well-defined map

$$f = f_E : X \to \mathbb{P}(E) = \mathbb{P}^2,$$

of degree $n = L \cdot L$ (intersection-number of $\mathcal{L}$ on X). We call E generic if the following hold:

(a) E generates $\mathcal{L}$;
(b) the discriminant locus $Z \subset X$ of f_E is smooth;

(c) the restriction of f_E to Z is birational onto its image $D \subset P = \mathbb{P}(E)$;
(d) D has only cusps and simple double points as singularities.

We shall prove:

Proposition 1

(i) *Generic E's form a dense open subset G' of G.*
(ii) *For generic E let $Y \to X \to P$ denote the associated (normal) Galois covering. Then Y is smooth, Z is irreducible, and the covering group* $\mathrm{Aut}(Y/P)$ *is the full symmetric group S_n.*
(iii) *These Y's form a smooth projective scheme over G'.*

Proof We use the standard technique. Construct closed subsets of $G \times X$ consisting of pairs (E, x) such that the projection f_E has a bad property at x. If these subsets have codimension > 2 their projection to G is a proper closed subset which can be removed, and our bad property does not occur in the generic case. If the codimension is 2 we obtain generically a finite set of bad points. All our exceptional subsets will be bundles over X.

(a1) A subspace $E \subset \Gamma(X, \mathcal{L})$ does not generate $\mathcal{L}$ at x if it is contained in the subspace of sections vanishing at x. This subspace has codimension 1 as $\Gamma(X, \mathcal{L})$ generates $\mathcal{L}$, and the Grassmannian of this subspace has codimension 3 in G. The union over $x \in X$ gives a proper closed subset of G which we remove. So from now on assume that E generates $\mathcal{L}$, and f_E is defined everywhere. Then, in local coordinates at x, f_E is defined by a pair of functions f, g with $f(x) = g(x) = 0$. The condition on $\mathcal{L}$ means that we can describe freely the 3-jets of f and g. That means if we write them as power series the coefficients become local functions on the Grassmannians, and the coefficients of total degree ≤ 3 define locally a smooth map from the Grassmannian to affine space.

(a2) A point x lies in the discriminant locus of f_E if E the Jacobian $J(f, g)$ has a zero at x, a condition of codimension 1. It is a singular point on this locus iff the Jacobian vanishes to at least second order. This can happen if either all first derivatives of f and g vanish at x (codimension 4), or if some first derivative, say of f, does not vanish at x, but the gradient g is, up to order ≥ 2, a multiple of the gradient of f (codimension 3). Thus again we may assume that this does not happen, that is the discriminant is smooth.

(a3) Assume x lies in the discriminant. Then we can chose local coordinates (u, v) at x such that $f = u$, and g has order ≥ 2. If the coefficient of v^2 in g

does not vanish, the projection f_E is, in suitable coordinates, given by

$$f_E(u, v) = (u, v^2),$$

and the restriction to the discriminant is an immersion at x. Now assume that this coefficient vanishes. In more invariant form this means that, up to order ≥ 3, g is a multiple of f. This condition defines a set of codimension 2 in the product $G \times X$. If we require order ≥ 4 instead of order ≥ 3, we obtain codimension 3, thus may assume that this does not happen. This means that the coefficient of v^3 in g does not vanish, and, in suitable local coordinates, f_E is given by

$$f_E(u, v) = (u, 3uv - v^3).$$

The image of the discriminant then acquires a cusp. Here we have used characteristic 0, or better that the characteristic is neither 2 nor 3.

(a4) Assume two different point $x, y \in X$ both lie in the discriminant locus of f_E and have the same image in P. We are 'free' to choose the 2-jets of f_E at x and at y, by the condition on $\mathcal{L}$. Only using 1-jets we see that our condition defines a subset of codimension 4 in $G \times X \times X$. If in addition the restriction of f_E to the discriminant is not an immersion at x or y, we have to fulfill another condition (using 2-jets), so we may assume that this does not happen. Finally first note that the tangent directions of the projection of the discriminant are determined by 2-jets. That they are equal means another condition, so again this does not happen generically. Second, having three different points (on the discriminant) with the same image in P happens only on a proper subvariety of G, and does not happen generically. This finishes the proof of (a).

For (b) we first note that Z corresponds to a section of $\mathcal{K}_X \otimes \mathcal{L}^{\otimes 3}$ which is ample, so that it is connected and non-empty. Furthermore f_E is an etale covering over $P - D$, from which one constructs a canonical Galois cover with group S_n. Its fibre over a point p of $P - D$ classifies orderings of the n-points of the fibre $f_E^{-1}(p)$. It extends to a normal ramified covering $Y \to X \to P$. That Y itself is smooth amounts to a local calculation around points of D.

For a smooth point p of D the fibre $f_E^{-1}(p)$ contains one double point in which the map f_E looks in local coordinates like

$$f_E(u, v) = (u, v^2).$$

Locally Y is étale over X, and the pullback of D to Y has one irreducible component of multiplicity 2. In local coordinates it is defined by $v = 0$.

For a cusp p the fibre $f_E^{-1}(p)$ contains a triple point where f_E looks in local coordinates like

$$f_E(u, v) = (u, 3uv - v^3).$$

To obtain Y we have to adjoin the three roots of the equation

$$T^3 - 3uT = v^3 - 3uv.$$

One of them is v, and the others are

$$-v/2 \pm w \text{ with } w^2 = 3u - 3/4 \cdot v^2.$$

Thus v and w form local coordinates in Y which is smooth. We also note that the pullback of D to Y has three irreducible components through the point $v = w = 0$. They have multiplicity 2, are smooth and meet with different tangents. In our local coordinates they are given by $w = 0$ respectively $w = \pm 3/2 \cdot v$.

Finally if p is a simple double point its preimage contains two points where f_E is locally defined by

$$\begin{aligned} f_E(u_1, v_1) &= (u_1, {v_1}^2), \\ f_E(u_2, v_2) &= ({u_2}^2, v_2). \end{aligned}$$

Then Y admits local coordinates (v_1, u_2), and the pullback of D to Y has two irreducible components of multiplicity 2. These meet transversally, and their local equations are $v_1 = 0$, $u_2 = 0$. □

As these choices of local coordinates can be done in families it follows that the universal Y over G' is itself smooth over G'. For irreducibility we use that a two-fold transitive subgroup of S_n which contains a transposition (ij) must contain all transpositions and thus be equal to S_n. In our case the decomposition group of any connected component contains a transposition, namely the inertia at a generic point of D (if D is empty the covering $X \to P$ is trivial because P is simply connected. This contradicts the assumptions on $\mathcal{L}$). That the group is twofold transitive means that the normalisation of $X \times_P X$ has two irreducible components. Now we use the theorem of Fulton & Hansen (1979). Namely the subscheme $X \times_P X \subset X \times X$ is the preimage of the diagonal in $P \times P$ and thus connected. One of its irreducible components is the diagonal X. We claim that it is smooth away from the discriminant curve, $Z \subset X$ which is diagonal, and that at each point of Z its formal completion has two irreducible components. This calculation can be done in local coordinates. First of all it is clear that away from $Z \times Z$ locally $X \times X$ is étale over one of its factors and thus smooth. Next $Z \times_P Z$ consists of Z and pairs (x, y) of points of Z mapping to the same double point in D. Near such pairs the maps have local equations

$$\begin{aligned} f_E(u_1, v_1) &= (u_1, {v_1}^2), \\ f_E(u_2, v_2) &= ({u_2}^2, v_2), \end{aligned}$$

thus $X \times_P X$ is defined by

$$u_1 = u_2{}^2, v_2 = v_1{}^2$$

and is smooth. Next, for a general point on the diagonal Z, we have local equations

$$f_E(u, v) = (u, v^2),$$

thus $X \times_P X$ is defined by

$$u_1 = u_2, v_1 = \pm v_2$$

and has two smooth local components meeting in Z. At a cusp point with

$$f_E(u, v) = (u, 3uv - v^3)$$

the local equations become

$$u_1 = u_2 = u, 3 \cdot (v_1 - v_2) \cdot (v_1{}^2 + v_1 v_2 + v_2{}^2 - u) = 0.$$

Thus two smooth irreducible components again meet in Z. It follows that two irreducible components of $X \times_P X$ can only meet along Z. As Z is irreducible this can only happen for the diagonal X and one other irreducible component which provides the other local component at each point of Z. This other component is smooth and must be the quotient X_2 of Y under S_{n-2}. We denote by $Z_2 \subset X_2$ the image of Z.

Remark 7 An alternative proof for this could be done as follows.

Define an equivalence relation on the set $\{1, \ldots, n\}$ by the rule that i is equivalent to j if the transposition (ij) lies in G. This relation is G-invariant and non-trivial as G contains at least one transposition. The quotient by the relation defines a factorisation $X \to X' \to P$. The whole preimage of the ramification locus of X'/P is contained in Z. As the restriction of f_E to Z is generically injective this means that X' is unramified over P, and thus $X' = P$ as P is simply connected. It follows that there is only one equivalence class, and $G = S_n$.

We also need some geometric facts established above. In particular denote by $X_2 \subset X \times_P X$ the other irreducible component, besides the diagonal. It is the quotient of Y by S_{n-2}, and we have seen that it is smooth. Its intersection with the diagonal is the smooth curve Z_2 (isomorphic to Z) imbedded diagonally. Denote its preimage in Y by Z_{12}. It is the set of fixed points of the transposition (12). By the local calculations Z_{12} is smooth: this is obvious away from fixed points of the group S_n. What remains is either fixed by some S_3 and corresponds to a cusp in D, or by $S_2 \times S_2$ corresponding to an ordinary double point. In the cusp case we have on, Y, local coordinates (v, w), and the map to $X \times_P X$ has two components $(u, v_\pm)$ with

$$\begin{aligned} 3u &= w^2 + 3/4 \cdot v^2, \\ v_\pm &= -v/2 \pm w. \end{aligned}$$

Also the $v_\pm$ are local coordinates on X_2. The curves Z on the factors X are defined by $u = v_\pm^2$, the diagonal by $v_+ = v_-$, and the preimage of the diagonal Z is $w = 0$. Thus the projection $Y \to X_2$ is locally étale. Finally for the other projection $X_2 \to X$ the induced map on normal bundles of Z has a simple zero at such points.

For double points the calculation is even easier. There the local map $Z_{12} \to Z$ has ramification of order 2, but the normal bundle of Z_{12} in Y is again the pullback of the normal bundle of Z in X_2 or of Z in X. We define $Z_{ij} \subset Y$ as the transforms of Z_{12} under S_{n-2}. These are smooth divisors. At S_3-fixed-points (corresponding to cusps), three of them intersect with different tangents. At $S_2 \times S_2$-fixed-points, two of them intersect transversally. Otherwise there are no intersections. We shall need a criterion for whether Z_{ij} is connected (hence irreducible).

Lemma 1 *Suppose D contains a double point. Then all Z_{ij} are connected (thus irreducible).*

Proof We show that the covering group of Z_{12}/Z is all of S_{n-2}. For each ordinary double point of D it contains a transposition. To show that it is two-fold transitive we consider its quotient under S_{n-4}. It is a subscheme of X^4, and generically is the closure of the set of quadruples (x, y, z, z) with $\{x, y\}$ elements of $X - Z$, $z \in Z$, and all three are different but have the same image in P. To study this set consider for any triple $(x, y, z) \in X_3$ the set of $E \in G'$ such that they satisfy the conditions above for f_E, that is they all have the same image, Z lies in the different but x and y do not. As we may freely prescribe the 2-jets of f_E in these three points one easily sees that our conditions define an irreducible subscheme of G', and, varying (x, y, z), we obtain an irreducible subscheme of $G' \times X^3$. The fibre of this subscheme over the generic point η of G is again irreducible. If we first work over the function field $k(G) = k(\eta)$ it follows that the covering group over this field is two-fold transitive, thus equal to S_{n-2}. Over the algebraic closure $\{\overline{\eta}\}$ we obtain a normal subgroup which still contains a transposition, hence the desired result over the geometric generic fibre. But as the Z_{ij} form a smooth projective family over G' the number of geometric connected components is constant on G'. □

3 Dimensions and expectation values

Our aim is to prove finiteness for integral points on $P - D$. Thus choose a number field K, a finite set of places of K (containing all infinite places), and a model for $P - D$ over the S-integers $\mathcal{O}_S$ of K. Also $\mathcal{L}$ should satisfies the ampleness conditions in the beginning of the section.

Theorem 1 *Assume $D - \alpha Z$ is ample on X, for some $\alpha > 12$. Then $P - D$ has only finitely many $\mathcal{O}_S$-points.*

As $D = dL$ and $Z = K + 3L$ the condition on α means that $(d/\alpha - 3)L - K$ is ample. For example for $X = \mathbb{P}^1 \times \mathbb{P}^1$ and $\mathcal{L} = \mathcal{O}(a, b)$ this holds if $d \geq 3\alpha$, and if $a, b \geq 3$ we have $d \geq 42$ and may use $\alpha = 14$.

Proof By the usual lift over unramified extensions we can reduce this to the corresponding problem on the complement in Y of the union of the divisors Z_{ij}. We also use the divisors A_i which are the sums of the Z_{ij}, over all $j \neq i$. They are the pullbacks of Z under the ith projection $Y \to X$. As we required, for some rational $\alpha > 12$, the divisor $D - \alpha Z$ on X to be ample, it follows that $dL - \alpha A_i$ is ample on Y, and that some multiple is globally generated. Choose generators F_i for it.

Next, for some big integer N, consider on $\Gamma(Y, \mathcal{O}(N \cdot L))$ the filtration by order of vanishing along A_i. As explained in Faltings & Wüstholz (1994) this filtration defines a probability measure on the real line which converges as $N \to \infty$ and whose most important invariant is its expectation value (see Faltings & Wüstholz 1994). In our case this value turns out to be α/d. Recall that these expectation values are used to construct sections F of $\mathcal{O}(d \cdot d_1, \ldots, d \cdot d_r)$ on Y^r which have high index on all subschemes $A_i{}^r$. By the theory this is possible as long as the indices are less than one third of the expectation values above, that is less than $\alpha \cdot r/3$ (in the definition of the index the νth coordinate has weight $1/d_\nu$). Furthermore one then can apply Siegel's lemma (as in Faltings & Wüstholz 1994) to obtain suitable bounds on the size of this section. Namely F can be written (in various ways) as sum of monomials in the F_i, with good bounds on the coefficients. Here the number r must be big and can be estimated from the law of large numbers. Next, for an integral point y, its L-height, Y, is an Arakelov-type intersection number which can be written as a sum of local contributions. Namely the pullback of D is on the one hand equal to $d \cdot L$, on the other hand represented by

$$2 \cdot \sum Z_{ij} = \sum A_i .$$

Thus $d \cdot h(y)$ is a sum of local intersection numbers $h_v(y)$ of y with the divisors

A_i, indexed by the places v at infinity and the indices i, or even intersection numbers with the Z_{ij}. If we divide them by their sum $h(y)$ we obtain finitely many reals lying in a fixed compact subset. If there exist infinitely many ys we can also find infinitely many for which these ratio lies in a small neighbourhood of a given point in this compact set. Also we find an r-tuple $(y_1, \ldots, y_r)$ of them with strongly increasing heights, as in the introduction. It then follows from our numerical assumptions that F must have high index at $(y_1, \ldots, y_r)$.

We check that F vanishes at this point. For higher derivatives similar reasonings apply. Namely if $F(y_1, \ldots, y_r) \neq 0$ we can use its local v-norms to calculate the height of this point. By choosing local trivialisations of $\mathcal{L}$, we see that F lies in the intersection of certain ideals J_i which define the index condition along ${A_i}^r$. If f_i denote local equations for A_i then J_i is generated by all monomials

$$\prod pr_\mu^*(f_i)^{n_{i,\mu}}$$

with

$$\sum n_{i,\mu}/d_\mu \geq \alpha \cdot r/3.$$

Similarly we use local equations f_{ij} for Z_{ij} to define ideals I_{ij} as generated by monomials

$$\prod {pr_\mu}^*(f_{ij})^{n_{ij,\mu}}$$

with

$$\sum n_{ij,\mu}/d_\mu \geq \alpha \cdot r/3.$$

We define the ideal I as the product of the I_{ij}, and claim that $J^4 \subset I^2$. To see this first note that, near $(y_1, \ldots, y_r)$, I_{ij} is locally the unit ideal unless all y_μ lie in Z_{ij}. In general this can happen only for one pair (ij), unless all y_μ are either S_3 or $S_2 \times S_2$ fixed points where we have to consider three, respectively, two, pairs. Let us check the claimed identity in the typical cases. If the y_μ lie only in Z_{12} only $f_{12,\mu}$, $f_{1,\mu}$ and $f_{2,\mu}$ are not units, and they have (for fixed μ) the same divisor. Thus $J_1 = J_2 = J = I_{12}$. If the y_μ lie in Z_{12}, Z_{13} and Z_{23}, we have $J^4 \subset J_1 \cdot J_2 \cdot J_3$. The latter ideal is generated by all monomials

$$\prod {pr_\lambda}^*(f_{12} \cdot f_{13})^{n_{1,\lambda}} \cdot {pr_\mu}^*(f_{12} \cdot f_{23})^{n_{2,\mu}} {pr_\nu}^*(f_{13} \cdot f_{23})^{n_{3,\nu}}$$

with

$$\sum n_{i,\lambda}/d_\lambda \geq \alpha \cdot r.$$

All these monomials occur in the product of ${I_{ij}}^2$, using monomials with indices

$m_{ij,\lambda} = n_{i,\lambda}$ or $n_{j,\lambda}$. Finally if only Z_{12} and Z_{34} matter we have, up to units,

$$\begin{aligned} f_{1,\mu} &= f_{2,\mu} = f_{12,\mu}, \\ f_{3,\mu} &= f_{4,\mu} = f_{34,\mu}, \end{aligned}$$

the rest being units.

Hence $J_1 = J_2 = I_{12}$ and $J_3 = J_4 = I_{34}$. Also in this case the elements $f_{12,\mu}$ and $f_{34,\mu}$ form a regular sequence near $(y_1, \ldots, y_r)$ because Z_{12} and Z_{34} meet transversally. Thus $J^2 \subset I_{12} \cdot I_{34} = I$.

We deduce that F^4 is a global section of $I^2 \cdot \mathcal{O}(4d \cdot d_1, \ldots, 4d \cdot d_r)$. Furthermore we can find a covering of (an integral model) of Y^r by Zariski-open sets on which F is a sum of generators of I^2, with good bounds for the size of the coefficients. Hence the logarithms of its v-norms at $(y_1, \ldots, y_r)$ can be estimated from the local intersection numbers of y_μ with Z_{ij}s. Adding up we obtain that

$$4rd \cdot h(y_1)$$

is bounded by

$$2\alpha rd/3 \cdot h(y_1) + \text{constant}.$$

As $\alpha > 6$ this cannot happen if $h(y_1)$ is big.

Thus by the product theorem we get that one of the y_μ is contained in a proper subscheme of Y, and we know bounds for its degree and height. If this subscheme is a point we are done. Thus assume that we have a curve $C \subset Y$. Here we repeat our previous considerations with Y^r replaced by a product of Y's and various C's. Again we find a section F of index $\geq \alpha \cdot r/3$ along the ${A_i}^r$, and F^4 is locally contained in the product of the ideals ${I_{ij}}^2$. This allows the induction to proceed until we reach a contradiction.

□

4 Intersection numbers

To show that our main theorem does not easily reduce to known results we need to investigate the intersection products on Y of the curves Z_{ij}. Because of S_n-invariance the product $Z_{ij} \cdot Z_{kl}$ can take three different values, depending on how equal the indices are. Denote by γ, respectively δ, the number of cusps and double points on the discriminant curve D. These are given by some Pluecker formula which we shall recall below. If $\{ij\}$ and $\{kl\}$ are disjoint, the cycles Z_{ij} and Z_{kl} meet only over the double points of D. Over each such double point lie $n!/4$ points of Y (the inertia at each such point is $S_2 \times S_2$), and $2 \cdot (n-4)!$ of them contribute to $Z_{ij} \cdot Z_{kl}$. Hence for disjoint $\{ij\}$ and $\{kl\}$ we have

$$Z_{ij} \cdot Z_{kl} = 2 \cdot (n-4)! \cdot \delta.$$

Next, if one index coincides, the cycles meet over the curps of D. The preimage of each such cusp contains $n!/6$ points, from which $(n-3)!$ contribute to the intersection. Hence

$$Z_{ij} \cdot Z_{ik} = (n-3)! \cdot \gamma.$$

It remains to compute the self-intersection $Z_{ij} \cdot Z_{ij}$. We do this in two steps.

First, the normal bundle of Z_{ij} in Y is the pullback of the normal bundle of Z_2 in X_2, so it suffices to compute the degree of the latter. The conormal bundle of Z_2 in X_2 is a quotient of the conormal bundle of the diagonal in $X \times X$, that is Ω_X. In fact the differential of f_E defines an inclusion $f_E{}^*(\Omega_P) \subset \Omega_X$, with the quotient a line bundle on Z. By local calculations in generic points of Z one checks that this quotient is the desired normal bundle: in local coordinates (u, v) near a generic point of Z we see f_E is given by (u, v^2), the image of Ω_P is generated by d_u, and $u_1 - u_2$ vanishes on

$$X_2 \subset X \times_P X \subset X \times X.$$

Hence the self-intersection number of Z_2 on X_2 is $-\deg(\Omega_X / f_E{}^*(\Omega_P))$, and

$$Z_{ij} \cdot Z_{ij} = -(n-2)! \cdot \deg(\Omega_X / f_E{}^*(\Omega_P)).$$

Second, the projection from X_2 to X induces on normal bundles of Z a map with γ zeroes. Thus

$$Z_{ij} \cdot Z_{ij} = (n-2)! \cdot (Z \cdot Z - \gamma).$$

We further compute these numbers using the cohomology of X. In degree 2 it contains classes K, L and Z corresponding, respectively, to the canonical sheaf of X, to $\mathcal{L}$, and finally to the divisor Z. Furthermore in degree 4 there is the second Chern class c_2 equal to the Euler characteristic of X. The Chern character takes values

$$\begin{aligned} \mathrm{ch}(\Omega_X) &= 2 + K + K^2/2 - c_2, \\ \mathrm{ch}(f_E{}^*(\Omega_P)) &= 3 \cdot e^{-L} - 1, \end{aligned}$$

and finally

$$\begin{aligned} \deg(\Omega_X / f_E{}^*(\Omega_P)) &= \gamma - Z^2 \\ &= \text{terms of degree four in} \\ &\qquad 2 + K + K^2/2 - c_2 + 1 - 3 \cdot e^{-L} - 1 + e^{-Z} \end{aligned}$$

$$= K^2/2 - c_2 - 3 \cdot L^2/2 + Z^2/2$$
$$= K^2 + 3KL + 3L^2 - c_2.$$

We deduce that

$$\begin{aligned} \gamma &= (K+3L)^2 + K^2 + 3KL + 3L^2 - c_2 \\ &= 2K^2 + 9KL + 12L^2 - c_2 = 2K^2 - c_2 + 9d - 15n \end{aligned}$$

with

$$d = ZL = KL + 3L^2 = KL + 3n \text{ the degree of } D = f_E(Z).$$

Finally $\gamma + \delta$ is the difference between arithmetic genera of D and Z, that is,

$$\gamma + \delta = d(d-3)/2 - Z(Z+K)/2 = d^2/2 - 6d + 9n - K^2,$$

hence

$$\delta = d^2/2 - 15d + 24n - 3K^2 + c_2.$$

For example, for $X = \mathbb{P}^1 \times \mathbb{P}^1$, $\mathcal{L} = \mathcal{O}(a, b)$ we obtain

$$\begin{aligned} n &= 2ab, \\ d &= 6ab - 2(a+b), \\ \gamma &= 24ab - 16(a+b) + 12, \\ \delta &= 18a^2b^2 - 12ab(a+b) + 2(a+b)^2 - 42ab + 30(a+b) - 20. \end{aligned}$$

If $a, b \geq 3$ these numbers γ and δ never vanish ($\delta \neq 0$ was needed for irreducibility of Z_{ij}).

5 Nondegeneracy

We want to know whether the intersection form on Y applied to the cycles Z_{ij} is nondegenerate. It defines a symmetric S_n-invariant bilinear form on the vector space V which is the S_n-representation induced from the trivial representation of $S_2 \times S_{n-2}$. The dimension of its ring of S_n-endomorphisms is equal to that of its $S_2 \times S_{n-2}$-invariants, which is 3. It thus must have three (distinct) irreducible components all generated by their $S_2 \times S_{n-2}$-invariants, and the intersection form is nondegenerate if it is so on these three invariants. In terms of cycles these are spanned by the pullbacks to Y of D, the sum $A_i + A_j$ of two pullbacks of $Z \subset X$, and the diagonal $Z \subset X_2$. Namely D pulls back to twice the sum of all Z_{ij}, and $A_i + A_j - Z_{ij}$ is the sum of those Z_{ij} where one of the indices is 1 or 2, and pulls back Z_2 to Z_{12}. Of course instead of D we could also use L.

Naturally D^2 is positive as D is ample. Denote by $A^\circ{}_i$ and $Z^\circ{}_{ij}$ the projections to the perpendicular space of D. Then $A^\circ{}_i = A_i - D/n, Z^\circ{}_{ij} = Z_{ij} - D/n(n-1)$, as the projections of Z, respectively Z_2, to Y are D. Moreover the projection of Z_2 to X is equal to Z, and thus $Z_{ij} - A_i/(n-1)$ is perpendicular to A_i. and D. The self-intersection was computed above and is equal to $-\gamma \cdot (n-2)!$. Furthermore $Z^\circ{}_{ij} \cdot A^\circ{}_i = -A^\circ{}_i{}^2/(n-1)$. Finally D is the sum of all A_i, thus the sum of the $A^\circ{}_i$ vanishes, and also $A^\circ{}_i \cdot A^\circ{}_j = -A^\circ{}_i{}^2/(n-1)(i \neq j)$. So finally $A^\circ{}_1 + A^\circ{}_2$ and $Z^\circ{}_{12} + (A^\circ{}_1 + A^\circ{}_2)/(n-2)$ constitute an orthogonal basis for the perpendicular to D, and their squares are

$$\begin{aligned}(A^\circ{}_1 + A^\circ{}_2)^2 &= 2 \cdot A^\circ{}_1{}^2 + 2 \cdot A^\circ{}_1 \cdot A^\circ{}_2 \\ &= 2(n-2)/(n-1) \cdot A^\circ{}_1{}^2,\end{aligned}$$

and

$$\begin{aligned}\left(Z^\circ{}_{12} + \frac{A^\circ{}_1 + A^\circ{}_2}{n-2}\right)^2 &= Z^\circ{}_{12}{}^2 - \left(\frac{A^\circ{}_1 + A^\circ{}_2}{n-2}\right)^2 \\ &= Z^\circ{}_{12}{}^2 - \frac{2 \cdot A^\circ{}_1{}^2}{(n-1)\cdot(n-2)} \\ &= \left(Z_{12} - \frac{A_1}{n-1}\right)^2 - \frac{n \cdot A^\circ{}_1{}^2}{(n-1)^2 \cdot (n-2)}.\end{aligned}$$

Now

$$A^\circ{}_1{}^2 = (n-1)! \cdot (K + (3 - d/n)L)^2 \text{ (intersection on } X),$$

hence

Proposition 2 *The pairing is nondegenerate if and only if*

$$\left(\frac{d}{n-3} \cdot L - K\right)^2 \neq 0, -\frac{\gamma \cdot (n-1) \cdot (n-2)}{n}.$$

The first inequality means that L and Z, or K and L, are numerically independent over X, that is $n \cdot K^2 \neq (d - 3n)^2$. For example this can never happen for $X = \mathbb{P}^2$.

For $X = \mathbb{P}^1 \times \mathbb{P}^1$ and $\mathcal{L} = \mathcal{O}(a, b)$ it means $a \neq b$. In fact here

$$\left(\frac{d}{n-3} \cdot L - K\right)^2 = -\frac{2(a-b)^2}{ab}.$$

Also

$$\begin{aligned}\gamma &= 24ab - 18(a+b) + 12 \\ &= \frac{3}{2} \cdot \big((4a-3) \cdot (4b-3) - 1\big)\end{aligned}$$

is (if, say, $a, b \geq 2$) never equal to

$$-\frac{n \cdot ((d/n - 3) \cdot L - K)^2}{(n-1) \cdot (n-2)} = \frac{4(a-b)^2}{(2ab-1)(2ab-2)}.$$

As an application we treat the question whether our result might be reduced to known facts about subvarieties of semiabelian group schemes, by mapping $Y - \cup Z_{ij}$ to such a scheme. This is not possible if the affine space has a finite abelianised fundamental group. If the Z_{ij} are numerically independent this is equivalent to Y having a finite abelianised fundamental group. By results of Moishezon & Teicher (see Moishzon & Teicher 1987) this holds if $X = \mathbb{P}^1 \times \mathbb{P}^1$, $\mathcal{L} = \mathcal{O}(a, b)$, with a and b coprime. However it is open whether one could use instead an étale covering of $Y - Z_{ij}$.

References

Faltings, G. (1991), Diophantine approximation on abelian varieties, *Ann. of Math.* **133**, 549–576.

Faltings, G. & G. Wüstholz (1994), Diophantine approximations on projective spaces, *Invent. Math.* **116** (1994), 109–138.

Fulton, W. (1980), On the fundamental group of the complement of a node curve, *Ann. of Math.* **111**, 407–409.

Fulton, W. & A. Hansen (1979), A connectedness theorem for projective varieties with applications to intersections and singularities of mappings, *Ann. of Math.* **110**, 159–166.

Moishzon, B. & M. Teicher (1987), Simply-connected algebraic surfaces with positive index, *Invent. Math.* **89**, 601–643.

Vojta, P. (1991), Siegel's theorem in the compact case, *Ann. of Math.* **133**, 509–548.

Vojta, P. (1996), Integral points on subvarieties of semiabelian varieties I, *Invent. Math.* **126**, 133–181.

Vojta, P. (1999), Integral points on subvarieties of semiabelian varieties II, *Amer. J. of Math.* **121**, 283–313

16

Search Bounds for Diophantine Equations

D.W. Masser

Thanks for the Method

In this article we discuss diophantine equations from the point of view of 'search bounds'. We focus mainly on single quadratic equations, and we give an account of some work of the last decade culminating in the very recent results of Rainer Dietmann. On the way we pose some related open problems, and at the end we briefly mention some prospects for the future. We start with some general background.

Consider the equation

$$x^4 - 2y^4 + xy + x = 2000. \tag{1}$$

One can ask two basic questions:

(Q1) Are there solutions (x, y) in $\mathbf{Z}^2$?
(Q2) If so, how can we find one?

One can also ask many more questions, but we confine our discussion to just these.

It may well be that the first question cannot be effectively answered at present for (1); the associated curve has genus 3 but is not hyperelliptic or superelliptic or with 'separated variables', and the current effective results on rational approximations to $2^{1/4}$ do not appear to be sufficiently sharp.

Of course the situation gets better for genus 0. Consider the Pell equation

$$x^2 - 631y^2 = 1. \tag{2}$$

Here (Q1) and (Q2) happen to be easy because of $(x, y) = (\pm 1, 0)$, but we could modify these questions to exclude such trivial solutions. Now well-known classical results imply that there is a solution (x, y) in $\mathbf{Z}^2$ with $y \neq 0$; so (Q1) is answered without effort. However (Q2) does need some effort, as the solution with smallest $y > 0$ is (u_0, v_0) with

$$u_0 = 48961575312998650035560, \quad v_0 = 1949129537575151036427. \tag{3}$$

If we alter (2) slightly, to say

$$x^2 - 631y^2 = 2000, \tag{4}$$

then (Q1) and (Q2) both become far less easy to answer, although we will see in a moment that an effective algorithm does exist.

Naturally (1), (2) and (4) are all special cases of a system

$$P_1(x_1, \dots, x_N) = \cdots = P_M(x_1, \dots, x_N) \ = \ 0 \tag{5}$$

for positive integers M and N and polynomials $P_1(X_1, \dots, X_N), \dots, P_M(X_1, \dots, X_N)$ in $\mathbf{Z}[X_1, \dots, X_N]$. Then (Q1) and (Q2) can be similarly formulated for solutions $(x_1, \dots, x_N)$ in $\mathbf{Z}^N$. For the purposes of this article we want to answer both questions in one stroke by producing a 'search bound' with the property that there is always a solution of (5) with

$$\max\{|x_1|, \dots, |x_N|\} \ \le \ B \tag{6}$$

provided a solution exists at all. Of course B should be effectively computable in some sense, and preferably given explicitly as a function of some upper bound $H \ge 1$ for the absolute values of the coefficients in (5); it is easy to find artificial definitions of B which satisfy tautologically the above property.

We may also allow 'restricted search bounds' to take into account side conditions such as $x_1 \ne 0$ or $x_1, \dots, x_N$ not all zero, like $y \ne 0$ in (2).

For example suppose that $P_1, \dots, P_M$ in (5) are all polynomials of total degree at most 1, and $H \ge 1$ measures the coefficients as above. Then $B = (NH)^{\min\{M,N\}}$ is a search bound. The proof is a simple application of Cramer's Rule. We may call this a 'polynomial' search bound, because it is a polynomial in H for fixed M and N. Of course in practice there are more efficient ways of answering (Q1) and (Q2) in this case (for example through Gaussian elimination). But this search bound has a certain theoretical interest, and in fact it is essentially best possible as $H \to \infty$ for any fixed M and N. An extremal system is

$$x_1 = Hx_2, \ \dots, \ x_{k-1} = Hx_k, \ x_k = H \tag{7}$$

with $k = \min\{M, N\}$. This forces $x_1 = H^k$ and so the exponent $\min\{M, N\}$ is sharp.

Actually almost the same bound B is valid for relations $P_m \sim 0$ ($1 \le m \le M$), where '$\sim$' is chosen from any of the symbols $=, \ne, <, >, \le, \ge$. See for example Flahive (1989) and the references therein.

This bound $(NH)^{\min\{M,N\}}$ can be considerably improved if $P_1, \dots, P_M$ are homogeneous and the problem is restricted to exclude the trivial solution $(0, \dots, 0)$. In this case it is reasonable to guarantee a non-trivial solution by imposing $M < N$; and then $(NH)^{M/(N-M)}$ is a restricted search bound. This is just the Siegel Lemma, and the exponent $M/(N-M)$ is also sharp for any fixed M and N (see for example Schmidt 1991, p. 2).

So much for degree 1. We now assume that $P_1, \ldots, P_M$ have total degree at most 2. The main part of this article concerns the case $M = 1$ and so we are dealing with a single quadratic equation

$$P(x_1, \ldots, x_N) = 0 \tag{8}$$

with $P(X_1, \ldots, X_N)$ in $\mathbf{Z}[X_1, \ldots, X_N]$; again suppose $H \geq 1$ is an upper bound for the absolute values of the coefficients.

Siegel (1972) used Hermite reduction theory to derive an effective decision procedure. See also the interesting paper of Grunewald & Segal (1981) for a variant of his algorithm. In terms of H one can deduce a search bound $\exp(CH^{\kappa})$ with C and κ depending only on N, although this does not seem to be explicitly in the literature. Schinzel (1972) gave the bound $(3H)^{300H^3}$ for $N = 2$. This suffices for equations like (4).

This latter bound is not polynomial in H, and in principle it cannot be; in fact Schinzel himself and Lagarias (1980) gave examples showing that functions like $\exp(\sqrt{H})$ cannot be avoided. One such example (not the simplest) starts off with the equation

$$25x^2 - 631.9z^2 = 1. \tag{9}$$

This amounts to (2) with harmless congruence conditions; and the theory of Pell's equation shows that the positive solutions of (9) are given by

$$5x + \sqrt{631}.3z = (u_0 + \sqrt{631}.v_0)^s \quad (s = 1, 3, 5, \ldots)$$

with u_0 and v_0 as in (3). Here the congruence conditions induce a simple congruence condition on s.

Now pick a large positive integer n, and add the not so harmless condition $z \equiv 0 \pmod{3^n}$. This too induces a condition on s, which turns out to be $s \equiv 0 \pmod{3^n}$. It follows that the smallest positive integer solution to

$$25x^2 - 631.3^{2n+2}y^2 = 1 \tag{10}$$

is given by

$$5x + \sqrt{631}.3^{n+1}y = (u_0 + \sqrt{631}.v_0)^s \quad (s = 3^n).$$

Thus as $n \to \infty$ we see that $\max\{|x|, |y|\}$ must be at least exponential in 3^n, which is itself at least of order $\sqrt{H}$ in (10).

Kornhauser (1990a) improved the upper bound to $(14H)^{5H}$, and also proved that it must exceed $2^{H/5}$ for infinitely many H. The lower bound comes out of (10) after adding yet another congruence condition $x \equiv 0 \pmod{5^m}$; and the upper bounds are deduced from classical estimates (see for example Hua 1942) for the smallest non-trivial solution of Pell's equation $x^2 - dy^2 = 1$.

In a second paper Kornhauser (1990b) treated the case $N \geq 5$ and established the search bound $(N^3 H)^{51N}$ provided the homogeneous quadratic part of P in (8) is non-singular. Some such proviso is necessary to exclude equations like (10) with $0x_3{}^2 + 0x_4{}^2 + 0x_5{}^2$ added to the left-hand side. And simple examples in the style of (7) show that the exponent $51N$ cannot be replaced by anything less than $N/2$. The proof of the upper bound uses an elementary but complicated method of Watson which is ultimately based on p-adic considerations.

Thus for $N \geq 5$ we have polynomial search bounds, whereas for $N = 2$ they must grow exponentially.

The case $N = 4$ was treated by Dietmann (1997). In that Diplomarbeit (Master's Thesis) he obtained a polynomial bound CH^κ with absolute constants C, κ. The proof uses the Circle Method. It is well-known that this method becomes more efficient with more variables, and in this way he has recently (Dietmann 2001) improved Kornhauser's bounds for $N \geq 5$ to CH^{5N+93}, where C is now an effective but not yet explicit function of N.

The final case $N = 3$ was settled very recently also by Dietmann (2001), and this case of ternary quadratic equations $P(x, y, z) = 0$ seems to be the most difficult of all.

Here is a summary of these recent results.

Theorem (Dietmann) *For any $N \geq 3$ there exist $C = C(N)$ and $\kappa = \kappa(N)$, depending only on N, with the following property. Suppose that P is a quadratic polynomial in $\mathbf{Z}[X_1, \ldots, X_N]$, whose homogeneous quadratic part is non-singular, such that the equation $P(x_1, \ldots, x_N) = 0$ has a solution in $\mathbf{Z}^N$. Then there is one with $\max\{|x_1|, \ldots, |x_N|\} \leq CH^\kappa$, where $H \geq 1$ is an upper bound for the absolute values of the coefficients in P. Further one may take*

$$\kappa(3) = 3000, \quad \kappa(4) = 100, \quad \kappa(N) = 5N + 93$$

for any $N \geq 5$.

Before we sketch the proof of this result, let us quote an elegant result of Cassels (1978) for the homogeneous case. It says that if P is a quadratic form and the equation (8) has a non-zero solution in $\mathbf{Z}^N$ then it has one with

$$\max\{|x_1|, \ldots, |x_N|\} \leq (6N^2 H)^{(N-1)/2}. \tag{11}$$

So this is a search bound that is restricted in the sense described above; and once again examples based on (7) show that the exponent is sharp. The proof is an ingenious application of the geometry of numbers.

Let us now return to the general case (8). The first step in Dietmann's proof, as in all other work, is to get rid of the linear terms by completing the square. For example, we multiply $3x^2 + x = y^2$ by 12 and write $z = 6x + 1$ to yield $z^2 = 12y^2 + 1$ for $z \equiv 1 \pmod 6$. In this way we reduce (8) to an equation

$$F(x_1, \ldots, x_N) = d \tag{12}$$

for a quadratic form $F(X_1, \ldots, X_N)$ in $\mathbf{Z}[X_1, \ldots, X_N]$ and d in $\mathbf{Z}$, together with congruence conditions

$$x_1 \equiv d_1, \ \ldots, \ x_N \equiv d_N \pmod D \tag{13}$$

for $d_1, \ldots, d_N$ and D in $\mathbf{Z}$.

If $N = 2$ then already (12) is very close to a Pell equation, and we argue as in Kornhauser (1990a).

If $N = 4$ then all depends on the quantity d. If $d = 0$ in (12) then we are in the homogeneous case; but we have the extra conditions (13) to watch, so we cannot just apply Cassels' result. Instead Dietmann shows how to modify the proof (or rather a variant due to Davenport 1957) to obtain a appropriate search bound that is polynomial in D as well as H. The argument is by no means self-evident and it also raises interesting side questions; for example, can one obtain such a bound of the shape $C(N, D)H^{(N-1)/2}$ with Cassels' exponent?

If $N = 4$ and $d \neq 0$ then the Circle Method is applicable, and in fact Dietmann uses a version based on the Poisson summation formula as in Heath-Brown (1983). This also works if $N \geq 5$ (independent of d); but the boundary nature of $N = 4$ is well-known.

Therefore if $N = 3$ one must expect other methods to be necessary, and in fact Dietmann proceeds using ideas of equivalence and automorphisms that are quite special to the theory of quadratic forms.

Recall that two quadratic forms $F = F(X) = F(X_1, \ldots, X_N)$ and G are said to be $\mathbf{Z}$-equivalent if $G(X) = F(UX)$ for some matrix U in the general linear group $GL_N(\mathbf{Z})$; for brevity we write $G = F[U]$. It is not easy to determine whether two given forms are $\mathbf{Z}$-equivalent or not; for example Cassels (1978), p. 132, remarks that neither L.E. Dickson nor A.E. Ross ("both enthusiastic calculators") could decide if the forms

$$x^2 - 3y^2 - 2yz - 23z^2, \quad x^2 - 7y^2 - 6yz - 11z^2 \tag{14}$$

are $\mathbf{Z}$-equivalent or not. Of course for positive definite binary forms this problem is classical, and one finds in a matter of seconds that the parts in (14) involving y and z are not **Z**-equivalent; naturally enough, otherwise the equivalence of (14) themselves would follow immediately. But in the event, nothing follows about (14).

Siegel (1972) gave an effective decision procedure, also based on Hermite reduction theory, to solve this general problem, but nothing as definite as search bounds. Using Siegel's argument Dietmann establishes a more precise connexion as follows. If we have polynomial search bounds for $N = 3$ (in a natural sense soon to be described) for this equivalence problem then we can deduce polynomial search bounds for the equations (12) with $N = 3$ and $d \neq 0$.

For $N = 2$ this connexion goes as follows. If

$$F(x, y) = d \tag{15}$$

is solvable with coprime x, y in **Z**, then F is **Z**-equivalent to a form involving dX^2. Shifting X by an integral multiple of Y we can find an equivalent form G whose coefficient of XY is at most $|d|$ in absolute value; and then the remaining coefficient of G is bounded polynomially in terms of d and the discriminant $\det F$. So there is a similarly bounded U_0 in $GL_2(\mathbf{Z})$ with $G = F[U_0]$; and now the first column of U_0 provides the required solution of (15).

The extension to $N = 3$ uses the classical reduction theory of binary forms. Furthermore if we adjoin suitable congruence conditions throughout then we can deal in the same way with (12) and (13) together.

So this reduces everything to the **Z**-equivalence problem. Some search bounds (without congruence conditions) were worked out for general N in the Diplomarbeit of Straumann (1999), but they are not polynomial; indeed for $N = 2$ they cannot be, in view of the above connexion and the negative results of Schinzel, Lagarias and Kornhauser. It is plausible that polynomial search bounds exist for each $N \geq 3$ but this seems to be very hard to prove. Here is a precise version (without congruence conditions).

Conjecture (Polynomial Bounds for **Z**-Equivalence) *For any $N \geq 3$ there exist C and κ, depending only on N, with the following property. Suppose that F and G are non-singular quadratic forms in $\mathbf{Z}[X_1, \ldots, X_N]$ which are* **Z***-equivalent. Then there is U_0 in $GL_N(\mathbf{Z})$ with $G = F[U_0]$ and*

$$||U_0|| \quad \leq \quad C(||F|| + ||G||)^{\kappa}. \tag{16}$$

Here the norms can be chosen in any crude sense; thus for example $||U_0||$ is the maximum of the absolute values of the entries of U_0, and $||F||$ and $||G||$ are the maximum of the absolute values of the coefficients of F and G respectively.

Dietmann succeeds in establishing this result for $N = 3$ (even with extra congruence conditions), and his Theorem for $N = 3$ follows as we have described. The Conjecture is not yet proved for any $N \geq 4$.

Let us now sketch Dietmann's amazing proof of the Conjecture with $N = 3$.

Many aspects of quadratic forms are easier to handle over fields (a simple example is the estimate in Masser (1998) for rational solutions of (8)), and equivalence is no exception; thus we define **Q**-equivalence with matrices V in $GL_N(\mathbf{Q})$. Dietmann starts by obtaining polynomial search bounds fo the **Q**-equivalence problem for ternary forms. The proof has elements in common with the proof of Cassel' result (11), and it again raises an interesting side problem: what is the sharp exponent in the inequality analogous to (16) for V_0 in $GL_N(\mathbf{Q})$? Now one should define $||V_0||$ more arithmetically to take denominators into account; and probably one should estimate $\max\{||V_0||, ||V_0^{-1}||\}$ on grounds of symmetry.

How does this effective **Q**-equivalence help with **Z**-equivalence?

Take **Z**-equivalent F and G as in the Conjecture. Then $G = F[U]$ for some unknown U in $GL_N(\mathbf{Z})$. Of course F and G are also **Q**-equivalent, and so (if $N = 3$) $G = F[V_0]$ for some V_0 in $GL_N(\mathbf{Q})$ which is polynomially bounded in terms of $||F||$ and $||G||$; for brevity let us refer to V_0 just as 'small'.

Now eliminating G gives

$$F = F[\Omega] \tag{17}$$

for a matrix $\Omega = UV_0{}^{-1}$ in $GL_N(\mathbf{Q})$ with 'small denominator'.

We are suddenly in a different world, that of rational automorphisms of a single form F, and we digress to describe the landscape. The automorphisms form an algebraic group Aut_F, and this is the object that causes the basic trouble. It is known classically that the dimension is $N(N-1)/2$ (for example $N = 3$ and $F = x_1{}^2 + x_2{}^2 + x_3{}^2$ gives the orthogonal group O_3 with three dimensions – two for the axis of rotation and one for the angle of rotation). And in fact the component $\mathrm{Aut}_F{}^+$ consisting of matrices with determinant $+1$ was parametrized by Cayley using the expressions $\phi(S) = (F - S)^{-1}(F + S)$ for varying skew-symmetric matrices S; here we are using F also to denote the symmetric matrix associated to the form F. However the rational map ϕ taking S to $\phi(S)$ is undefined outside a large set corresponding to $\det(F - S) = 0$; and this fact causes some complications in the theory for general N (see for example Weyl 1946, p. 56 or Watson 1960, pp. 132–133).

A substitute for $N = 3$ was given by Hermite in the shape of a rational map ω from projective space $\mathbf{P}_3$ to $\mathrm{Aut}_F{}^+$ in the affine space $\mathbf{A}^9$. See Theorem 58 (p. 96) of Watson (1960). It is *a priori* defined only on the quasiprojective variety Y obtained by omitting the quadric surface in $\mathbf{P}_3$ defined by the vanishing of a polynomial

$$Q(T) = Q(T_1, T_2, T_3, T_4) = (\det F)F(T_1, T_2, T_3) + 4T_4{}^2.$$

Jones & Watson (1956) proved that ω is injective, and also that it is surjective even between the corresponding sets of rational points $Y(\mathbf{Q})$ and $\mathrm{Aut}_F{}^+(\mathbf{Q})$ (and this remains true over any field of zero characteristic). Its coordinates in $\mathbf{A}^9$ are the matrix entries

$$\omega_{ij}(t) = \omega_{ij}(t_1, t_2, t_3, t_4) = Q_{ij}(t)/Q(t) \quad (i, j = 1, 2, 3)$$

with $Q_{ij}(T)$ like $Q(T)$ quadratic forms whose coefficients are explicit functions of F.

More surprisingly, if we embed $\mathbf{A}^9$ in $\mathbf{P}_9$ it turns out that ω defines a projective morphism from $\mathbf{P}_3$ to $\mathbf{P}_9$, so that the above forms Q_{ij} and Q have no common projective zero.

Finishing the digression we return to $F = F[\Omega]$ as in (17), with Ω in the set $\mathrm{Aut}_F(\mathbf{Q})$ of rational automorphisms. Changing Ω to $\pm\Omega$ we can get into $\mathrm{Aut}_F{}^+$, and it follows that $\Omega = \omega(\tau)$ for some τ in $\mathbf{P}_3(\mathbf{Q})$. We can assume that its coordinates $\tau_1, \tau_2, \tau_3, \tau_4$ are in $\mathbf{Z}$.

Now it is well-known (see for example Silverman 1986, Theorem 5.6 on p. 208) that projective morphisms behave well with respect to heights. It follows that the 'formal denominator' $Q(\tau)$ of $\Omega = \omega(\tau)$ is not much bigger than the 'actual denominator'. Now $\Omega = UV_0{}^{-1}$ has small denominator. So $Q(\tau) = q$ (say) is small. But this equation

$$(\det F)F(\tau_1, \tau_2, \tau_3) + 4\tau_4{}^2 = q$$

has exactly the shape (12) with $N = 4$!

We can therefore apply Dietmann's Theorem for $N = 4$ to find small τ_0 in $\mathbf{P}_3(\mathbf{Q})$ with integral coordinates also satisfying $Q(\tau_0) = q$. Furthermore by the congruence version we can assume that τ_0 and τ are congruent to some suitable modulus r. Now $\Omega_0 = \omega(\tau_0)$ satisfies $F = F[\Omega_0]$ and is small not just in denominator but in numerator too. Combining this with $G = F[V_0]$ we find $G = F[U_0]$ with $U_0 = \Omega_0 V_0$ also small. At first sight it looks as if U_0 is only rational and so no great advance on V_0; but choosing the modulus r sufficiently divisible allows us to deduce that

$$U_0 - U = (\Omega_0 - \Omega)V_0 = (\omega(\tau_0) - \omega(\tau))V_0$$

is integral. Because U was integral, so is U_0, and this is what is needed in the Conjecture for $N = 3$.

This completes our discussion of Dietmann's Theorem on single quadratic equations. It supplements the decision procedure of Siegel and of Grunewald–Segal by something much more mechanical, and furthermore with polynomial bounds. As a vague question one could ask for similar bounds in the more extensive situation of the other paper of Grunewald & Segal (1980), which contains full details of their procedure. Even more ambitiously, one could try to do everything over rings of integers of algebraic number fields. But is there a good analogue of Hua's estimate (Hua 1942)?

Let us now briefly discuss the further prospects for general systems (5).

These don't look too good. The famous negative results of Matijasevich imply that the basic question (Q1) is undecidable, so there must exist a system (5) with no search bound at all in the sense we have been using. However the logicians can reduce any system to a system of quadratics; for example

$$x^3 + y^3 + z^3 = 3 \tag{18}$$

is the same as

$$xu + yv + zw = 3, \quad u = x^2, \ v = y^2, \ w = z^2.$$

So the general system of quadratic polynomials (5) is undecidable. Only the case

$$P(x, y, z) = Q(x, y, z) = 0 \tag{19}$$

with $M = 2$, $N = 3$ looks hopeful, because it probably defines a curve of genus $g \leq 1$, which could then be handled by Baker's theory of linear forms in logarithms (see Baker 1966, 1967a, 1967b, 1968a, 1972, 1973a, 1973b, 1974, 1977, Baker & Stark 1971, Baker & Wüstholz 1993).

And even the homogeneous case of (19), much easier for a single equation, presents severe difficulties; for example the system

$$x^2 - az^2 = ct^2, \quad y^2 - bz^2 = dt^2$$

with $M = 2$, $N = 4$ arises in the theory of 2-descents of elliptic curves (for example Silverman 1986, p. 281), and is responsible for the notorious ineffectiveness of the Mordell–Weil Theorem.

So much for quadratics. A single cubic equation (8) is also probably hopeless as soon as $N \geq 3$. Indeed no-one knows, even after extensive computer calculations as well as heuristic theroretical considerations, if the equation (18) has a solution with $x \neq -5, 1, 4$. Only the case

$$P(x, y) = 0 \tag{20}$$

with $N = 2$ can be dealt with at present using the full power of linear forms in logarithms; indeed one of the early successes of Baker's theory was the result (Baker & Coates 1970) that if P is absolutely irreducible of degree δ and (20) defines a curve of genus $g = 1$ then all solutions (x, y) in $\mathbf{Z}^2$ satisfy

$$\max\{|x|, |y|\} \leq \exp\exp\exp\{(2H)^{\Delta}\} \qquad (\Delta = 10^{\delta^{10}}) \tag{21}$$

If $\delta = 3$ then necessarily $g = 0$ or 1, so if $g = 1$ we obtain

$$B = \exp\exp\exp\{(2H)^{\Delta}\} \qquad (\Delta = 10^{59049})$$

as a bound for all solutions, and without doubt the estimates Kornhauser (1990a) on $g = 0$ (together with effective Riemann–Roch) allow this B to be taken as a search bound in (20) for any cubic $P(X, Y)$ whatsoever.

The bound in (21) has since been reduced by Schmidt (1992), Theorem 4 p. 35, to a single exponential $\exp\{C(\delta)H^{\Delta}\}$ with $\Delta = (4\delta)^{13}$ but this is still far from polynomial. And despite the enormous progress (see for example Baker 1964a, 1964b, 1967c, 1968b, 1968c, 1969, Baker & Davenport 1969, Baker & Stewart 1988) on binary diophantine equations in general we still do not have polynomial search bounds for Mordell's $y^2 = x^3 + k$ or even Thue's $x^3 - ay^3 = 2$.

In the homogeneous case a cubic form $P(X_1, \ldots, X_N)$ in $\mathbf{Z}[X_1, \ldots, X_N]$ always has a non-trivial zero in $\mathbf{Z}^N$ if $N \geq 16$, see Davenport 1963; and there is even a corresponding restricted search bound $C(\epsilon)H^{\epsilon}$ if N is sufficiently large with respect to $\epsilon > 0$, Schmidt (1980), (and similar results for homogeneous systems (5) of odd degree). See also Schmidt (1985). These are obtained using the Circle Method, the boundary of which appears to be $N = 10$ or 9, see Heath-Brown (1983), Hooley (1988). And the case $N = 3$ leads again to the Mordell–Weil Theorem for elliptic curves.

Finally we really do have to stop at quartic polynomials; if $P_1, \ldots, P_M$ are quadratic polynomials with no search bound for (5) then another logician's trick shows that the quartic equation $P_1{}^2 + \ldots + P_M{}^2 = 0$ has no search bound. So a single quartic is undecidable.

And even $M = 1$, $N = 2$ is hopeless, because we can return to our starting point (1) with genus 3, for which Siegel's Theorem after more than 70 years remains as ineffective as ever.

Acknowledgement I wish to thank Jörg Brüdern and Rainer Dietmann for their comments on an earlier version of this article.

References

Baker, A. (1964a), Rational approximations to certain algebraic numbers, *Proc. London Math. Soc.* **4**, 385–398.

Baker, A. (1964b), Rational approximations to $\sqrt[3]{2}$ and other algebraic numbers, *Quart. J. Math. Oxford* **15**, 375–383.

Baker, A. (1966), Linear forms in the logarithms of algebraic numbers I, *Mathematika* **13**, 204–216.

Baker, A. (1967a), Linear forms in the logarithms of algebraic numbers II, *Mathematika* **14**, 102–107.

Baker, A. (1967b), Linear forms in the logarithms of algebraic numbers III, *Mathematika* **14**, 220–228.

Baker, A. (1967c), Simultaneous rational approximations to certain algebraic numbers, *Proc. Camb. Phil. Soc.* **63**, 693–702.

Baker, A. (1968a), Linear forms in the logarithms of algebraic numbers IV, *Mathematika* **15**, 204–216.

Baker, A. (1968b), Contributions to the theory of Diophantine equations: I On the representation of integers by binary forms, II The Diophantine equation $y^2 = x^3 + k$, *Phil. Trans. Roy. Soc. London* **A263**, 173–208.

Baker, A. (1968c), The Diophantine equation $y^2 = ax^3 + bx^2 + cx + d$, *J. London Math. Soc.* **43**, 1–9.

Baker, A. (1969), Bounds for the solutions of the hyperelliptic equation, *Proc. Camb. Phil. Soc.* **65**, 439–444.

Baker, A. (1972), A sharpening of the bounds for linear forms in logarithms I, *Acta Arithmetica* **21**, 117–129.

Baker, A. (1973a), A sharpening of the bounds for linear forms in logarithms II, *Acta Arithmetica* **24**, 33–36.

Baker, A. (1973b), A central theorem in transcendence theory. In *Diophantine Approximation and its Applications*, Academic Press, 1–23.

Baker, A. (1974), A sharpening of the bounds for linear forms in logarithms III, *Acta Arithmetica* **27**, 247–252.

Baker, A. (1977), The theory of linear forms in logarithms. In *Transcendence Theory: Advances and Applications*, A. Baker & D.W. Masser (eds.), Academic Press, 1–27.

Baker, A. & J. Coates (1970), Integer points on curves of genus 1, *Proc. Camb. Phil. Soc.* **67**, 595–602.

Baker, A. & H. Davenport (1969), The equations $3x^2 - 2 = y^2$ and $8x^2 - 7 = z^2$, *Quart. J. Math. Oxford* **20**, 129–137.

Baker, A. & H.M. Stark (1971), On a fundamental inequality in number theory, *Ann. Math.* **94**, 190–199.

Baker, A. & C.L. Stewart (1988), On effective approximations to cubic irrationals. In *New Advances in Transcendence Theory*, A. Baker (ed.), Cambridge University Press, 1–24.

Baker, A. & G. Wüstholz (1993), Logarithmic forms and group varieties, *J. Reine Angew. Math.* **442**, 19–62.

Cassels, J.W.S. (1978), *Rational Quadratic Forms*, Academic Press.

Davenport, H. (1957), Note on a theorem of Cassels, *Proc. Camb. Phil. Soc.* **53**, 539–540.

Davenport, H. (1963), Cubic forms in sixteen variables, *Proc. Roy. Soc. London* **A272**, 285–303.

Dietmann, R. (1997), *Kleine Lösungen quadratischer diophantischer Gleichungen in vier Veränderlichen*, Diplomarbeit Stuttgart (53 pages).

Dietmann, R. (2001), Small solutions of quadratic diophantine equations, *Proc. London Math. Soc.*, to appear.

Flahive, M. (1989), Integral solutions of linear systems. In *Théorie des Nombres – Number Theory*, de Gruyter, 213–219.

Grunewald, F. & G. Segal (1980), Some general algorithms. I: Arithmetic groups, *Ann. Math.* **112**, 531–583.

Grunewald, F. & G. Segal (1981), How to solve a quadratic equation in integers, *Math. Proc. Camb. Phil. Soc.* **89**, 1–5.

Heath-Brown, D.R. (1983), Cubic forms in ten variables, *Proc. London Math. Soc.* **47**, 225–257.

Hooley, C. (1988), On nonary cubic forms, *J. Reine Angew. Math.* **386**, 32–98.

Hua, L.K. (1942), On the least solution of Pell's equation, *Bull Amer. Math. Soc.* **48**, 731–735.

Jones, B.W. & G.L. Watson (1956), On indefinite ternary quadratic forms, *Canadian J. Math.* **8**, 592–608.

Kornhauser, D. (1990a), On the smallest solution to the general binary quadratic equation, *Acta Arith.* **55**, 83–94.

Kornhauser, D. (1990b), On small solutions of the general nonsingular quadratic Diophantine equation in five and more unknowns, *Math. Proc. Camb. Phil. Soc.* **107**, 197–211.

Lagarias, J, (1980), On the computational complexity of determining the solvability or unsolvability of the equation $x^2 - dy^2 = -1$, *Trans. Amer. Math. Soc.* **260**, 485–508.

Masser, D.W. (1998), How to solve a quadratic equation in rationals, *Bull. London Math. Soc.* **30**, 24–28.

Schinzel, A. (1972), Integer points on conics, Ann. Soc. Math. Polon., Ser I: *Comment. Math.* **16**, 133–135; Errata, *ibid*, **17** (1973), 305.

Schmidt, W.M. (1980), Diophantine inequalities for forms of odd degree, *Adv. in Math.* **38**, 128–151.

Schmidt, W.M. (1985), The density of integer points on homogeneous varieties, *Acta Math.* **154**, 243–296.

Schmidt, W.M. (1991), *Diophantine Approximation and Diophantine Equations*, Lecture Notes in Math. **1467**, Springer.

Schmidt, W.M. (1992), Integer points on curves of genus 1, *Compositio Math.* **81**, 33–59.

Siegel, C.L. (1972), Zur Theorie der quadratischen Formen, *Nachr. Akad. Wiss. Göttingen Math.-Phys. Kl.* **II**, 21–46; also in *Ges. Abh.*, **IV**, 224–249.

Silverman, J.H. (1986), *The Arithmetic of Elliptic Curves*, Springer.

Straumann, S. (1999), *Das Äquivalenzproblem ganzer quadratischer Formen: Einige explizite Resultate*, Diplomarbeit Basel, (32 pages).

Watson, G.L. (1960), *Integral Quadratic Forms*, Cambridge University Press.

Weyl, H. (1946), *The Classical Groups: their Invariants and Representations*, Princeton University Press.

17

Regular Systems, Ubiquity and Diophantine Approximation

V.V. Beresnevich, V.I. Bernik & M.M. Dodson

1 Introduction

Approximation of real and complex numbers by rationals and algebraic numbers appeared first in papers by Dirichlet, Liouville and Hermite on Diophantine approximation and the theory of transcendental numbers. During the first three decades of the 20th century, E. Borel and A. Khintchine introduced the so-called metric (or measure theoretic) approach in which one considers approximation to any number which does not belong to an exceptional null set (i.e., a set of measure zero). Neglecting such exceptional sets can lead to strikingly simple and general theorems, such as Khintchine's theorem (see below). The exceptional sets can be analysed more deeply by using Hausdorff dimension, which can distinguish between different null sets.

This article gives an account of results, methods and ideas connected with Lebesgue measure and Hausdorff dimension of such exceptional sets. We will be concerned mainly with the lower bound of the Hausdorff dimension. Although determining the correct lower bound for the Hausdorff dimension of a set is often (though by no means always) harder than determining the correct upper bound, recent developments indicate that for many problems, the correct lower bound can be established using information associated with the upper bound. There are some exceptions to this principle. For example, convergence in the Khintchine–Groshev type theorem (for terminology see Bernik & Dodson 1999) for the parabola is related to the upper bound which was proved in Bernik (1979). Nevertheless the divergence case is still unsettled.

For the most part, lower bounds are proved using methods which involve a knowledge of the distribution of some special sets. These sets are very close (or equal) to the solution sets for the Diophantine inequalities under consideration. Originally these methods were developed for sets consisting of points with a distribution described in terms of regular systems. Ubiquitous systems, a multidimensional and more geometrical generalization of regular systems, were introduced in order to investigate more complicated Diophantine approximation, such as on manifolds. Regular and ubiquitous systems have proved to be very effective techniques for obtaining lower bounds for the Hausdorff dimension but in rather different directions.

The development of these ideas has resulted in extensive generalisations of two classical theorems: one due to A.I. Khintchine (see Cassels 1957, Chapter VII, or Khintchine 1924) and the other to V. Jarník (1929) and A.S. Besicovich (1934). Some notation is needed at this point. As usual $|A|$ and $\dim A$ will denote, respectively, the Lebesgue measure and the Hausdorff dimension of the set A. Throughout this article, unless otherwise stated, the function $\psi : \mathbb{N} \to \mathbb{R}^+$ ($\mathbb{N}$ is the set of positive integers) will be monotonically decreasing. A number x will be called *ψ-approximable* if the inequality

$$|qx - p| < \psi(q) \tag{1}$$

holds for infinitely many $(p, q) \in \mathbb{Z} \times \mathbb{N}$. This definition will be carried over to more general and sometimes different situations.

The set of ψ-approximable numbers will be denoted by $\mathcal{K}^1(\psi)$. Note that $\mathcal{K}^1(\psi)$ can be expressed as a general kind of 'lim-sup' set:

$$\mathcal{K}^1(\psi) = \bigcap_{N=1}^{\infty} \bigcup_{q=N}^{\infty} \bigcup_{p \in \mathbb{Z}} \left(\frac{p}{q} - \frac{\psi(q)}{q}, \frac{p}{q} + \frac{\psi(q)}{q} \right). \tag{2}$$

The first result shows how the size of $\mathcal{K}^1$ in terms of Lebesgue measure depends on the convergence properties of ψ.

Theorem 1 (Khintchine) *Let $q\psi(q)$ be monotonically decreasing. Then, for any finite interval $I \subset \mathbb{R}$,*

$$|\mathcal{K}^1(\psi) \cap I| = \begin{cases} 0, & \text{if } \sum_{q=1}^{\infty} \psi(q) < \infty, \\ |I|, & \text{if } \sum_{q=1}^{\infty} \psi(q) = \infty. \end{cases} \tag{3}$$

The second gives the Hausdorff dimension of the exceptional set $\mathcal{K}^1_v$ of very well approximable points corresponding to $\psi(q) = q^{-v}$, where $v > 1$, in (1).

Theorem 2 (Jarník–Besicovitch) *For any $v \geqslant 1$,*

$$\dim \mathcal{K}^1_v = \frac{2}{v+1}. \tag{4}$$

The convergence case of Khintchine's theorem and the correct upper bound for the Hausdorff dimension in the Jarník–Besicovitch theorem are quite straightforward. Indeed, there are natural covers of the sets $\mathcal{K}^1(\psi)$ and $\mathcal{K}_v^1$ arising from (2) of intervals defined by (1). Applying the Borel–Cantelli Lemma and the Hausdorff–Cantelli Lemma (the Hausdorff dimension analogue of the former, see Bernik & Dodson 1999, p. 67) to these covers gives the desired results. It is worth repeating that the main difficulty lies in the complementary cases of these theorems.

Regular systems and ubiquitous systems are introduced separately and then some applications are discussed.

2 Regular systems

Mahler (1932) gave a classification of real and complex numbers and raised a problem about the approximation type of almost all real numbers. For each $n \in \mathbb{N}$ and $v \in \mathbb{R}$, let $\mathfrak{M}_v^{(n)}$ denote the set of $x \in \mathbb{R}$ such that there are infinitely many integer polynomials P of degree at most n satisfying the inequality

$$|P(x)| < H(P)^{-v}, \tag{5}$$

where $H(P)$ is the height of P (essentially, $\mathfrak{M}_v^{(1)}$ is $\mathcal{K}_v^1$). Mahler conjectured that for any $v > n$, the set $\mathfrak{M}_v^{(n)}$ is of measure zero. This was solved by V.G. Sprindžuk (1964).

Theorem 3 (Sprindžuk) *Let $n \in \mathbb{N}$ and let $v > n$. Then $|\mathfrak{M}_v^{(n)}| = 0$.*

The Hausdorff dimension of the null set $\mathfrak{M}_v^{(n)}$ naturally became of interest. Some upper bounds for $\dim \mathfrak{M}_v^{(n)}$ had been obtained before 1964 but no lower bound was known. Baker & Schmidt (1970) introduced a very powerful method for obtaining lower bounds for Hausdorff dimension and used it to establish the correct lower bound:

$$\dim \mathfrak{M}_v^{(n)} \geqslant \frac{n+1}{v+1}. \tag{6}$$

Let us now explain the basic ideas of their method. Let $P \in \mathbb{Z}[x]$, $\deg P \leqslant n$ and let α be a real root of P. By the continuity of P, the closer x is to α, the smaller $|P(x)|$. Thus it is very natural to consider approximation of real numbers by real algebraic numbers in (1) with $\psi(q) = q^{-v}$.

Let $\mathbb{A}^{(n)}$ denote the set of real algebraic numbers of degree at most n. Given $v \in \mathbb{R}$, let $\mathbb{A}_v^{(n)}$ be the set of $x \in \mathbb{R}$ such that there are infinitely many $\alpha \in \mathbb{A}^{(n)}$ satisfying

$$|x - \alpha| < H(\alpha)^{-v-1}, \tag{7}$$

where $H(\alpha)$ is the height of α. It is not difficult to see that if $n < w < v$, then

$$\mathbb{A}_v^{(n)} \subset \mathfrak{M}_w^{(n)}. \tag{8}$$

Thus, since $A \subset B$ implies that $\dim A \leqslant \dim B$, a lower bound for $\dim \mathbb{A}_v^{(n)}$ is also a lower bound for $\dim \mathfrak{M}_w^{(n)}$. Note that the inequality

$$\dim \mathbb{A}_v^{(n)} \leqslant (n+1)/(v+1)$$

can be easily obtained by the Hausdorff–Cantelli Lemma exactly as in the case $n = 1$, which has already been discussed.

Now let us consider the rationals again. They are dense in $\mathbb{R}$ and also uniformly distributed. Moreover, in a certain sense any two different rational numbers are well separated. This can be described as follows. Let T be a large positive number. Let $\mathbb{Q}_T$ be the set of rationals with height (the modulus of the denominator) less than or equal to T. The number of such rationals in an interval I, $\mathrm{card}(\mathbb{Q}_T \cap I)$, is $O(T^2|I|)$. It is easily seen that the average of the distances between two consecutive rationals in $\mathbb{Q}_T \cap I$ is asymptotically $|I|/\mathrm{card}(\mathbb{Q}_T \cap I) = O(T^{-2})$. Also the distance between two consecutive rationals in $\mathbb{Q}_T \cap I$ is at least T^{-2}. Thus on average the points of $\mathbb{Q}_T \cap I$ are separated as they are individually.

This is not the case for algebraic numbers of higher degree. However, it can be shown that a positive proportion of points in $\mathbb{A}^{(n)}(T)$, the set of algebraic numbers of degree at most n and height at most T, consists of well separated points. Thus the set $\mathbb{A}^{(n)}(T)$ can be refined so that we will have a system of points with a distribution similar to the rationals. This fact, first established by Baker & Schmidt (1970), is described using the concept of a regular system of points.

Definition 1 Let Γ be a countable set of real numbers and let $N : \Gamma \to \mathbb{R}$ be a positive function. The pair (Γ, N) is called *a regular system of points* if there exists a constant $C_1 = C_1(\Gamma, N) > 0$ such that for any finite interval I there exists a sufficiently large number $T_0 = T_0(\Gamma, N, I) > 0$ such that for any integer $T \geqslant T_0$ there exists a collection

$$\gamma_1, \ldots, \gamma_t \in \Gamma \cap I \tag{9}$$

such that $N(\gamma_i) \leqslant T \quad (1 \leqslant i \leqslant t)$, $|\gamma_i - \gamma_j| \geqslant T^{-1} \quad (1 \leqslant i < j \leqslant t)$, and $t \geqslant C_1|I|T$.

Example 2 It is readily verified that the set of all rational numbers together with the function $N(p/q) = q^2$, where p and q are relatively prime, is a regular system.

As usual, $\{x\}$ denotes the fractional part of the real number x and

$$\|x\| = \min\{|x - k| : k \in \mathbb{Z}\}.$$

A number α is badly approximable if $\inf\{n\|n\alpha\| : n \in \mathbb{N}\} > 0$.

Example 3 When $\alpha \in \mathbb{R}$ is a badly approximable number, the pair (Γ, N), where $\Gamma = \{\{\alpha n\} : n \in \mathbb{N}\}$ and $N(\{\alpha n\}) = n$, is a regular system.

The following non-trivial example of a regular system was given by Baker & Schmidt (1970).

Example 4 Let $\Gamma = \mathbb{A}^{(n)}$ and $N(\gamma) = H(\gamma)^{n+1}(\log H(\gamma))^{-3n(n+1)}$ for $\gamma \in \Gamma$. Then (Γ, N) is a regular system.

The next lemma is the key point of the Baker–Schmidt method for obtaining lower bounds for Hausdorff dimension (see Baker & Schmidt 1970, Rynne 1992).

Lemma 1 *Suppose that* $\psi : \mathbb{R} \to \mathbb{R}^+$ *is decreasing with* $x\psi(x) \leqslant 1/2$ *for large* x. *If* (Γ, N) *is a regular system then*

$$\dim \Lambda(\Gamma, N; \psi) \geqslant s_0 = \sup\{s : \lim_{x\to\infty} x\psi(x)^s = \infty\},$$

where $\Lambda(\Gamma, N, \psi)$ *is the set of all real numbers* x *for which the inequality* $|x - \gamma| < \psi(N(\gamma))$ *holds for infinitely many* $\gamma \in \Gamma$.

Baker & Schmidt (1970) applied this lemma to Example 4 and used inequality (8) to establish the correct lower bounds for $\dim \mathbb{A}_v^{(n)}$ and $\dim \mathfrak{M}_v^{(n)}$. The correct upper bound for $\dim \mathbb{A}_v^{(n)}$ can be easily obtained by Hausdorff–Cantelli Lemma in the same way as in the case $n = 1$ already discussed. Determining the correct upper bound for $\dim \mathfrak{M}_v^{(n)}$ is much harder and involves different arguments based on careful and complicated analysis of the distribution of all the algebraic numbers, not only regularly distributed ones, since any subclass of $\mathbb{A}_v^{(n)}$ may contribute to $\dim \mathfrak{M}_v^{(n)}$ (Bernik 1983).

Theorem 4 *For any* $v > n$,

$$\dim \mathbb{A}_v^{(n)} = \dim \mathfrak{M}_v^{(n)} = (n + 1)/(v + 1).$$

Melián & Pestana (1993) have considered the hyperbolic space analogue of the Jarník–Besicovitch theorem. To obtain the correct lower bound for the Hausdorff dimension, they used 'well-distributed' systems, an extension of regular systems to higher dimensions.

3 Ubiquity

Ubiquitous systems were introduced in Dodson, Rynne & Vickers (1990) as another technique for obtaining a lower bound for the Hausdorff dimension of sets of 'very well approximable' points and of general 'exceptional' sets associated with questions of 'small denominators' which arise in normal forms and stability of dynamical systems (see Arnold 1963, Bernik & Dodson 1999, Chapter 7, and Dodson, Rynne & Vickers 1989, for more details). In one dimension, ubiquitous and regular systems, discussed above, are almost equivalent (regular systems are more general in one respect, and ubiquity is in another, but both have been extended to equivalent forms in Rynne 1992). Regular systems lend themselves to more refined simultaneous estimates in higher dimensions but ubiquitous systems deal with broader questions and yield a lower bound for the Hausdorff dimension more directly in terms of the geometry. In addition, ubiquity allows the approximation function q^{-v} to be replaced naturally by $\psi: \mathbb{N} \to \mathbb{R}^+$, where $\psi(q) \to 0$ monotonically as $q \to \infty$.

In the type of Diophantine approximation considered here, we are concerned with the set consisting of points x in Euclidean space which are, roughly speaking, a small distance from a member of a special class of subsets of the space infinitely often. The set is related to a general sort of 'lim-sup' of a sequence of neighbourhoods of special sets. In the Jarník–Besicovitch theorem described above, the special class of subsets is the set of rationals $\mathbb{Q}$ and the distance is less than q^{-v}. There is no loss of generality in confining attention to the (open or closed) unit interval or to hypercubes in higher dimensions. The hard part of this theorem is establishing the correct lower bound for $\dim \mathcal{K}_v^1$. As has been pointed out, this can be obtained using regular systems, which were introduced to establish the generalisation of the Jarník–Besicovitch theorem to approximation by real algebraic irrationals of given degree (Baker & Schmidt 1970). Ubiquity, however, was framed to deal with higher dimensional sets, such as the systems of linear forms arising in a general form of the Jarník–Besicovitch theorem established by Bovey & Dodson (1986). For each $\mathbf{x} = (x_1, \ldots, x_n) \in \mathbb{R}^n$, we let

$$|\mathbf{x}| = \max\{|x_1|, \ldots, |x_n|\} \text{ and } \|\mathbf{x}\| = \max\{\|x_1\|, \ldots, \|x_n\|\}$$

be the height of $\mathbf{x}$ and the distance of $\mathbf{x}$ from $\mathbb{Z}^n$ respectively. Let $\mathcal{K}^{m,n}(\psi)$ be the set of real $m \times n$ matrices $(a_{ij}) = A$, regarded as points in $\mathbb{R}^{mn}$, such that the system of inequalities

$$\|\mathbf{q}A\| = \max_{j=1,\ldots,n} \{\|q_1 a_{1j} + \cdots + q_m a_{mj}\|\} < \psi(|\mathbf{q}|)$$

holds for infinitely many $\mathbf{q} = (q_1, \ldots, q_m) \in \mathbb{Z}^m$. The set $\mathcal{K}^{m,n}(\psi)$ is related to a 'lim-sup' set of a sequence of neighbourhoods of finite unions of subsets of hyperplanes $R(\mathbf{q}) = \{A \in (0,1)^{mn} : \|\mathbf{q}A\| = 0\}$. For another more complicated example, see Dodson, Rynne & Vickers (1994). Following Arnol'd (1983), these sets $R(\mathbf{q})$ are called *resonant* because of the association with the physical phenomenon of resonance.

The definition of ubiquity was abstracted from Bovey & Dodson (1986) which involved systems of linear forms and used geometrical ideas based on those in Besicovitch (1934) combined with a mean and variance argument from Cassels (1957), Chapter 7. By design, resonant sets play a fundamental role in ubiquity which in essence ensures that they are in good supply. They can be thought of as generalisations of rational numbers and are of a relatively simple nature, being finite unions of points or parts of lines, curves, planes, surfaces and so on, which are solution sets of Diophantine equations.

Definition 2 Take U to be a non-empty open subset of $\mathbb{R}^m$. Let

$$\mathcal{R} = \{R_j \subset U : j \in J\} \tag{10}$$

be a family of resonant sets, indexed by J, where each $j \in J$ has a weight $\lfloor j \rfloor > 0$. The resonant set $R_j = R(\mathbf{q})$ with $\lfloor j \rfloor = |\mathbf{q}|$ in the above. Let the function $\rho : \mathbb{N} \to \mathbb{R}^+$ converge to 0 at ∞ and let $A(Q)$, $Q = 1, 2, \ldots$, be a sequence of subsets of U such that $\lim_{Q\to\infty} |A(Q)| = |U|$. Let

$$B(R_j; \delta) = \{u \in U : \mathrm{dist}_\infty(u, R_j) < \delta\},$$

where $\mathrm{dist}_\infty(u, R) = \inf\{|u - r| : r \in R\}$, the distance from u to R in the supremum norm. Suppose that there exists a constant $d \in [0, m]$ such that given any hypercube $H \subset U$ with $\ell(H) = \rho(Q)$ and such that $H/2$ intersects $A(Q)$, then there exists a $j \in J$ with $\lfloor j \rfloor \leqslant Q$ such that for all $\delta \in (0, \rho(Q)]$,

$$|H \cap B(R_j; \delta)| \gg \delta^{m-d}\, \ell(H)^d, \tag{11}$$

where $\gg$ and $\ll$ are the Vinogradov symbols ($b \gg a$ and $a \ll b$ mean that $a = O(b)$). Suppose further that for any other hypercube H' in U with $\ell(H') \leqslant \rho(Q)$,

$$|H' \cap H \cap B(R_j; \delta)| \ll \delta^{m-d}\, \ell(H')^d. \tag{12}$$

Then the pair $(\mathcal{R}, \lfloor\cdot\rfloor)$ is called a *ubiquitous system with respect to* ρ (reference to the weight is usually omitted).

The intersection estimates (11) and (12) have been used in preference to more geometrical descriptions of the intersections $H \cap R_j$ for generality. The first requires that the hypercube H and the resonant set R_j intersect substantially and that small hypercubes H' intersect $H \cap R_j$ as they 'should'. For resonant sets R_j with a reasonable structure, d will be the topological dimension of each R_j and the intersection conditions (11) and (12) will be satisfied more or less automatically. Indeed when the R_j are d-dimensional affine spaces in Euclidean space, we can take the approximating set $A(Q)$ to be a union of $\rho(Q)$-neighbourhoods of R_j, namely $A(Q) = \bigcup_{\lfloor j \rfloor \leqslant Q} B(R_j; \rho(Q))$. It is then readily verified that the intersection conditions (11) and (12) can be replaced by the single measure condition

$$\left| \bigcup_{\lfloor j \rfloor \leqslant Q} B(R_j; \rho(Q)) \right| \to |U| \text{ as } Q \to \infty.$$

This condition can be weakened to the limit of the left-hand side being at least $c|U|$ for some constant c, $0 < c \leqslant 1$ and all Q sufficiently large, see Rynne (1992). Ubiquity can be relatively simple to establish and in practice the function ρ emerges naturally. For instance Dirichlet's theorem implies that the set of rationals (with weight the modulus of the denominator) is ubiquitous with respect to a function comparable with $Q^{-2} \log Q$; in $\mathbb{R}^n$, the rational points $\mathbf{p}/q$, where $\mathbf{p} \in \mathbb{Z}^n$, $q \in \mathbb{Q}$, are ubiquitous with respect to a function comparable with $Q^{-1-1/n} \log Q$, see Dodson, Rynne & Vickers (1990).

A general lower bound

The distribution of the resonant sets in ubiquitous systems allows the determination of a general lower bound for the Hausdorff dimension of the lim-sup set

$$\Lambda(\mathcal{R}; \psi) = \{u \in U\colon \operatorname{dist}_\infty(u, R_j) < \psi(\lfloor j \rfloor) \text{ for infinitely many } j \in J\},$$

where $\psi\colon \mathbb{N} \to \mathbb{R}^+$ is a non-increasing function and the resonant sets have common dimension $d = \dim \mathcal{R}$, say, and codimension $m - d = \operatorname{codim}\mathcal{R}$.

Theorem 5 *Suppose $\mathcal{R}$ is a family of resonant sets which is ubiquitous with respect to ρ and that $\widetilde{\psi}\colon \mathbb{R}^+ \to \mathbb{R}^+$ is a non-increasing function satisfying $\widetilde{\psi}(Q) \leqslant \rho(Q)$ for Q sufficiently large. Then*

$$\dim \Lambda(\mathcal{R}; \widetilde{\psi}) \geqslant \dim \mathcal{R} + \gamma \operatorname{codim}\mathcal{R},$$

where $\gamma = \limsup_{Q \to \infty} \log \rho(Q) / \log \widetilde{\psi}(Q) \leqslant 1$.

This is proved in Dodson, Rynne & Vickers (1990) and another proof using the mass distribution principle is given in Bernik & Dodson (1999), Chapter 5. The constant γ is at most 1 since $\widetilde{\psi}(Q) \leqslant \rho(Q)$ for Q sufficiently large.

This result can be used to establish the correct lower bound for the Hausdorff dimension of ψ-approximable systems of linear forms. It can be shown using Minkowski's linear forms theorem that the resonant sets $R_{\mathbf{q}}$, where $\mathbf{q} \in \mathbb{Z}^m$ is non-zero, given by

$$R_{\mathbf{q}} = \{A \in [0,1]^{mn} : \mathbf{q}A \in \mathbb{Z}^n\}$$

are ubiquitous with respect $mQ^{-1-m/n}\log Q$ and 'most' matrices A in $[0,1]^{mn}$ are 'close' to a set $R_{\mathbf{q}}$ with weight $\lfloor \mathbf{q} \rfloor = |\mathbf{q}|$ not too large (see Dodson 1992, 1993). The correct upper bound for the Hausdorff dimension of $\mathcal{K}^{m,n}(\psi)$ can obtained using a straightforward covering argument. The complementary result follows from Theorem 5 with $\widetilde{\psi}(Q) \asymp Q\psi(Q)$ ($a \asymp b$ means that a and b are comparable, i.e., $a \ll b$ and $b \ll a$). The lower order $\lambda(f)$ of a function $f : \mathbb{N} \to \mathbb{R}^+$ is defined to be $\liminf_{Q\to\infty}(\log f(Q))/(\log Q)$.

Theorem 6 *Let $\psi : \mathbb{N} \to \mathbb{R}^+$ be a decreasing function and let λ be the lower order of $1/\psi$. Then*

$$\dim \mathcal{K}^{m,n}(\psi) = \begin{cases} (m-1)n + (m+n)/(\lambda+1) & \text{when } \lambda \geqslant m/n, \\ mn & \text{when } \lambda \leqslant m/n. \end{cases}$$

The Jarník–Besicovitch theorem corresponds to $m = n = 1$ and $\psi(q) = q^{-v}$. Dickinson & Velani (1997) have extended Jarník's Hausdorff measure analogue (Jarník 1929) of Khintchine's theorem for simultaneous Diophantine approximation to systems of linear forms, thus establishing the Hausdorff measure analogue of the Khintchine–Groshev theorem (Sprindžuk 1979). Instead of ubiquity, they work with an elaborate Cantor-type construction. There are some interesting applications to normal forms of pseudodifferential operators (Dickinson, Gramchev, & Yoshino 1995). The complete hyperbolic analogue of the Jarník–Besicovitch theorem was established by Hill & Velani (1998) using Cantor type subsets.

Inhomogeneous Diophantine approximation, which in one dimension concerns the size of $|qx - \alpha - p|$ for some fixed $\alpha \in \mathbb{R}$, differs somewhat from homogeneous Diophantine approximation, where $\alpha = 0$. It is easier in the doubly metric case where one considers the joint measure of the set of points (x, α) but harder in the singly metric case where α is given. Using ubiquity, a general form of the inhomogeneous version of the Jarník–Besicovitch theorem has been obtained for the doubly metric case (Dodson 1997) and Levesley

(1998) has established it in the more difficult singly metric case with the additional help of uniform distribution.

Dickinson has discussed the Hausdorff dimension for systems of linear forms which have small modulus and has shown that the set of matrices $A = (a_{ij}) \in \mathbb{R}^{mn}$ such that for infinitely many $\mathbf{q} \in \mathbb{Z}^m$,

$$|\mathbf{q}A| = \max_{j=1,\dots,n} \{|q_1 a_{1j} + \cdots + q_m a_{mj}|\} < |\mathbf{q}|^{-v} \tag{13}$$

has Hausdorff dimension $m(n-1) + m/(v+1)$ for $v \geqslant (m/n) - 1$ and mn otherwise (Dickinson 1993). The upper bound is obtained by the usual covering argument but ubiquity gives the correct lower only for $m > n$. In the complementary range where $m \leqslant n$, a diffeomorph of the set is decomposed into the cartesian product of an $(m-1)(n-m+1)$-cube and a space to which the arguments in the first range apply. For a more general approach, see Dickinson (1993) and Rynne (1998a,b). The p-adic version of the general Jarník–Besicovitch theorem is essentially of the form (13). The Hausdorff dimension of the corresponding set was obtained by Abercrombie (1995) for $m > n$ using Billingsley dimension; the dimension when $m \leqslant n$ has been determined in Dickinson, Dodson & Jin (1999) using the same approach as in Dickinson (1993).

4 Khintchine-type theorems on manifolds

Theorem 1 has been generalised to approximation by real algebraic numbers and to Diophantine approximation on manifolds in Euclidean space. The functional dependence between the coordinates in the latter case causes formidable technical problems but approximation on the rational normal curve

$$\mathcal{V} = \{(x, \dots, x^n) : x \in \mathbb{R}\} \tag{14}$$

is related to approximation by real algebraic numbers. In this connection, in 1966 Baker raised (with a slightly different notation) the question of the measure of the set $\mathfrak{M}^{(n)}(\psi)$ of $x \in \mathbb{R}$ such that the inequality

$$|P(x)| < \psi(H(P)) \tag{15}$$

has infinitely many solutions $P \in \mathbb{Z}[x]$ with $\deg P \leqslant n$. Baker proved that $|\mathfrak{M}^{(n)}(\psi)| = 0$ if ψ is monotonic and if $\sum_{q=1}^{\infty} \psi^{1/n}(q) < \infty$ (see Baker 1966). He further conjectured that the convergence condition can be replaced with $\sum_{q=1}^{\infty} q^{n-1}\psi(q) < \infty$; this was proved by Bernik (1989), see also Bernik & Dodson (1999).

Theorem 7 *Let* $\psi : \mathbb{N} \to \mathbb{R}^+$ *be monotonic. Then* $|\mathfrak{M}^{(n)}(\psi)| = 0$ *whenever the sum*

$$\sum_{q=1}^{\infty} q^{n-1}\psi(q) \tag{16}$$

converges.

The proof of this result was used to improve the regular system of algebraic numbers constructed by Baker & Schmidt.

Example 5 (see Bernik & Dodson 1999, p. 101). For each $n \in \mathbb{N}$, let

$$N(\gamma) = H(\gamma)^{n+1}(\log H(\gamma))^{-(n+1)}$$

for $\gamma \in \mathbb{A}^{(n)}$. Then $(\mathbb{A}^{(n)}, N)$ is a regular system.

Regular systems were used to establish a Khintchine–Groshev type theorem for $\mathfrak{M}^{(n)}(\psi)$ when the sum (16) diverges (Beresnevich 1999):

Theorem 8 *Let* $\psi : \mathbb{N} \to \mathbb{R}^+$ *be a monotonic sequence. Then for each* $n \in \mathbb{N}$*, the set* $\mathfrak{M}^{(n)}(\psi)$ *has full measure*† *whenever* $\sum_{q=1}^{\infty} q^{n-1}\psi(q) = \infty$.

Theorem 8 can be derived from Theorem 9 below using the following arguments. Let $\mathbb{A}^{(n)}(\psi)$ be the set consisting of $x \in \mathbb{R}$ such that there are infinitely many $\gamma \in \mathbb{A}^{(n)}$ satisfying

$$|x - \gamma| < \psi(H(\gamma)). \tag{17}$$

It can be verified easily that for any interval I_0, there is a sufficiently small positive constant c such that $\mathbb{A}^{(n)}(\tilde{\psi}) \cap I_0 \subset \mathfrak{M}^{(n)}(\psi) \cap I_0$ if $\tilde{\psi}(q) \leqslant c\psi(q)/q$ for all $q \in \mathbb{N}$. Thus the following suffices to prove Theorem 8, see Beresnevich (1999).

Theorem 9 *For any* $n \in \mathbb{N}$ *and monotonic sequence* $\psi : \mathbb{N} \to \mathbb{R}^+$*, the set* $\mathbb{A}^{(n)}(\psi)$ *has full measure whenever* $\sum_{q=1}^{\infty} q^n\psi(q) = \infty$.

Regular systems play the key role in the proof of Theorem 9. Indeed this Khintchine-type result requires 'optimal' knowledge about the distribution of real algebraic numbers. Given an interval I and a positive number T, the collection (9) is chosen from the set $\Gamma_N(I, T) = \{\gamma \in \Gamma \cap I : N(\gamma) \leqslant T\}$. This

† A set A has full measure if $|\mathbb{R} \setminus A| = 0$

set may contain many points which are not included in (9). For example, in the Baker–Schmidt regular system (see Example 4),

$$\frac{\mathrm{card}\Gamma_N(I,T)}{\text{the number of points satisfying (9)}} \asymp |I|(\log T)^{3n(n+1)}.$$

In Example 5 this ratio is $\asymp |I|(\log T)^{n+1}$, a little smaller. This suggests the following.

Definition 3 (see Beresnevich 2000) The regular system (Γ, N) will be called *optimal* if for any finite interval I

$$\sup_{T>0} \frac{\mathrm{card}\Gamma_N(I,T)}{T} < \infty. \tag{18}$$

The following example of a best possible regular system is given in Beresnevich (1999) where more details are given.

Example 6 For each $n \in \mathbb{N}$, let $N(\gamma) = H(\gamma)^{n+1}/(1+|\gamma|)^{n(n+1)}$. Then $(\mathbb{A}^{(n)}, N)$ is a regular system.

The proof of this example is based on measuring the solution sets for certain Diophantine inequalities efficiently (see Beresnevich 1999 for more details). The proof of Theorem 9 is based on the following generalised Borel–Cantelli lemma, also used in the proof of the Khintchine–Groshev theorem (see Sprindžuk 1979, Chapter 2,§ 2; Harman 1998, p. 35).

Lemma 2 *Let $E_i \subset \mathbb{R}$ be a sequence of measurable sets and the set E consist of points x belonging to infinitely many E_i. If all the sets E_i are uniformly bounded and the sum $\sum_{i=1}^{\infty} |E_i|$ diverges, then*

$$|E| \geqslant \limsup_{N\to\infty} \frac{\left(\sum_{i=1}^{N} |E_i|\right)^2}{\sum_{i=1}^{N}\sum_{j=1}^{N} |E_i \cap E_j|}. \tag{19}$$

The sets E_i are taken to be small neighbourhoods of points (9). The set $\mathbb{A}^{(n)}$ having an optimal regular system makes it possible to control the sum in both the numerator and the denominator of (19). As far as approximation by points of regular system is concerned, the following Khintchine-type result is proved in Beresnevich (2000).

Lemma 3 *Let* (Γ, N) *be a regular system,* $\psi : \mathbb{N} \to \mathbb{R}^+$ *be monotonic and* $\Lambda(\Gamma, N; \psi)$ *be the set defined in Lemma 1. Then, for any interval* $I \subset \mathbb{R}$

$$|\Lambda(\Gamma, N; \psi) \cap I| = \begin{cases} 0, & \text{if } \sum_{q=1}^{\infty} \psi(q) < \infty \text{ and } (\Gamma, N) \text{ is optimal,} \\ |I|, & \text{if } \sum_{q=1}^{\infty} \psi(q) = \infty. \end{cases}$$

Extending Khintchine-type theorems to a manifold M in $\mathbb{R}^n$ is difficult because of the functional relationships between the coordinates of points (the measure is of course the induced Lebesgue measure on M). A half-way house is to show that the set $\mathcal{L}(M; \psi)$ of points $\mathbf{x} \in M$ such that

$$\|\mathbf{x} \cdot \mathbf{q}\| < \psi(|\mathbf{q}|)$$

for infinitely many $\mathbf{q} \in \mathbb{Z}^n$ is null (in the induced measure on M) when $\psi(q) = q^{-v}$ for any $v > n$. The set $\mathcal{L}(M; \psi)$ is dual to the set $\mathcal{S}(M; \psi)$ of points $\mathbf{x} \in M$ satisfying

$$\max\{\|qx_i\| : i = 1, \ldots, n\} < \psi(q)$$

for infinitely many $q \in \mathbb{N}$. By Khintchine's Transference Principle, the set $\mathcal{S}_v(M)$, which is $\mathcal{S}(M; \psi)$ with $\psi(q) = q^{-v}$, is also null for $v > 1/n$ (see Bernik & Dodson 1999).

The first general result in the metrical theory of Diophantine approximation on manifolds was due to Schmidt (1964). He investigated C^3 planar curves of the form $\Gamma = \{(f_1(x), f_2(x)) : x \in I\}$, where I is an interval, $f_1, f_2 : I \to \mathbb{R}$ are C^3 functions such that the curvature $f_1'(s) f_2''(s) - f_1''(s) f_2'(s) \neq 0$ for almost all $s \in I$, and proved that for any such curve, the set of very well approximable points is relatively null, i.e., the curve is *extremal* (see Bernik & Dodson 1999 for terminology). Thus M is extremal if the set of simultaneously very well approximable points

$$\mathcal{S}_v(M) = \{\xi \in M : \|q\,\xi\| < q^{-v} \text{ for infinitely many } q \in \mathbb{N}\} \qquad (20)$$

is relatively null when $v > 1/n$ or equivalently if the set

$$\mathcal{L}_v(M) = \{\xi \in M : \|\mathbf{q} \cdot \xi\| < |\mathbf{q}|^{-v} \text{ for infinitely many } \mathbf{q} \in \mathbb{Z}^n\}$$

is null in M when $v > n$ respectively. The terminology reflects the fact that for almost all points on an extremal set, the exponents in Dirichlet's theorem are unimprovable, for more details see Bernik & Dodson (1999), Koksma (1936) p. 67. A manifold M is *strongly extremal* if given any $v > n$, the set of points $x = (x_1, \ldots, x_n) \in M$ satisfying

$$\|\mathbf{q} \cdot \mathbf{x}\| < \prod_{j=1}^{n} (|q_j| + 1)^{-v/n},$$

for infinitely many $\mathbf{q} \in \mathbb{Z}^n$ is null in M, so that a strongly extremal manifold is extremal. Baker conjectured that the rational normal curve $\mathcal{V}$ (see (14)) is strongly extremal in Baker (1975) and Sprindžuk extended the conjecture to any manifold M satisfying the conditions of H_1 (Conjecture H_2 in Sprindžuk 1980). The rational normal curve $\mathcal{V}$ was shown to be strongly extremal by Bernik & Borbat (1997) for $n = 4$.

Manifolds satisfying a variety of analytic, geometric and number theoretic conditions have been shown to be extremal (more details are in Bernik & Dodson 1999; Sprindžuk 1979, 1980; Vinogradov & Chudnovsky 1984). In an extension of Schmidt's theorem to higher dimensional manifolds, Kovalevskaya (1978) has shown that surfaces in $\mathbb{R}^3$ having non-zero Gaussian curvature almost everywhere are extremal (see also or Sprindžuk 1979, p. 149, Theorem 18) and, together with Bernik, later extended this result to m-dimensional surfaces in $\mathbb{R}^{2m}$, see Bernik & Kovalevskaya (1990). These are special cases of the more general result that smooth (C^3) manifolds of dimension at least 2 (so that M is at least a surface) and satisfying a curvature condition (which specialises to non-zero Gaussian curvature for surfaces in $\mathbb{R}^3$) are also extremal (see Dodson, Rynne & Vickers 1989, 1991, and the next section).

Schmidt's result has been extended to C^4 curves in $\mathbb{R}^3$ by Beresnevich & Bernik (1996). Recently Kleinbock & Margulis (1998) have proved that manifolds which are nondegenerate almost everywhere are strongly extremal. Nondegeneracy can be regarded as a generalisation of non-zero curvature and is defined as follows. For each $j \leqslant k$, the point $\mathbf{x} = \theta(u) \in M \subset \mathbb{R}^n$ is j-*nondegenerate* if the partial derivatives of θ at u up to order j span $\mathbb{R}^n$. The point $\mathbf{x}$ is *nondegenerate* if it is j-*nondegenerate* for some j. This result is best possible and implies both Sprindžuk's and the stronger Baker–Sprindžuk conjectures. The proof uses ideas from dynamical systems, particularly unipotent flows in homogeneous spaces of lattices. Their techniques are likely to lead to further progress and have led to a generalisation of Baker's result (Baker 1966).

In 1991, the following Khintchine–Groshev-type result was obtained for fairly general manifolds. Let M be a C^3 manifold embedded in $\mathbb{R}^n$ with dimension at least 2 and 2-convex almost everywhere (i.e., M has at least 2 principal curvatures with strictly positive product almost everywhere). Then $\mathcal{L}(M; \psi)$ is null if the sum (16) converges. If the sum diverges and M satisfies a stronger curvature condition, then $\mathcal{L}(M; \psi)$ is full (Dodson, Rynne & Vickers 1991, Theorem 1.1. If the sum $\sum_{q=1}^{\infty} \psi(q)^n$ converges, then $\mathcal{S}(M; \psi)$ is null. Khintchine–Groshev type analogues of Schmidt's theorem were obtained in Beresnevich, Bernik, Dodson & Dickinson (1999) and Bernik, Dodson &

Dickinson (1998). A slightly different notation has been adopted, reflecting the connection with $\mathfrak{M}^{(n)}(\psi)$.

Theorem 10 *Let I be a finite interval and $f_1, f_2 : I \to \mathbb{R}$ be C^3 functions such that $f_1'(x)f_2''(x) - f_1''(x)f_2'(x) \neq 0$ for almost all $x \in I$. Let $\psi : \mathbb{N} \to \mathbb{R}^+$ be monotonic and let $\mathcal{L}_{f_1,f_2}(\psi)$ be the set of $x \in I$ such that the inequality*

$$|a_2 f_2(x) + a_1 f_1(x) + a_0| < \psi(|\mathbf{a}|_\infty), \tag{21}$$

where $|\mathbf{a}|_\infty = \max\{|a_0|, |a_1|, |a_2|\}$, has infinitely many solutions $\mathbf{a} = (a_0, a_1, a_2) \in \mathbb{Z}^3$. Then

$$|\mathcal{L}_{f_1,f_2}(\psi)| = \begin{cases} 0, & \text{if } \sum_{q=1}^{\infty} q\psi(q) < \infty, \\ |I|, & \text{if } \sum_{q=1}^{\infty} q\psi(q) = \infty. \end{cases} \tag{22}$$

Analogues of Khintchine's theorem have been obtained recently for non-degenerate manifolds. This was done independently by Beresnevich (2001b) and Bernik, Kleinbock & Margulis (1999). In the former, a development of Sprindžuk's method of essential and inessential domains is applied to smooth curves with Wronskians which are non-zero almost everywhere and the result extended to nondegenerate manifolds. In the latter, the geometry of lattices in Euclidean spaces and flows on homogeneous spaces are used†. The complementary divergence case has also been established for various cases in Beresnevich (1999, 2000a, 2000b) and Bernik, Kleinbock & Margulis (1999), and the full result will appear in Beresnevich, Bernik, Kleinbock & Margulis (2002).

5 Hausdorff dimension on manifolds

Extremal results and the convergence case of Khintchine–Groshev type theorems give rise to null sets and so the question of their Hausdorff dimension arises naturally. A manifold M which is a C^3 planar curve with non-vanishing curvature everywhere except on a set with zero Hausdorff dimension is extremal by Schmidt (1964). By extending this and the results of Baker & Schmidt (1970), R.C. Baker proved that the Hausdorff dimension of $\mathcal{L}_v(M)$ is $3/(v+1)$ for $v \geqslant 2$ (R.C. Baker 1978). It is shown in Dodson, Rynne & Vickers (1989) that for manifolds M with dimension $m \geqslant 2$ and 2-curved (this specialises to non-zero Gaussian curvature for surfaces in $\mathbb{R}^3$) everywhere except on a set of Hausdorff dimension at most $m-1$,

$$\dim \mathcal{L}_v(M) = m - 1 + (n+1)/(v+1) \tag{23}$$

† A stronger multiplicative version is also proved. These results can be extended to manifolds which can be sliced into suitable curves, such as, for example, analytic manifolds.

for $v \geqslant n$. Note that this implies that M is extremal and that when ψ is decreasing, this result can be extended to

$$\mathcal{L}(M\,;\psi) = \{\mathbf{x} \in M \colon \|\mathbf{q}\cdot\mathbf{x}\| < \psi(|\mathbf{q}|) \text{ for infinitely many } \mathbf{q} \in \mathbb{Z}^n\}.$$

Ubiquity has been used to show that the right-hand side of (23) is a general lower bound for the Hausdorff dimension of $\mathcal{L}(M;\psi)$ when M is a C^1 extremal manifold in $\mathbb{R}^n$ (Dickinson & Dodson 2000).

Theorem 11 *Let $\psi\colon \mathbb{N} \to \mathbb{R}^+$ be decreasing with the lower order λ. Let M be a C^1 extremal manifold embedded in $\mathbb{R}^n$ and suppose $\lambda \geqslant n$. Then*

$$\dim \mathcal{L}(M;\psi) \geqslant m - 1 + (n+1)/(\lambda+1).$$

The proof uses the geometry of numbers and Fatou's lemma. The question of the correct upper bound is more difficult but we conjecture that equality holds for nondegenerate manifolds. Some of the results and methods discussed above have been extended to Diophantine approximation of complex and p-adic numbers (see Abercrombie 1995; Dickinson, Dodson, & Jin 1999).

Simultaneous Diophantine approximation on manifolds

Determining the Hausdorff dimension of the set $\mathcal{S}_v(M)$ defined in (20) of simultaneously v-approximable points on manifolds can be more difficult than the dual case. When M is the circle $\mathbb{S}^1$ and $v > 1$, the natural number q is part of the Pythagorean triple (p, r, q). Melnichuk (1979) exploited this to obtain the correct upper bound and (with regular systems) an estimate for the lower bound. In fact using either ubiquity or regular systems, it can be shown that

$$\dim \mathcal{S}_v(\mathbb{S}^1) = 1/(v+1)$$

for $v > 1$ (Dickinson & Dodson 2001). Exponential sums can be combined with regular systems to obtain estimates for the Hausdorff dimension of $\mathcal{S}_v(M)$ for certain manifolds M. The argument involves the distribution of rational points near the manifold. Further details on this and other aspects of the theory can be found in Bernik & Dodson (1999).

Acknowledgment The second author is grateful to ETH, Zürich for its hospitality and to G. Wüstholz for support and the opportunity to give a shorter version of this article at the international meeting to mark Alan Baker's 60th birthday. We are also grateful to H. Dickinson for helpful discussions and for assistance with the preparation of this article.

References

Abercrombie, A.G. (1995), The Hausdorff dimension of some exceptional sets of p-adic matrices, *J. Number Th.* **53**, 311–341.

Arnol'd, V.I. (1963), Small denominators and problems of stability of motion in classical and celestial mechanics, *Usp. Mat. Nauk* **18**, 91–192; English translation in *Russian Math. Surveys,* **18** (1963), 85–191.

Arnol'd, V.I. (1983), *Geometrical Methods in Ordinary Differential Equations*, Springer-Verlag.

Baker, A. (1966), On a theorem of Sprindžuk, *Proc. Roy. Soc. Series A* **292**, 92–104.

Baker, A. (1975), *Transcendental Number Theory*, Cambridge University Press; second edition (1979).

Baker, A. & W.M. Schmidt (1970), Diophantine approximation and Hausdorff dimension, *Proc. Lond. Math. Soc.* **21**, 1–11.

Baker, R.C. (1978), Dirichlet's theorem on Diophantine approximation, *Math. Proc. Cam. Phil. Soc.* **83**, 37–59.

Beresnevich, V.V. (1999), On approximation of real numbers by real algebraic numbers, *Acta Arith.* **90**, 97–112.

Beresnevich, V.V. (2000a), Application of the concept of regular system of points in metric number theory, *Vesti NAN Belarusi. Phys.-Mat. Ser.*, (1), 35–39.

Beresnevich, V.V. (2000b), On proving analogues of Khintchine's theorems for curves, *Vesti NAN Belarusi, Phys.-Mat. Ser.*, in Russian, (3), 35–40.

Beresnevich, V.V. & V.I. Bernik (1996), On a metrical theorem of W. Schmidt, *Acta Arith.* **75**, 219–233.

Beresnevich, V.V., V.I. Bernik, H. Dickinson & M.M. Dodson (1999), The Khintchine–Groshev theorem for planar curves, *Proc. Roy. Soc. Lond. A* **455**, 3053–3063.

Beresnevich, V.V., V.I. Bernik, D.Y. Kleinbock & G.A. Margulis (2002), Metric Diophantine approximation: the Khintchine–Groshev theorem for non-degenerate manifolds, *Moscow Mathematical Journal*, to appear.

Bernik, V.I. (1979), On the exact order of approximation of almost all points on the parabola, *Mat. Zametki* **26**, 657–665.

Bernik, V.I. (1983), An application of Hausdorff dimension in the theory of Diophantine approximation, *Acta Arith.* **42**, 219–253; English translation in American Mathematical Society Translations **140** (1988), 15–44.

Bernik, V.I. (1989), On the exact order of approximation of zero by values of integral polynomials, *Acta Arith.* **53**, 17–28 (in Russian).

Bernik, V.I. & V. N. Borbat (1997), Polynomials with coefficients of different modulus and A. Baker's conjecture, *Vesti Akad, Navuk Belarus. Ser. Fiz.-Mat. Navuk.* **3**, 5–8 (in Russian).

Bernik, V.I.& M.M. Dodson (1999), *Metric Diophantine Approximation on Manifolds*, Cambridge University Press.

Bernik, V.I. & E.I. Kovalevskaya (1990), Diophantine approximation on n-dimensional manifolds in $\mathbb{R}^{2n}$, *Dokl. Bel.* **12**, 1061–1064.

Bernik, V.I., H. Dickinson & M.M. Dodson (1998), A Khintchine-type version of Schmidt's theorem for planar curves, *Proc. Roy. Soc. Lond. A* **454**, 179–185.

Bernik, V.I., D.Y. Kleinbock & G.A. Margulis (1999), Khintchine-type theorems for manifolds: convergence case for standard and multiplicative versions, Preprint 99–092, Bielefeld. Submitted to *International Math. Research Notices*.

Besicovitch, A.S. (1934), Sets of fractional dimensions (IV): on rational approximation to real numbers, *J. Lond. Math. Soc.* **9**, 126–131.

Bovey, J.D. & M.M. Dodson (1986), The Hausdorff dimension of systems of linear forms, *Acta Arith.* **45**, 337–358.

Cassels, J.W.S. (1957), *An Introduction to Diophantine Approximation*, Cambridge University Press.

Dickinson, H. (1993), The Hausdorff dimension of systems of simultaneously small linear forms, *Mathematika* **40**, 367–374.

Dickinson, H. (1994), The Hausdorff dimension of sets arising in Diophantine approximation, *Acta Arith* **53**, 133–140.

Dickinson, H. & M.M. Dodson (2000), Extremal manifolds and Hausdorff dimension, *Duke Math. J.* **101**, 337–347.

Dickinson, H. & M.M. Dodson (2001), Diophantine approximation and Hausdorff dimension on the circle, *Math. Proc. Cam. Philos. Soc.* **130** (2001), 515–522.

Dickinson, H., M.M. Dodson, & Jin Yuan (1999), Hausdorff dimension and p-adic Diophantine approximation, *Indag. Mathem., N.S.* **10**, 337–347.

Dickinson, H., T. Gramchev, & M. Yoshino (1995), First order pseudodifferential operators on the torus: normal forms, Diophantine approximation and global hypoellipticity, *Ann. Univ. Ferrara, Sez. VII – Sc. Mat.* **61**, 51–64.

Dickinson, H. & S.L. Velani (1997), Hausdorff measure and linear forms, *J. Reine Angew. Math.* **490**, 1–36.

Dodson, M.M. (1992), Hausdorff dimension, lower order and Khintchine's theorem in metric Diophantine approximation, *J. Reine Angew. Math.* **432** (1992), 69–76.

Dodson, M.M. (1993), Geometric and probabilistic ideas in the metrical theory of Diophantine approximation, *Usp. Mat. Nauk* **48**, 77–106; English translation in *Russian Math. Surveys* **48** (1993), 73–102.

Dodson, M.M. (1997), A note on metric inhomogeneous Diophantine approximation, *J. Austral. Math. Soc. (Series A)* **62**, 175–185.

Dodson, M.M., B.P. Rynne, & J.A.G. Vickers (1989), Metric Diophantine approximation and Hausdorff dimension on manifolds, *Math. Proc. Cam. Phil. Soc.* **105**, 547–558.

Dodson, M.M., B.P. Rynne, & J.A.G. Vickers (1990), Diophantine approximation and a lower bound for Hausdorff dimension, *Mathematika* **37**, 59–73.

Dodson, M.M., B.P. Rynne, & J.A.G. Vickers (1991), Khintchine-type theorems on manifolds, *Acta Arith.* **57**, 115–130.

Dodson, M.M., B.P. Rynne, & J.A.G. Vickers (1994), The Hausdorff dimension of exceptional sets associated with normal forms, *J. Lond. Math. Soc.* **49**, 614–624.

Harman, G. (1998), *Metric Number Theory*, Clarendon Press.

Hill, R. & S.L. Velani (1998), The Jarník–Besicovitch theorem for geometrically finite Kleinian groups, *Proc. Lond. Math. Soc.* **77**, 524–550.

Jarník, V. (1929), Diophantischen Approximationen und Hausdorffsches Mass, *Mat. Sbornik* **36**, 371–382.

Khintchine, A.I. (1924), Einige Sätze über Kettenbruche, mit Anwendungen auf die Theorie der Diophantischen Approximationen, *Math. Ann.* **92**, 115–125.

Kleinbock, D.Y. & G.A. Margulis (1998), Flows on homogeneous spaces and Diophantine approximation on manifolds, *Ann. Math.* **148** (1998), 339–360.

Koksma, J.F. (1936), *Diophantische Approximationen*, Springer-Verlag.

Kovalevskaya, E.I. (1978), A geometric property of extremal surfaces, *Mat. Zametki* **23**, 99–101.

Levesley, J. (1998), A general inhomogeneous Jarník–Besicovitch theorem, *J. Number Th.* **71**, 65–80.

Mahler, K. (1932), Über das Mass der Menge aller S-Zahlen, *Math. Ann.* **106**, 131–139.

Melián, M.V. & D. Pestana (1993), Geodesic excursions into cusps in finite volume hyperbolic manifolds, *Mich. Math. J.* **40**, 77–93.

Melnichuk, Y.V. (1979), Diophantine approximations on a circle and Hausdorff dimension, *Mat. Zametki* **26**, 347–354; English translation in *Math. Notes* **26** (1980), 666–670.

Rynne, B.P. (1992), Regular and ubiquitous systems, and $\mathcal{M}^s_\infty$-dense sequences, *Mathematika* **39** (1992), 234–243.

Rynne, B.P. (1998a), Hausdorff dimension and generalised simultaneous Diophantine approximation, *Bull. Lond. Math. Soc.* **30**, 365–376.

Rynne, B.P. (1998b), The Hausdorff dimension of sets arising from Diophantine approximation with a general error function, *J. Number Th.* **71**, 166–177.

Schmidt, W.M. (1964), Metrische Sätze über simultane Approximation abhängiger Grössen, *Monatsh. Math.* **68**, 154–166.

Sprindžuk, V.G. (1969), *Mahler's Problem in Metric Number Theory*, American Mathematical Society (1969), translated by B. Volkmann.

Sprindžuk, V.G. (1979), *Metric theory of Diophantine Approximations*, Wiley.

Sprindžuk, V.G. (1980), Achievements and problems in Diophantine approximation theory, *Usp. Mat. Nauk* **35**, 3–68; English translation in *Russian Math. Surveys,* **35**, (1980), 1–80.

Vinogradov, A.I. & G.V. Chudnovsky (1984), The proof of extremality of certain manifolds. In *Contributions to the Theory of Transcendental Numbers*, G.V. Chudnovsky (ed.), American Mathematical Society, 421–447.

18

Diophantine Approximation, Lattices and Flows on Homogeneous Spaces

Gregory Margulis

Introduction

During the last 15–20 years it has been realized that certain problems in Diophantine approximation and number theory can be solved using geometry of the space of lattices and methods from the theory of flows on homogeneous spaces. The purpose of this survey is to demonstrate this approach on several examples. We will start with Diophantine approximation on manifolds where we will briefly describe the proof of Baker–Sprindžuk conjectures and some Khintchine-type theorems. The next topic is the Oppenheim conjecture proved in the mid-1980s and the Littlewood conjecture, still not settled. After that we will go to quantitative generalizations of the Oppenheim conjecture and to counting lattice points on homogeneous varieties. In the last part we will discuss results on unipotent flows on homogeneous spaces which play, directly or indirectly, the most essential role in the solution of the above-mentioned problems. Most of those results on unipotent flows are proved using ergodic theorems and also notions such as minimal sets and invariant measures. These theorems and notions have no effective analogs and because of that the homogeneous space approach is not effective in a certain sense. We will briefly discuss the problem of the effectivization at the very end of the paper.

The author would like to thank A. Eskin and D. Kleinbock for their comments on a preliminary version of this article.

1 Diophantine approximation on manifolds

We start by recalling some standard notation and terminology. For $\mathbf{x}, \mathbf{y} \in \mathbf{R}^n$ we let

$$\mathbf{x} \cdot \mathbf{y} = \sum_{i=1}^{n} x_i y_i, \qquad \|\mathbf{x}\| = \max_{1 \le i \le n} |x_i|,$$

$$\Pi(\mathbf{x}) = \prod_{i=1}^{n} |x_i| \quad \text{and} \quad \Pi_+(\mathbf{x}) = \prod_{i=1}^{n} |x_i|_+,$$

where $|x|_+$ stands for $\max(|x|, 1)$. A vector $\mathbf{y} \in \mathbf{R}^n$ is called *very well approximable*, to be abbreviated as VWA, if the following two equivalent conditions are satisfied:

(V1) for some $\varepsilon > 0$ there are infinitely many $\mathbf{q} \in \mathbf{Z}^n$ such that

$$|\mathbf{q} \cdot \mathbf{y} + p| \cdot \|\mathbf{q}\|^n \le \|\mathbf{q}\|^{-n\varepsilon}$$

for some $p \in \mathbf{Z}$;

(V2) for some $\varepsilon > 0$ there are infinitely many $q \in \mathbf{Z}$ such that

$$\|q\mathbf{y} + \mathbf{p}\|^n \cdot |q| \le |q|^{-\varepsilon}$$

for some $\mathbf{p} \in \mathbf{Z}^n$.

A vector $\mathbf{y} \in \mathbf{R}^n$ is called *very well multiplicatively approximable*, to be abbreviated as VWMA, if the following two equivalent conditions are satisfied:

(VM1) for some $\varepsilon > 0$ there are infinitely many $\mathbf{q} \in \mathbf{Z}^n$ such that

$$|\mathbf{q} \cdot \mathbf{y} + p| \cdot \Pi_+(\mathbf{q}) \le \Pi_+(\mathbf{q})^{-\varepsilon}$$

for some $p \in \mathbf{Z}$;

(VM2) for some $\varepsilon > 0$ there are infinitely many $q \in \mathbf{Z}$ such that

$$\Pi(q\mathbf{y} + \mathbf{p}) \cdot |q| \le |q|^{-\varepsilon}$$

for some $\mathbf{p} \in \mathbf{Z}^n$.

Remark It is clear that if a vector is VWA then it is also VWMA. The equivalence of (V1) and (V2) (resp. the equivalence of (VM1) and (VM2)) follows from (resp. from a modification of) Khintchine's transference principle.

The just-introduced definitions can be generalized in the following way. Let ψ be a positive non-increasing function defined on the set of positive integers, and let $\lfloor x \rfloor$ denote the distance between $x \in \mathbf{R}$ and the closest integer. We say that a vector $\mathbf{y} \in \mathbf{R}^n$ is *ψ-approximable*, to be abbreviated as ψ-A (resp. *ψ-multiplicatively approximable*, to be abbreviated as ψ-MA), if there are infinitely many $\mathbf{q} \in \mathbf{Z}^n$ such that

$$\lfloor \mathbf{q} \cdot \mathbf{y} \rfloor \le \psi(\|\mathbf{q}\|)^n \quad \text{(resp.} \quad \lfloor \mathbf{q} \cdot \mathbf{y} \rfloor \le \psi(\Pi_+(\mathbf{q}))).$$

Remark If a vector is ψ-A then it is also ψ-MA. A vector is VWA (resp. VWMA) if it is ψ_ε-A (resp. ψ_ε-MA) for some positive ε, where $\psi_\varepsilon(k) \doteq k^{-(1+\varepsilon)}$.

It easily follows from (the simple part of) the Borel–Cantelli lemma that almost all $\mathbf{y} \in \mathbf{R}^n$ are not ψ-A if

$$\sum_{k=1}^{\infty} \psi(k) < \infty, \tag{1}$$

and that almost all $\mathbf{y} \in \mathbf{R}^n$ are not ψ-MA if

$$\sum_{k=1}^{\infty} (\log k)^{n-1} \psi(k) < \infty. \tag{2}$$

Conversely, according to the Khintchine–Groshev theorem and its multiplicative version (see Schmidt 1960 and Sprindžuk 1979, Chapter I, Theorem 12), almost all $\mathbf{y} \in \mathbf{R}^n$ are ψ-A (resp. ψ-MA) if the series $\sum_{k=1}^{\infty} \psi(k)$ diverges (resp. the series $\sum_{k=1}^{\infty} (\log k)^{n-1} \psi(k)$ diverges); these two statements are usually referred to as the convergence and the divergence parts of the theorem. As an example we get almost all $\mathbf{y} \in \mathbf{R}^n$ are not VWMA and hence are not VWA either. Much more difficult questions arise if one considers almost all points on a submanifold M of $\mathbf{R}^n$. Mtivated by his theory of types of transcendental numbers, Mahler (1932) conjectured that almost all points of the curve

$$M = \{(x, x^2, \ldots, x^n) \mid x \in \mathbf{R}\} \tag{3}$$

are not VWA. In 1964 V. Sprindžuk proved this conjecture (see Sprindžuk 1964, 1969) using what is later became known as 'the method of essential and inessential domains'. Also that year, Schmidt (1964) proved that if a smooth planar curve $\{(f_1(x), f_2(x)) \mid x \in \mathbf{R}\}$ has non-zero curvature for almost all x then almost all points of this curve are not VWA. Sprindžuk's result was improved by Baker (1966): he showed that if ψ is a positive non-increasing function such that

$$\sum_{k=1}^{\infty} \frac{\psi(k)^{1/n}}{k^{1-1/n}} < \infty, \tag{4}$$

then almost all points of the curve (3) are not ψ-A. The above and some other results obtained in mid-1960s eventually led to the development of a new branch of metric number theory, usually referred to as 'Diophantine approximation with dependent quantities' or 'Diophantine approximation on manifolds'.

Baker conjectured that (4) could be replaced by the optimal condition (1); this conjecture was proved by Bernik (1984). And recently Beresnevich (2001) proved the complementary divergence case for the curve (3). That is, the complete analogue of the Khintchine–Groshev Theorem for the curve (3) has been established.

Baker (1975) stated, in his book, another conjecture. This conjecture is about the multiplicative approximation, and it says that almost all points of the curve (3) are not VWMA. Later it was generalized by Sprindžuk (1980) in a survey paper:

Conjecture 1 *Let $\mathbf{f} = (f_1, \ldots, f_n)$ be an n-tuple of real analytic functions on a domain V in $\mathbf{R}^d$ which together with 1 are linearly independent over $\mathbf{R}$. Then for almost all $\mathbf{x} \in V$ the vector $\mathbf{f}(\mathbf{x})$ is not VWMA.*

Remark In the same survey Sprindžuk (1980) stated also a weaker version of Conjecture 1 where VWMA is replaced by VWA.

Conjecture 1 was proved by Kleinbock & Margulis (1998) not only for analytic but also for smooth functions. To state our main result we have to introduce the following definition: if V is an open subset of $\mathbf{R}^d$ and $l \leq k$, an n-tuple $\mathbf{f} = (f_1, \ldots, f_n)$ of C^k functions $V \mapsto \mathbf{R}$ is called *l-nondegenerate* at $\mathbf{x} \in V$ if the space $\mathbf{R}^n$ is spanned by partial derivatives of $\mathbf{f}$ at $\mathbf{x}$ of order up to l. The n-tuple $\mathbf{f}$ is *nondegenerate* at $\mathbf{x}$ if it is l-nondegenerate at $\mathbf{x}$ for some l. We say that $\mathbf{f} : V \mapsto \mathbf{R}^n$ is *nondegenerate* if it is nondegenerate at almost every point of V. Note that if the functions $f_1, \ldots, f_n$ are analytic and V is connected, the nondegeneracy of $\mathbf{f}$ is equivalent to the linear independence of $1, f_1, \ldots, f_n$ over $\mathbf{R}$.

Theorem 1 (Kleinbock & Margulis 1998, Theorem A) *Let $\mathbf{f} : V \mapsto \mathbf{R}^n$ be a nondegenerate C^k map of an open subset V of $\mathbf{R}^d$ into $\mathbf{R}^n$. Then $\mathbf{f}(\mathbf{x})$ is not VWMA (hence not VWA either) for almost every point $\mathbf{x}$ of V.*

If $M \subset \mathbf{R}^n$ is a d-dimensional C^k submanifold, we say that M is *nondegenerate* at $\mathbf{y} \in M$ if any (equivalently some) diffeomorphism $\mathbf{f}$ between an open subset V of $\mathbf{R}^d$ and a neighborhood $\mathbf{y}$ in M is nondegenerate at $\mathbf{f}^{-1}(\mathbf{y})$. We say that M is *nondegenerate* if it is nondegenerate at almost every point of M (in the sense of the natural measure class on M). A connected analytic submanifold $M \subset \mathbf{R}^n$ is nondegenerate if and only if it is not contained in any hyperplane in $\mathbf{R}^n$. Now we can reformulate Theorem 1.

Corollary 1 *Let M be a nondegenerate C^k submanifold of $\mathbf{R}^n$. Then almost all points of M are not VWMA (hence not VWA either).*

The proof of Theorem 1 in Kleinbock & Margulis (1998) was based on a new method which used the correspondence (cf. Dani 1985; Kleinbock 1996, 1998) between approximation properties of vectors $\mathbf{y} = (y_1, \ldots, y_n) \in \mathbf{R}^n$ and the behavior of certain orbits in the space of unimodular lattices in $\mathbf{R}^{n+1}$. More precisely, let

$$U_{\mathbf{y}} = \begin{pmatrix} 1 & y_1 & y_2 & \cdots & y_n \\ 0 & 1 & 0 & \cdots & 0 \\ 0 & 0 & 1 & \cdots & 0 \\ \cdots & \cdots & \cdots & \cdots & \cdots \\ 0 & 0 & 0 & \cdots & 1 \end{pmatrix} \in SL(n+1, \mathbf{R}).$$

Thus $U_{\mathbf{y}}$ is a unipotent matrix with all rows, except the first one, the same as in the identity matrix. Note that

$$U_{\mathbf{y}} \begin{pmatrix} p \\ \mathbf{q} \end{pmatrix} = \begin{pmatrix} \mathbf{q} \cdot \mathbf{y} + p \\ \mathbf{q} \end{pmatrix}, \qquad p \in \mathbf{Z}, \quad \mathbf{q} \in \mathbf{Z}^n. \tag{5}$$

We also have to introduce some diagonal matrices. Let

$$g_s = \begin{pmatrix} e^{ns} & 0 & \cdots & 0 \\ 0 & e^{-s} & \cdots & 0 \\ \cdots & \cdots & \cdots & \cdots \\ 0 & 0 & \cdots & e^{-s} \end{pmatrix} \in SL(n+1, \mathbf{R}), \qquad s \geq 0,$$

and

$$g_{\mathbf{t}} = \begin{pmatrix} e^{t} & 0 & \cdots & 0 \\ 0 & e^{-t_1} & 0 & 0 \\ \cdots & \cdots & \cdots & \cdots \\ 0 & 0 & \cdots & e^{-t_n} \end{pmatrix} \in SL(n+1, \mathbf{R}),$$

$$\mathbf{t} = (t_1, \ldots, t_n), \quad t_i \geq 0, \quad t = \sum_{i=1}^{n} t_i;$$

(it is clear that $g_s = g_{(s,\ldots,s)}$). Next define a function δ on the space of lattices by

$$\delta(\Lambda) \doteq \inf_{\mathbf{v} \in \Lambda \setminus \{0\}} \|\mathbf{v}\|.$$

Note that the ratio of $1 + \log(1/\delta(\Lambda))$ and $1 + \mathrm{dist}(\Lambda, \mathbf{Z}^{n+1})$ is bounded between two positive constants for any $SL(n+1, \mathbf{R})$-invariant metric 'dist' on the space of lattices Λ in $\mathbf{R}^{n+1}$. The equality (5) immediately implies that

$$\delta(g_{\mathbf{t}} U_{\mathbf{y}} \mathbf{Z}^{n+1}) = \inf_{(p,\mathbf{q}) \in \mathbf{Z}^{n+1} \setminus \{0\}} \min\{\mathbf{q} \cdot \mathbf{y} + p, e^{-t_1} q_1, \ldots, e^{-t_n} q_n\} \tag{6}$$

where $\mathbf{q} = (q_1, \ldots, q_n)$. It is a rather easy consequence of (6) that a vector $y \in \mathbf{R}^n$ is VWA (resp. VWMA) if and only if there exists $\gamma > 0$ and infinitely many $t \in \mathbf{Z}_+$ (resp. infinitely many $\mathbf{t} \in \mathbf{Z}_+^n$) such that

$$\delta(g_t U_{\mathbf{y}} \mathbf{Z}^{n+1}) \leq e^{-\gamma t} \qquad (\text{resp.} \quad \delta(g_{\mathbf{t}} U_{\mathbf{y}} \mathbf{Z}^{n+1}) \leq e^{-\gamma \|\mathbf{t}\|}); \tag{7}$$

in other words, $\mathbf{y}$ is not VWA (resp. not VWMA) if and only if $\operatorname{dist}(g_t U_{\mathbf{y}} \mathbf{Z}^{n+1}, \mathbf{Z}^{n+1})$ (resp. $\operatorname{dist}(g_{\mathbf{t}} U_{\mathbf{y}} \mathbf{Z}^{n+1}, \mathbf{Z}^{n+1})$), as a function of $t \in \mathbf{Z}_+$ (resp. as a function of $\mathbf{t} \in \mathbf{Z}_+^n$), grows slower than any linear function. Thus Theorem 1 is equivalent to the statement that for almost all $\mathbf{x} \in V$ and any $\gamma > 0$, there are at most finitely many $\mathbf{t} \in \mathbf{Z}_+^n$ such that (7) holds for $\mathbf{y} = \mathbf{f}(\mathbf{x})$. In view of the Borel–Cantelli lemma, this statement can be proved by estimating the measure of the sets

$$E_{\mathbf{t}} \doteq \{\mathbf{x} \in V \mid \delta(g_{\mathbf{t}} U_{\mathbf{f}(\mathbf{x})} \mathbf{Z}^{n+1}) \leq e^{-\gamma \|\mathbf{t}\|}\}$$

for any given $\mathbf{t} \in \mathbf{Z}_+^n$, so that

$$\sum_{\mathbf{t} \in \mathbf{Z}_+^n} |E_{\mathbf{t}}| < \infty$$

(here and hereafter $|\cdot|$ stands for the Lebesgue measure). Such estimates are easily deduced from the following

Proposition 1 (Kleinbock & Margulis 1998, Proposition 2.3) *Let $\mathbf{f} : V \mapsto \mathbf{R}^n$ be a C^k map of an open subset V of $\mathbf{R}^d$ into $\mathbf{R}^n$, and let $\mathbf{x}_0 \in V$ be such that $\mathbf{R}^n$ is spanned by partial derivatives of $\mathbf{f}$ at $\mathbf{x}_0$ of order up to k. Then there exists a ball $B \subset U$ centered at $\mathbf{x}_0$, and positive constants D and ρ, such that for any $\mathbf{t} \in \mathbf{R}_+^n$ and $0 \leq \varepsilon \leq \rho$ one has*

$$|\{\mathbf{x} \in B \mid \delta(g_{\mathbf{t}} U_{\mathbf{f}(\mathbf{x})} \mathbf{Z}^{n+1}) \leq \varepsilon\}| \leq D \left(\frac{\varepsilon}{\rho}\right)^{1/dk} |B|.$$

Proposition 1 is deduced in Kleinbock & Margulis (1998) from a more general theorem (Theorem 2 below). To state that theorem we need some terminology. Let V be a subset of $\mathbf{R}^d$ and f a continuous function on V. We write $\|f\|_B \doteq \sup_{\mathbf{x} \in B} |f(\mathbf{x})|$ for a subset B of V. For positive numbers C and α, say that f is (C, α)-*good on* V if for any open ball $B \subset V$ and any $\varepsilon > 0$ one has

$$|\{\mathbf{x} \in B \mid |f(\mathbf{x})| < \varepsilon\}| \leq C \cdot \left(\frac{\varepsilon}{\|f\|_B}\right)^{\alpha} \cdot |B|.$$

A model example of good functions are polynomials: for any $k \in \mathbf{R}$, any polynomial $f \in \mathbf{R}[x]$ of degree not greater than k is $(2k(k+1)^{1/k}, 1/k)$-good on $\mathbf{R}$.

We fix $k \in \mathbf{N}$ and a basis $\mathbf{e}_1, \dots, \mathbf{e}_k$ of $\mathbf{R}^k$, and for $I = \{i_1, \dots, i_j\} \subset \{1, \dots, k\}$, $i_1 < i_2 < \cdots < i_j$, we let $\mathbf{e}_I \doteq \mathbf{e}_{i_1} \wedge \cdots \wedge \mathbf{e}_{i_j} \subset \bigwedge^j(\mathbf{R}^k)$. We extend the norm $\|\cdot\|$ from $\mathbf{R}^k$ to the exterior algebra $\bigwedge(\mathbf{R}^k)$ by $\|\sum_{I\subset\{1,\dots,k\}} w_I \mathbf{e}_I\| = \max_{I\subset\{1,\dots,k\}} |w_I|$. For a discrete nonzero subgroup Γ of $\mathbf{R}^k$, we define the norm of Γ by $\|\Gamma\| \doteq \|\mathbf{w}\|$, where $\mathbf{w} = \mathbf{v}_1 \wedge \cdots \wedge \mathbf{v}_j$ and $\mathbf{v}_1, \cdots, \mathbf{v}_j$ is a basis of Γ (note that $\|\Gamma\|$ is correctly defined because $\mathbf{w}$ is defined up to a sign).

Let Λ be a discrete subgroup of $\mathbf{R}^k$. We say that a subgroup Γ of Λ is *primitive* (in Λ) if $\Gamma = \Gamma_{\mathbf{R}} \cap \Lambda$ where $\Gamma_{\mathbf{R}}$ denotes the minimal linear subspace of $\mathbf{R}^k$ containing Γ. Let us denote by $\mathcal{L}(\Lambda)$ the set of all nonzero primitive subgroups of Λ. Another piece of notation which we need is $B(\mathbf{x}, r)$ which will stand for the open ball of radius $r > 0$ centered in $\mathbf{x}$.

Theorem 2 (Kleinbock & Margulis 1998, Theorem 5.2) *Let $d, k \in \mathbf{N}$, $C, \alpha > 0$, $0 < \rho < 1/k$ and let a ball $B = B(\mathbf{x}_0, r_0) \subset \mathbf{R}^d$ and a map $h : \tilde{B} \to GL(k, \mathbf{R})$ be given, where $\tilde{B}$ stands for $B(\mathbf{x}_0, 3^k r_0)$. For any $\Gamma \in \mathcal{L}(\mathbf{Z}^k)$, denote by ψ_Γ the function $\psi_\Gamma(\mathbf{x}) \doteq \|h(\mathbf{x})\Gamma\|$, $\mathbf{x} \in \tilde{B}$. Assume that for any $\Gamma \in \mathcal{L}(\mathbf{Z}^k)$,*

(i) *ψ_Γ is (C, α)-good on $\tilde{B}$;*
(ii) *$\|\psi_\Gamma\|_B \geq \rho$.*

Then for any positive $\varepsilon \leq \rho$ one has

$$\left|\{\mathbf{x} \in B \mid \delta(h(\mathbf{x})\mathbf{Z}^k) < \varepsilon\}\right| \leq kC(3^d N_d)^k \left(\frac{\varepsilon}{\rho}\right)^\alpha |B|,$$

where N_d is an integer (from Besicovitch's Covering Theorem) depending only on d.

In Kleinbock & Margulis (1998) the method of proof of Theorem 2 is based, with some technical changes, on the argument from Margulis (1975) and its modification in Dani (1986) where it was applied to prove some results on nondivergence of unipotent flows in the space of lattices. We will discuss these results in Section 5. Let us state now a slight generalization of the $d = 1$ case of Theorem 2.

Theorem 3 (Bernik, Kleinbock & Margulis 1999, Theorem 5.1) *Let $k \in \mathbf{N}$, $C, \alpha > 0$, $0 < \rho < 1/k$, an interval $B \subset \mathbf{R}$ and a continuous map $h : B \to GL(k, \mathbf{R})$ be given. Take $\Lambda \in \mathcal{L}(\mathbf{Z}^k)$ and, for $\varepsilon > 0$, denote by $B_{h,\Lambda,\varepsilon}$ the set*

$$B_{h,\Lambda,\varepsilon} \doteq \{x \in B \mid \|h(x)v\| < \varepsilon \quad \textit{for some} \quad \mathbf{v} \in \Lambda \setminus \{0\}\}.$$

Assume that for any $\gamma \in \mathcal{L}(\Lambda)$,

(i) *the function* $x \mapsto \|h(x)\Gamma\|$ *is* (C, α)*-good on* B, *and*
(ii) *there exists* $x \in B$ *such that* $\|h(x)\Gamma\| \geq \rho$.

Then for any positive $\varepsilon \leq \rho$ *one has*

$$|B_{h,\Lambda,\varepsilon}| \leq C \dim(\Lambda_{\mathbf{R}}) 2^{\dim(\Lambda_{\mathbf{R}})} \left(\frac{\varepsilon}{\rho}\right)^{\alpha} |B|.$$

Theorem 3 is used to prove the following theorem about ψ-approximability and ψ-multiplicative approximability.

Theorem 4 (Bernik, Kleinbock & Margulis 1999, Theorem 1.1) *Let* $B \subset \mathbf{R}$ *be an interval,* $\mathbf{f} = (f_1, \ldots, f_n)$ *a nondegenerate* n*-tuple of* C^n *functions on* B, *and* $\psi : \mathbf{N} \to (0, \infty)$ *a non-increasing function. Then*

(S) *assuming* (1), $\mathbf{f}(x)$ *is not* ψ*-A for almost all* $x \in B$;
(M) *assuming* (2), $\mathbf{f}(x)$ *is not* ψ*-MA for almost all* $x \in B$.

Using a standard 'foliation' technique, one can deduce the following corollary from Theorem 4.

Corollary 2 *Let* ψ *be as in Theorem 4, and let* M *be a* C^n *submanifold of* $\mathbf{R}^n$ *such that for almost all* $\mathbf{y} \in M$ *there exists a curve nondegenerate at* $\mathbf{y}$ *and contained in* M. *Then*

(S) *assuming* (1), *almost all points of* M *are not* ψ*-A;*
(M) *assuming* (2), *almost all points of* M *are not* ψ*-MA.*

In particular the statements (S) *and* (M) *hold for nondegenerate analytic submanifolds.*

By means of straightforward measure computations, Theorem 4 is deduced in Bernik, Kleinbock & Margulis (1999) from the following two propositions.

Proposition 2 *Let an interval* $B \subset \mathbf{R}$ *and functions* $\mathbf{f} = (f_1, \ldots, f_n) \in C^2(B)$ *be given. Fix* $\delta > 0$ *and define*

$$L = \max_{1 \leq j \leq n, x \in B} |f_j''(x)|.$$

Then for every $\mathbf{q} \in \mathbf{Z}^n$ *such that*

$$\|\mathbf{q}\| \geq \frac{1}{4nL|B|^2},$$

the set of solutions $x \in B$ of the inequalities

$$\begin{cases} \lfloor \mathbf{q} \cdot \mathbf{f}(x) \rfloor < \delta \\ |\mathbf{q} \cdot \mathbf{f}'(x)| \geq \sqrt{nL}\|\mathbf{q}\| \end{cases}$$

has measure at most $32\delta|B|$.

Proposition 3 *Let $V \subset \mathbf{R}$ be an interval, $x_0 \in V$, and let $\mathbf{f} = (f_1, \ldots, f_n)$ be an n-tuple of C^n functions on V which is nondegenerate at x_0. Then there exists a subinterval $B \subset V$ containing x_0 and constants $E > 0$ and $0 < \rho < 1$ such that for any choice of $\tau \geq n$, $\delta, K > 0$ and $Q_1, \ldots, Q_n \geq 1$ subject to the constraint*

$$\delta^\tau \leq \frac{K Q_1 \cdot \ldots \cdot Q_n}{\max_i Q_i} \leq \frac{\rho^{\tau+1}}{\delta},$$

the set

$$\begin{aligned} \Omega \doteq \; & \{x \in B \mid \exists \mathbf{q} \in \mathbf{Z}^n \setminus \{0\} \quad \text{such that} \\ & \lfloor \mathbf{q} \cdot \mathbf{f}(x) \rfloor < \delta; |\mathbf{q} \cdot \mathbf{f}'(x)| < K; |q_i| < Q_i, \quad i = 1, \ldots, n\} \end{aligned}$$

has measure at most

$$E \left(\frac{\delta K Q_1 \cdot \ldots \cdot Q_n}{\max_i Q_i} \right)^{\frac{1}{(\tau+1)(2n-1)}} |B|.$$

Proposition 2, roughly speaking, says that a function with big first derivative and not very big second derivative cannot have values very close to integers on a set of a big measure. It is proved in Bernik, Kleinbock & Margulis (1999) using an argument which is apparently due to Bernik, and is implicitly contained in one of the steps of Bernik, Dickinson & Dodson (1998).

As for Proposition 3, it is deduced from Theorem 3 in the following way. Let Λ denote the subgroup of $\mathbf{Z}^{n+2}$ consisting of integer vectors with zero second coordinate; that is,

$$\Lambda = \left\{ \begin{pmatrix} p \\ 0 \\ \mathbf{q} \end{pmatrix} \middle| \, p \in \mathbf{Z}, \; \mathbf{q} \in \mathbf{Z}^n \right\}.$$

Take $\delta, K, Q_1, \ldots, Q_n$ the same as in Proposition 3, fix $\varepsilon > 0$ and denote

$$d_0 = \frac{\delta}{\varepsilon}, \; d_* = \frac{K}{\varepsilon}, \; d_i = \frac{Q_i}{\varepsilon}, \quad i = 1, \ldots, n.$$

Now we can define a map $h : B \to GL(n+2, \mathbf{R})$ by

$$h(x) = DU_x, \quad x \in B,$$

where D and U_x denote the following diagonal and unipotent matrices:

$$D \doteq \begin{pmatrix} d_0^{-1} & 0 & 0 & \cdots & 0 \\ 0 & d_*^{-1} & 0 & \cdots & 0 \\ 0 & 0 & d_1^{-1} & \cdots & 0 \\ \cdots & \cdots & \cdots & \cdots & \cdots \\ 0 & 0 & 0 & \cdots & d_n^{-1} \end{pmatrix},$$

$$U_x \doteq \begin{pmatrix} 1 & 0 & f_1(x) & f_2(x) & \cdots & f_n(x) \\ 0 & 1 & f_1'(x) & f_2'(x) & \cdots & f_n'(x) \\ 0 & 0 & 1 & 0 & \cdots & 0 \\ \cdots & \cdots & \cdots & \cdots & \cdots & \cdots \\ 0 & 0 & 0 & 0 & \cdots & 1 \end{pmatrix}.$$

Proposition 3 is proved in Bernik, Kleinbock & Margulis (1999) by applying Theorem 3 to just defined Λ and h. The main difficulty in doing this is to find a neighborhood B of x_0 and constants C, α, ρ such that the conditions (i) and (ii) from Theorem 3 hold. After that it remains to notice that, as it can be directly checked, the set Ω from Proposition 3 is exactly equal to the set $B_{h,\Lambda,\varepsilon}$ from Theorem 3.

Remark 1 Theorem 4, which is the main result of Bernik, Kleinbock & Margulis (1999), was proved already in 1998, but only in the case when the functions $f_1, \ldots, f_n$ are analytic (in this case it is much easier to check the condition (i) from Theorem 3). Let us also mention that in 1999, Beresnevich (2001), improving Sprindžuk's method of essential and inessential domains, proved the statement (S) (but not (M)) of Corollary 2 for a C^k nondegenerate submanifold $M \subset \mathbf{R}^n$ without assuming that $k \geq n$.

Remark 2 (added on August 21, 2000). A modified version of the preprint Bernik, Kleinbock & Margulis (1999) was recently submitted for publication.† It contains the proof of both statements (S) and (M) of Corollary 2 under the same conditions as in Corollary 1 and in Beresnevich (2001),i.e. for a C^k nondegenerate submanifold $M \subset \mathbf{R}^n$ without assuming that $k \geq n$.

2 Values of quadratic forms and of products of linear forms at integral points

We will say that a real quadratic form is *rational* if it is a multiple of a form with rational coefficients and *irrational* otherwise.

† Published in *Int. Math. Res. Notes* (2001), No. 9, 453–486.

Theorem 5 *Let Q be a real irrational indefinite nondegenerate quadratic form in $n \geq 3$ variables. Then for any $\varepsilon > 0$ there exist integers $x_1, \ldots, x_n$ not all equal to 0 such that $|Q(x_1, \ldots, x_n)| < \varepsilon$.*

Theorem 5 was conjectured by A. Oppenheim in 1929 and proved by the author in 1986 (see Margulis 1997 and references therein). Oppenheim was motivated by Meyer's theorem that if Q is a rational quadratic form in $n \geq 5$ variables, then Q represents zero over $\mathbf{Z}$ nontrivially, i.e. there exists $\mathbf{x} \in \mathbf{Z}^n$, $\mathbf{x} \neq 0$ such that $Q(\mathbf{x}) = 0$. Because of that he originally stated the conjecture only for $n \geq 5$. Let us also note that the condition '$n \geq 3$' cannot be replaced by the condition '$n \geq 2$'. To see this, consider the form $x_1^2 - \lambda x_2^2$ where λ is an irrational positive number such that $\sqrt{\lambda}$ has a continued fraction development with bounded partial quotients; for example $\lambda = (1 + \sqrt{3})^2 = 4 + 2\sqrt{3}$.

It is a standard and simple fact that if Q is a real irrational indefinite nondegenerate quadratic form in n variables and $2 \leq m < n$ then $\mathbf{R}^n$ contains a rational subspace L of dimension m such that the restriction of B to L is irrational, indefinite and nondegenerate. Hence if the Oppenheim conjecture (Theorem 5) is proved for some n_0, then it is proved for all $n \geq n_0$. As a consequence of this remark, we see that it is enough to prove the conjecture for $n = 3$.

Before the Oppenheim conjecture was proved it was extensively studied mostly using analytic number theory methods (see Section 1 of Margulis 1997 mentioned above). In particular it had been proved for diagonal forms in five or more variables and for $n \geq 21$. Bentkus & Götze (1999) handled the case $n \geq 9$ and also proved, under the same assumption $n \geq 9$, the Davenport–Lewis conjecture about gaps between values of positive definite quadratic forms at integral points (see Corollary 4 below). But it seems that the methods of analytic number theory are not sufficient to prove the Oppenheim conjecture for general quadratic forms in a small number of variables.

Theorem 5 was proved by studying orbits of the orthogonal group $SO(2, 1)$ on the space of unimodular lattices in $\mathbf{R}^3$. It turns out that this theorem is equivalent to the following:

Theorem 6 *Let $G = SL(3, \mathbf{R})$ and $\Gamma = SL(3, \mathbf{Z})$. Let us denote by H the group of elements of G preserving the form $2x_1x_3 - x_2^2$ and by $\Psi_3 = G/\Gamma$ the space of lattices in $\mathbf{R}^3$ having determinant 1. Let G_y denote the stabilizer $\{g \in G \mid gy = y\}$ of $y \in \Psi_3$. If $z \in \Psi_3 = G/\Gamma$ and the orbit Hz is relatively compact in Ψ_3, then the quotient space $H/H \cap G_z$ is compact.*

For any real quadratic form B in n variables, let us denote the special orthogonal group

$$SO(B) = \{g \in SL(n, \mathbf{R}) \mid gB = B\} \quad \text{by} \quad H_B$$

and

$$\inf\{|B(x)| \mid x \in \mathbf{Z}^n, x \neq 0\} \quad \text{by} \quad m(B).$$

As was already noted, it is enough to prove Theorem 5 for $n = 3$. Then the equivalence of Theorems 5 and 6 is a consequence of the assertions (i) and (ii) below. The assertion (i) easily follows from Mahler's compactness criterion. As for (ii), it was essentially proved in Cassels & Swinnerton-Dyer (1995) by some rather elementary considerations; (ii) can be also deduced from Borel's density theorem.

(i) *Let B be a real quadratic form in n variables. Then the orbit $H_B\mathbf{Z}^n$ is relatively compact in the space $\Psi_n = SL(n, \mathbf{R})/SL(n, \mathbf{Z})$ of unimodular lattices in $\mathbf{R}^n$ if and only if $m(B) > 0$.*
(ii) *Let B be a real irrational indefinite nondegenerate quadratic form in 3 variables. Then $H_B/H_B \cap SL(3, \mathbf{Z})$ is compact if and only if the form is rational and anisotropic over $\mathbf{Q}$.*

Remark In implicit form the equivalence of Theorem 5 and 6 appears already in Section 10 of the just-mentioned paper of Cassels & Swinnerton-Dyer. In the mid-1970s, Raghunathan rediscovered this equivalence and noticed that the Oppenheim conjecture would follow from a conjecture about closures of orbits of unipotent subgroups; the Raghunathan conjecture was proved in 1991 by Ratner (see Theorem 25 below). Raghunathan's observations inspired the author's work on the homogeneous space approach to the Oppenheim conjecture.

Theorem 6 was also used to prove the following stronger statement:

Theorem 7 *If Q and n are the same as in Theorem 5, then $Q(\mathbf{Z}^n)$ is dense in $\mathbf{R}$ or, in other words, for any $a < b$ there exist integers $x_1, \ldots, x_n$ such that*

$$a < Q(x_1, \ldots, x_n) < b.$$

An integer vector $x \in \mathbf{Z}^n$ is called *primitive* if $x \neq ky$ for any $y \in \mathbf{Z}^n$ and $k \in \mathbf{Z}$ with $|k| \geq 2$. the set of all primitive vectors in $\mathbf{Z}^n$ will be denoted by $\mathcal{P}(\mathbf{Z}^n)$. A subset $(x_1, \ldots, x_m)$ of $\mathbf{Z}^n$ is said to be *primitive* if it is a part of a basis of $\mathbf{Z}^n$. We can state now a strengthening of Theorem 7.

Theorem 8 *Let Q be a real irrational indefinite nondegenerate quadratic form in $n \geq 3$ variables, and let B be the corresponding bilinear form defined by $B(v, w) = \frac{1}{4}\{B(v+w) - B(v-w)\}$, $v, w \in \mathbf{R}^n$.*

(a) (Dani & Margulis 1989, Theorem 1.) *The set $\{Q(x) \mid x \in \mathcal{P}(\mathbf{Z}^n)\}$ is dense in $\mathbf{R}$.*

(b) (Borel & Prasad 1992, Corollary 7.8; see also Dani & Margulis 1989, Theorem 1) for $m \leq 2$). *Let $m < n$ and $y_1, \ldots, y_m$ be elements of $\mathbf{R}^n$. Then there exists a sequence $(x_{j,1}, \ldots, x_{j,m})$ $(j = 1, \ldots)$ of primitive subsets of $\mathbf{R}^n$ such that*

$$B(y_a, y_b) = \lim_{j\to\infty} B(x_{j,a}, x_{j,b}) \ (1 \leq a, b \leq m).$$

(c) (see Borel & Prasad 1992, Corollary 7.9, or Borel 1995, Theorem 2.) *Let $c_i \in \mathbf{R}$ $(i = 1, \ldots, n-1)$. Then there exists a sequence $(x_{j,1}, \ldots, x_{j,n-1})$ $(j = 1, \ldots)$ of primitive subsets of $\mathbf{Z}^n$ such that*

$$\lim_{j\to\infty} Q(x_{j,i}) = c_i \quad (i = 1, \ldots, n-1).$$

Theorem 8 is deduced from the following Theorem 9 by an extension of an argument reducing Theorems 5 and 7 to Theorem 6. Theorem 6 is in fact an easy consequence of Ratner's orbit closure theorem (Theorem 25 below). However for $n = 3$ this theorem had been earlier proved in Dani & Margulis (1989). The proof given there uses techniques which involve, as in the original proof of Theorem 6, finding orbits of larger subgroups inside closed sets invariant under unipotent subgroups.

Theorem 9 *Let Q be a real irrational indefinite nondegenerate quadratic form in $n \geq 3$ variables. Let us denote by H the special orthogonal group $SO(B)$. Then any orbit of H in $SL(n, \mathbf{R})/SL(n, \mathbf{Z})$ either is closed and carries an H-invariant probability measure or is dense.*

Remark Borel & Prasad generalized Theorems 5, 7 and 8 for a family $\{Q_s\}$ where $s \in S$, S is a finite set of places of a number field k containing the set S_∞ of archimedian places, Q_s is a quadratic form on k_s^n, and k_s is the completion of k at s (see Borel 1995 and Borel & Prasad 1992). To prove these generalizations they use S-arithmetic analogs of Theorems 6 and 9.

Let us go now from quadratic forms to a different topic. About 1930 Littlewood stated the following:

Conjecture 2 *As before, let $\lfloor x \rfloor$ denote the distance between $x \in \mathbf{R}$ and the closest integer. Then*

$$\liminf_{n\to\infty} n \lfloor n\alpha \rfloor \lfloor n\beta \rfloor = 0$$

for any real numbers α and β.

A slightly stronger conjecture is

Conjecture 3 *Let $\{x\}$ denote the fractional part of x. Then*

$$\liminf_{n\to\infty} n\{n\alpha\}\{n\beta\} = 0$$

for any real numbers α and β.

We already mentioned the paper of Cassels & Swinnerton-Dyer in connection with earlier discussion about quadratic forms. But they also consider another type of form, namely products of three linear forms in 3 variables. In particular, they show (see also Section 2 in Margulis 1997) that the Littlewood conjecture will be proved if the following conjecture is proved for $n = 3$.

Conjecture 4 *Let L be the product of n linear forms on $\mathbf{R}^n$. Suppose that $n \geq 3$ and L is not a multiple of a form with rational coefficients. Then for any $\varepsilon > 0$ there exist integers $x_1, \ldots, x_n$ not all equal to 0 such that $|L(x_1, \ldots, x_n)| < \varepsilon$.*

Conjecture 4 can be considered as an analogue of Theorem 5. As in Theorem 5 and because of the same example, in this conjecture the condition $n \geq 3$ cannot be replaced by the condition $n \geq 2$. Conjecture 4 turns out to be equivalent to the following conjecture about orbits of the diagonal subgroup in $SL(n, \mathbf{R})/SL(n, \mathbf{Z})$. For $n = 3$ this equivalence was essentially noticed in Cassels & Swinnerton-Dyer (1955) and is explained in Margulis (1997); a similar argument can be applied for arbitrary n.

Conjecture 5 *Let $n \geq 3$, $G = SL(n, \mathbf{R})$, $\Gamma = SL(n, \mathbf{Z})$ and let A denote the group of all positive diagonal matrices in G. If $z \in G/\Gamma$ and the orbit Az is relatively compact in G/Γ then Az is closed.*

By now the strongest evidence for the truth of Conjecture 5 is provided by some recent results of Lindenstrauss & Weiss (2001):

Theorem 10 *Let n, G, Γ and A be as in Conjecture 5. Let $y \in G/\Gamma$, and let F denote the closure $\overline{Ay}$ of the orbit Ay. Assume that F contains a compact orbit* *of A. Then there are integers k and d with $n = kd$ and a permutation matrix P such that $F = Hy$, where*

$$H = \left\{ P \begin{pmatrix} B_1 & 0 & \cdots & 0 \\ 0 & B_2 & \cdots & 0 \\ \cdots & \cdots & \cdots & \cdots \\ 0 & 0 & \cdots & B_d \end{pmatrix} P^{-1} : B_i \in GL(k, \mathbf{R}) \right\} \cap G;$$

here the 0s stand for the zero matrices in $M_k(\mathbf{R})$. Moreover, if $F \neq Ay$ then F is not compact.

Corollary 3 *Assume in addition in the hypotheses of Theorem 10 that n is prime. Then Ay is either compact or dense.*

Remark From Corollary 3, Lindenstrauss & Weiss draw a generalization of an isolation theorem of Cassels & Swinnerton-Dyer.

3 A quantitative version of the Oppenheim conjecture

In connection with Theorem 7, it is natural to study some quantitative problems related to the distribution of values of Q at integral points (a quantitative version of the Oppenheim conjecture).

Let ν be a continuous function on the sphere $\{v \in \mathbf{R}^n \mid \|v\| = 1\}$ and let $\Omega = \{v \in \mathbf{R}^n \mid \|v\| < \nu(v/\|v\|)\}$. We denote by $T\Omega$ the dilate of Ω by T. Let Q be a real indefinite nondegenerate quadratic form in $n \geq 3$ variables. Let us denote by $N_{Q,\Omega}(a, b, T)$ the cardinality of the set

$$\{x \in \mathbf{Z}^n \mid x \in T\Omega \quad \text{and} \quad a < Q(x) < b\}$$

and by $V_{Q,\Omega}(a, b, T)$ the volume of the set

$$\{x \in \mathbf{R}^n \mid x \in T\Omega \quad \text{and} \quad a < Q(x) < b\}.$$

It is easy to verify that

$$V_{Q,\Omega}(a, b, T) \sim \lambda_{Q,\Omega}(b - a)T^{n-2}, \tag{8}$$

where

$$\lambda_{Q,\Omega} = \int_{L\cap\Omega} \frac{dA}{\|\nabla Q\|}, \tag{9}$$

L is the light cone $Q = 0$ and dA is the area element on L.

Let $\mathcal{O}(p, q)$ denote the space of quadratic forms of signature (p, q) and discriminant ± 1, let (a, b) be an interval. In combination with (8) and (9), the assertion (I) of the following theorem gives the asymptotically exact lower bound for $N_{Q,\Omega}(a, b, T)$.

Theorem 11 (Dani & Margulis 1993, Corollary 5) (I) *Let* $p \geq 2$ *and* $q \geq 1$. *Then for any irrational* $Q \in \mathcal{O}(p, q)$ *and any interval* (a, b)

$$\liminf_{T\to\infty} \frac{N_{Q,\Omega}(a, b, T)}{V_{Q,\Omega}(a, b, T)} \geq 1. \tag{10}$$

Moreover, the bound (10) *is uniform over compact sets of forms: if* $\mathcal{K}$ *is a compact subset of* $\mathcal{O}(p, q)$ *which consists of irrational forms, then*

$$\liminf_{T\to\infty} \inf_{Q\in\mathcal{K}} \frac{N_{Q,\Omega}(a, b, T)}{V_{Q,\Omega}(a, b, T)} \geq 1.$$

(II) *If* $p > 0, q > 0$ *and* $n = p + q \geq 5$, *then for any* $\varepsilon > 0$ *and any compact subset* $\mathcal{K}$ *of* $\mathcal{O}(p, q)$ *there exists* $c = c(\varepsilon, \mathcal{K})$ *such that for all* $Q \in \mathcal{K}$ *and* $T > 0$

$$N_{Q,\Omega}(a, b, T) \geq c\, V_{Q,\Omega}(a, b, T).$$

The situation with the asymptotics and upper bounds for $N_{Q,\Omega}(a, b, T)$ is more delicate. Rather surprisingly, here the answer depends on the signature of Q.

Theorem 12 (Eskin, Margulis & Mozes 1998, Theorem 2.1) *If* $p \geq 3, q \geq 1$ *and* $n = p + q$ *then, as* $T \to \infty$

$$N_{Q,\Omega}(a, b, T) \sim \lambda_{Q,\Omega}(b - a)T^{n-2} \tag{11}$$

for any irrational form $Q \in \mathcal{O}(p, q)$ *where* $\lambda_{Q,\Omega}$ *is as in* (9).

Remark There is a 'uniform' version of Theorem 12 (see Eskin, Margulis & Mozes 1998, Theorem 2.5). Let us also note that Corollary 5 in Dani & Margulis (1993) contains a slightly stronger statement than the 'uniform' version of (10) from Theorem 11(I).

If the signature of Q is (2, 1) or (2, 2), then no universal formula like (11) holds. In fact, we have the following theorem.

Theorem 13 (Eskin, Margulis & Mozes 1998, Theorem 2.2) *Let* Ω *be the unit ball, and let* $q = 1$ *or* 2. *Then for every* $\varepsilon > 0$ *and every interval* (a, b) *there exists an irrational form* $Q \in \mathcal{O}(2, q)$ *and a constant* $c > 0$ *such that for an infinite sequence* $T_j \to \infty$

$$N_{Q,\Omega}(a, b, T) > cT_j^q (\log T_j)^{1-\varepsilon}.$$

The case $q = 1$, $b \leq 0$ of this theorem was noticed by Sarnak and worked out in detail in Brennan (1994). The quadratic forms constructed are of the type $x_1^2 + x_2^2 - \alpha x_3^2$, or $x_1^2 + x_2^2 - \alpha(x_3^2 + x_4^2)$, where α is extremely well approximated by squares of rational numbers. Another point is that in the statement of Theorem 9, $(\log T)^{1-\varepsilon}$ can be replaced by $\log T/\nu(T)$ where $\nu(T)$ is any unbounded increasing function.

However in the (2, 1) and (2, 2) cases, there is an upper bound of the form $cT^q \log T$. This upper bound is effective, and is uniform over compact subsets of $\mathcal{O}(p, q)$. There is also an effective upper bound for the case $p \geq 3$.

Theorem 14 (Eskin, Margulis & Mozes 1998, Theorem 2.3) *Let $\mathcal{K}$ be a compact subset of $\mathcal{O}(p, q)$ and $n = p + q$. Then, if $p \geq 3$ and $q \geq 1$ there exists a constant $c = c(\mathcal{K}, a, b, \Omega)$ such that for any $Q \in \mathcal{K}$ and all $T > 1$*

$$N_{Q,\Omega}(a, b, T) < cT^{n-2}.$$

If $p = 2$ and $q = 1$ or $q = 2$, then there exists a constant $c = c(\mathcal{K}, a, b, \Omega)$ such that for any $Q \in \mathcal{K}$ and all $T > 2$

$$N_{Q,\Omega}(a, b, T) < cT^{n-2} \log T.$$

Also, for the (2, 1) and (2, 2) cases, the following ‘almost everywhere’ result is true:

Theorem 15 (Eskin, Margulis & Mozes 1998, Theorem 2.5) *The asymptotic formula* (11) *holds for almost all quadratic forms of signature* (2, 1) *or* (2, 2).

We will now briefly describe how Theorems 11–15 are proved. Let $G = SL(n, \mathbf{R})$, $\Gamma = SL(n, \mathbf{Z})$, $\Psi_n = G/\Gamma$={the space of unimodular lattices in $\mathbf{R}^n$ with determinant 1}. One can associate to an integrable function f on $\mathbf{R}^n$, a function $\tilde{f}$ on Ψ_n by setting

$$\tilde{f}(\Delta) = \sum_{v \in \Delta, v \neq 0} f(v), \quad \Delta \in \Psi_n. \tag{12}$$

According to a theorem of Siegel

$$\int_{\mathbf{R}^n} f\, dm^n = \int_{\Psi_n} \tilde{f}\, d\mu, \tag{13}$$

where m^n is the Lebesgue measure on $\mathbf{R}^n$ and μ is the G-invariant probability measure on $\Psi_n = G/\Gamma$. In Dani & Margulis (1993), the proof of Theorem 11 is based on the following identity which is immediate from the definitions:

$$\int_T^{lT} \int_F \sum_{v \in g\mathbf{Z}^n} f(u_t k v)\, d\sigma(k)\, dt = \int_T^{lT} \int_F \tilde{f}(u_t k g \Gamma)\, d\sigma(k)\, dt, \tag{14}$$

where $\{u_t\}$ is a certain one-parameter unipotent subgroup of $SO(p,q)$, F is a Borel subset of the maximal compact subgroup K of $SO(p,q)$, σ is the normalized Haar measure on K, and f is a continuous function on $\mathbf{R}^n \setminus \{0\}$ with compact support. The number $N_{Q,\Omega}(a,b,T)$ can be approximated by the sum over m of the integrals on the left-hand side of (14) for an appropriate choice of g, $f = f_i$, $F = F_i$, $1 \le i \le m$. The-right hand side of (14) can be estimated, uniformly when $g\Gamma$ belongs to certain compact subsets of G/Γ, using (13) and Theorem 3 from Dani & Margulis (1993) which is a refined version of Ratner's uniform distribution theorem (Theorem 24 below). To prove the assertion (II) of Theorem 11, we have to use the following fact which is essentially equivalent to Meyer's theorem: if $n \ge 5$ then any closed orbit of $SO(p,q)$ in G/Γ is unbounded.

The just-mentioned Theorem 3 from Dani & Margulis (1993) and Ratner's uniform distribution theorem are proved (and in general true) only for bounded continuous functions. However, the function $\tilde{f}$ defined by (12) is unbounded for any continuous nonzero nonnegative function f on $\mathbf{R}^n$. Therefore we cannot use these theorems and (14) to get the asymptotic and upper bounds for $N_{Q,\Omega}(a,b,T)$. On the other hand, as was done by Dani & Margulis (1993), one can get lower bounds by considering bounded continuous functions $h \le \tilde{f}$ and applying their Theorem 3 to h.

Let $p \ge 2$, $p \ge q$, $q \ge 1$. We denote $p+q$ by n. Let $\{e_1, \dots, e_n\}$ be the standard basis of $\mathbf{R}^n$. Let Q_0 be the quadratic form of signature (p,q) defined by

$$Q_0\left(\sum_{i=1}^{n} v_i e_i\right) = 2v_1 v_n + \sum_{i=2}^{p} v_i^2 - \sum_{i=p+1}^{n-1} v_i^2 \qquad \text{for all} \quad v_1, \dots, v_n \in \mathbf{R}.$$

Let $H = SO(Q_0)$. For $t \in \mathbf{R}$, let a_t be the linear map so that $a_t e_1 = e^{-t} e_1$, $a_t e_n = e^t e_n$, and $a_t e_i = e_i$, $2 \le i \le n-1$. Let $\hat{K}$ be the subgroup of G consisting of orthogonal matrices, and let $K = H \cap \hat{K}$. It is easy to check that K is a maximal compact subgroup of H, and consists of all $h \in H$ leaving invariant the subspace spanned by $\{e_1 + e_n, e_2, \dots, e_n\}$. For technical reasons, we consider in Eskin, Margulis & Mozes (1998) not the identity (14) but another:

$$\int_K \sum_{v \in g\mathbf{Z}^n} f(a_t k v)\nu(k)\, d\sigma(k) = \int_K \tilde{f}(a_t k g\Gamma)\nu(k)\, d\sigma(k), \tag{15}$$

where ν is a bounded measurable function on K.

Let Δ be a lattice in $\mathbf{R}^n$. We say that a subspace L of $\mathbf{R}^n$ is Δ-*rational* if $L \cap \Delta$ is a lattice in L. For any Δ-rational subspace L, we denote by $d(L)$ the volume of $L/L \cap \Delta$. Note that the ratio of $d(L)$ and the norm $\|L \cap \Delta\|$ of the discrete subgroup $L \cap \Delta$, defined in the Section 1 after Proposition 1, is bounded between two positive constants. If $L = 0$ we write $d(L) = 1$. Let

$$\alpha(\Delta) = \sup \left\{ \frac{1}{d(L)} \,\middle|\, L \text{ is a } \Delta\text{-rational subspace} \right\}.$$

According to the 'Lipshitz Principle' from the geometry of numbers, for any bounded function f on $\mathbf{R}^n$ vanishing outside a compact subset, there exists a positive constant $c = c(f)$ such that

$$\tilde{f}(\Delta) < c\,\alpha(\Delta) \qquad \text{for any lattice} \quad \Delta \quad \text{in} \quad \mathbf{R}^n. \tag{16}$$

Theorems 12–14 are proved by combining the abovementioned Theorem 3 from Dani & Margul,is (1993) with the identity (15), the inequality (16) and the following integrability estimates:

Theorem 16 (Eskin, Margulis & Mozes 1998, Theorems 3.2 and 3.3) (a) *If* $p \geq 3, q \geq 1$ *and* $0 < s < 2$, *or if* $p = 2, q \geq 1$ *and* $0 < s < 1$, *then for any lattice* Δ *in* $\mathbf{R}^n$

$$\sup_{t>0} \int_K \alpha(a_t k \Delta)^s d\sigma(k) < \infty. \tag{17}$$

(b) *If* $p = 2$ *and* $q = 2$, *or if* $p = 2$ *and* $q = 1$, *then for any lattice* Δ *in* $\mathbf{R}^n$

$$\sup_{t>1} \frac{1}{t} \int_K \alpha(a_t k \Delta) d\sigma(k) < \infty. \tag{18}$$

The upper bounds (17) *and* (18) *are uniform as* Δ *varies over compact sets in the space of lattices.*

Theorem 15 is deduced from the identity (15), the inequality (16), Howe–Moore estimates for matrix coefficients of unitary representations, and from the fact that the function α on the space $\Psi_n = SL(n, \mathbf{R})/SL(n, \mathbf{Z})$ of unimodular lattices in $\mathbf{R}^n$ belongs to every L^μ, $1 \leq \mu < n$.

As was noticed before, there are quadratic forms Q of the type $x_1^2 + x_2^2 - \alpha x_3^2$, or $x_1^2 + x_2^2 - \alpha(x_3^2 + x_4^2)$, where α is extremely well approximated by squares of rational numbers, such that (11) does not hold. These examples can be generalized by considering irrational forms of the signature (2, 1) or (2, 2) which are extremely well approximated by split (over $\mathbf{Q}$) rational forms. As Theorem 17 below shows, this generalization is essentially the only method of constructing such forms in the (2, 2) case.

Fix a norm $\|\cdot\|$ on the space $\mathcal{O}(2, 2)$ of quadratic forms of signature (2, 2). We say that a quadratic form $Q \in \mathcal{O}(2, 2)$ is *extremely well approximable by split rational forms,* to be abbreviated as EWAS, if for any $N > 0$ there exist a split integral form Q' and a real number $\lambda > 2$ such that $\|\lambda Q - Q'\| \leq \lambda^{-N}$. It is clear that if the ratio of two nonzero coefficients of Q is Diophantine then Q is not EWAS; hence the set of EWAS forms has zero Hausdorff dimension as a subset of $\mathcal{O}(2, 2)$. (A real number x is called *Diophantine* if there exist $N > 0$ such that $|qx - p| > q^{-N}$ for any integers p and q. All algebraic numbers are Diophantine.)

Theorem 17 (Eskin, Margulis & Mozes 2001) *The asymptotic formula* (11) *holds if* $Q \in \mathcal{O}(2, 2)$ *is not EWAS and* $0 \notin (a, b)$.

Observe that whenever a form $Q \in \mathcal{O}(2, 2)$ has a rational 2-dimensional isotropic subspace L then $L \cap T\Omega$ contains of the order of T^2 integral points x for which $Q(x) = 0$, hence $N_{Q,\Omega}(-\varepsilon, \varepsilon) \geq cT^2$, independently of the choice of ε. This is exactly the reason why we assumed in Theorem 17 that $0 \notin (a, b)$. Let us also note that an irrational quadratic form Q of the signature $(2, 2)$ may have at most 4 rational isotropic subspaces and that if Q is such a form, then the number of points in the set $\{x \in \mathbf{Z}^n \mid Q(x) = 0,\ \|x\| < T,\ x$ is not contained in an isotropic (with respect to Q) subspace$\}$ grows not faster than a linear function of T.

Remark The proof of Theorem 17 in Eskin, Margulis & Mozes (2001) uses the approach developed in Dani & Margulis (1993) and Eskin, Margulis & Mozes (1998) but involves also a substantially more complicated analysis of the behavior of the function α on the sets $a_t K \Delta$. Though it seems that an analogue of Theorem 17 should be true for forms of signature (2, 1), it is not clear how the method of the proof of Theorem 17 can be extended to the (2, 1) case.

It was noted by Sarnak (1997) that the quantitative version of the Oppenheim conjecture in the (2, 2) case is related to the study of eigenvalue spacings of flat 2-tori. Let Δ be a lattice in $\mathbf{R}^2$ and let $M = \mathbf{R}^2/\Delta$ denote the associated flat torus. The eigenvalues of the Laplacian on M are the values of the binary quadratic form $q(m, n) = 4\pi^2\|mv_1 + nv_2\|^2$, where $\{v_1, v_2\}$ is a $\mathbf{Z}$–basis for the dual lattice Δ^*. We label these eigenvalues (with multiplicity) by

$$0 = \lambda_0(M) < \lambda_1(M) \leq \lambda_2(M) \cdots$$

It is easy to see that Weyl's law holds, i.e.

$$|\{j \mid \lambda_j(M) \leq T\}| \sim c_M T$$

where $c_M = (\text{area}M)/4\pi$. We are interested in the distribution of the local spacings $\lambda_j(M) - \lambda_k(M)$ and, in particular, in the so called *pair correlation*

$$R_M(a, b, T) = \frac{|\{(j, k) \mid \lambda_j(M) \le T, \lambda_k(M) \le T, j \ne k, a \le \lambda_j(M) - \lambda_k(M) \le b\}|}{T}.$$

Theorem 18 (Eskin, Margulis & Mozes 2001) *Let M be a flat 2-torus rescaled so that one of the coefficients in the associated binary quadratic form q is 1. Let A_1, A_2 denote the two other coefficients of q. Suppose that there exists $N > 0$ such that for all triples of integers (p_1, p_2, q),*

$$\max_{i=1,2} \left| A_i - \frac{p_i}{q} \right| > \frac{1}{q^N}.$$

Then, for any interval (a, b) which does not contain 0,

$$\lim_{T\to\infty} R_M(a, b, T) = c_M^2(b - a). \tag{19}$$

In particular, if one of the A_i is Diophantine, then (19) holds, and therefore the set of $(A_1, A_2) \subset \mathbf{R}^2$ for which (19) does not hold has zero Hausdorff dimension.

This theorem is proved by applying Theorem 17 to the form $Q(m_1, n_1, m_2, n_2) = q(m_1, n_1) - q(m_2, n_2)$. It is not difficult to give the asymptotics of $R_M(a, b, T)$ also in the case when $0 \in (a, b)$ and q is irrational. For this we have to study the multiplicity of eigenvalues $\lambda_i(M)$. This can be easily done if q is irrational, but it requires consideration of several cases. Note also that, for all $i > 0$, this multiplicity is at least 2.

The equality (19) is exactly what is predicted by the random number (Poisson) model. Sarnak (1997) showed that (19) holds on a set of full measure in the space of tori. But his method does not give any explicit example of such a torus. Let us also note that Theorem 18 is related to the Berry–Tabor conjecture that the distribution of the local spacings between eigenvalues of a completely integrable Hamiltonian is Poisson.

We finish this section with the formulation of some results of V. Bentkus and F. Götze on the distribution of values of a *positive definite* quadratic form at integral points.

Theorem 19 (Bentkus & Götze 1999, Theorem 1.1) *Let Q be a positive definite quadratic form in n variables, and let $N_{Q,a}(T)$ (resp. $V_{Q,a}(T)$), where $a \in \mathbf{R}^n$, denote the cardinality of the set $\{x \in \mathbf{Z}^n \mid Q(x - a) < T\}$ (resp. the volume of the set $\{x \in \mathbf{R}^n \mid Q(x - a) < T\}$). Assume that Q is irrational and $n \ge 9$.*

Then, as $T \to \infty$,

$$\sup_{a \in \mathbf{R}^n} \left| \frac{N_{Q,a}(T) - V_{Q,a}(T)}{V_{Q,a}(T)} \right| = o(T^{-1}). \tag{20}$$

It is conjectured that, for irrational Q, (20) is true if $n \geq 5$. But if n is 3 or 4, then (20) is not true for an arbitrary irrational Q. Theorem 19 easily implies the following

Corollary 4 (Bentkus & Götze 1999, Corollary 1.2) *Let Q be a positive definite quadratic form in n variables, let $a \in \mathbf{R}^n$, and let $d(T, Q, a)$ denote the maximal gap between successive values $Q(x - a)$, $x \in \mathbf{Z}^n$ in the interval $[T, \infty)$. Assume that Q is irrational and $n \geq 9$. Then* $\sup_{a \in \mathbf{R}^n} d(T, Q, a) \to 0$ *as* $T \to \infty$.

As we mentioned in Section 2, Corollary 4 proves the Davenport–Lewis conjecture in the case $n \geq 9$. This conjecture says that if Q is irrational and $n \geq 5$ then $d(T, Q, 0) \to 0$ as $T \to \infty$ (here probably the condition '$n \geq 5$' can be replaced by the condition '$n \geq 3$').

4 Counting lattice points on homogeneous varieties

Let V be a real algebraic subvariety of $\mathbf{R}^n$ defined over $\mathbf{Q}$, and let G be a reductive real algebraic subgroup of $GL(n, \mathbf{R})$ also defined over $\mathbf{Q}$. Suppose that V is invariant under G and that G acts transitively on V (or, more precisely, the complexification of G acts transitively on the complexification of V). Let $\|\cdot\|$ denote a Euclidean norm on $\mathbf{R}^n$. Let B_T denote the ball of radius T in $\mathbf{R}^n$ around the origin, and define

$$N(T, V) = |V \cap B_T \cap \mathbf{Z}^n|,$$

the number of integral points on V with norm less than T. In this section we are interested in the asymptotics of $N(T, V)$ as $T \to \infty$. Following Eskin (1998) and Eskin, Mozes & Shah (1996), let us describe how the homogeneous space approach can be applied to tackle this problem in some cases.

Let Γ denote $G(\mathbf{Z}) \doteq \{g \in G \mid g\mathbf{Z}^n = \mathbf{Z}^n\}$. By a theorem of Borel and Harish-Chandra, $V(\mathbf{Z})$ is a union of finitely many Γ-orbits. Therefore, to compute the asymptotics of $N(T, V)$, it is enough to consider each Γ-orbit, say $\mathcal{O}$, separately, and compute the asymptotics of

$$N(T, V, \mathcal{O}) = |\mathcal{O} \cap B_T|.$$

Of course, after that there is the problem, often non-trivial, of the summation over the set of Γ-orbits. This is essentially a problem from the theory of algebraic and arithmetic groups and is of completely different type than the computation of the asymptotics of $N(T, V, \mathcal{O})$.

Suppose that $\mathcal{O} = \Gamma v_0$ for some $v_0 \in V(\mathbf{Z})$. Then the stabilizer $H = \{g \in G \mid gv_0 = v_0\}$ is a reductive $\mathbf{Q}$-subgroup, and $V \cong G/H$. Define

$$R_T = \{gH \in G/H \mid gv_0 \in B_T\},$$

the pullback of the ball B_T to G/H.

Assume that G^0 and H^0 do not admit nontrivial $\mathbf{Q}$-characters, where G^0 (resp. H^0) denotes the connected component of identity in G (resp. in H). Then by a theorem of Borel and Harish-Chandra, G/Γ admits a G-invariant (Borel) probability measure, say μ_G, and $H/(\Gamma \cap H)$ admits an H-invariant probability measure, say μ_H. We can consider $H/(\Gamma \cap H)$ as a (closed) subset of G/Γ; then μ_H can be treated as a measure on G/Γ supported on $H/(\Gamma \cap H)$. Let $\lambda_{G/H}$ denote the (unique) G-invariant measure on G/H induced by the normalization of the Haar measures on G and H. We need the following definition:

Definition For a sequence $T_n \to \infty$, the sequence $\{R_{T_n}\}$ of open subsets in G/H is said to be *focused* if there exist a proper connected reductive $\mathbf{Q}$-subgroup L of G containing H^0, and a compact subset $C \subset G$ such that

$$\limsup_{n\to\infty} \frac{\lambda_{G/H}(q_H(CL(Z(H^0) \cap \Gamma)) \cap R_{T_n})}{\lambda_{G/H}(R_{T_n})} > 0,$$

where $q_H : G \to G/H$ is the natural quotient map and $Z(H^0)$ denotes the centralizer of H^0 in G. Now we can state the main result of Eskin, Mozes & Shah (1996).

Theorem 20 (Eskin, Mozes & Shah 1996, Theorem 1.16) *Suppose that H^0 is not contained in any proper $\mathbf{Q}$-parabolic subgroup of G^0 (equivalently, $Z(H^0)/(Z(H^0) \cap \Gamma$ is compact), and for some sequence $T_n \to \infty$ with bounded gaps, the sequence $\{R_{T_n}\}$ is not focused. Then, asymptotically as $T \to \infty$,*

$$N(T, V, \mathcal{O}) \sim \lambda_{G/H}(R_T). \tag{21}$$

The conditions of Theorem 20 can be checked in many cases. In particular, we have the following:

Theorem 21 (Eskin, Mozes & Shah 1996, Theorem 1.11) *The asymptotic formula* (21) *holds if H^0 is a maximal proper connected $\mathbf{Q}$-subgroup of G.*

Example Let $p(\lambda)$ be a monic polynomial of degree $n \geq 2$ with integer coefficients and irreducible over $\mathbf{Q}$. Let $M_n(\mathbf{Z})$ denote the set of $n \times n$ integer matrices, and put

$$V_p(\mathbf{Z}) = \{A \in M_n(\mathbf{Z}) \mid \det(\lambda I - A) = p(\lambda)\}.$$

Thus $V_p(\mathbf{Z})$ is the set of integer matrices with characteristic polynomial $p(\lambda)$. Consider the norm on $n \times n$ matrices given by $\|(x_{ij})\| = \sqrt{\sum_{ij} x_{ij}^2}$.

Corollary 5 (Eskin, Mozes & Shah 1996, Theorem 1.3) *Let $N(T, V_p)$ denote the number of elements of $V_p(\mathbf{Z})$ with norm less than T. Then asymptotically as $T \to \infty$,*

$$N(T, V_p) \sim c_p T^{n(n-1)/2}, \tag{22}$$

where $c_p > 0$ is an explicitly computable constant.

In the above example, the group G is $SL(n, \mathbf{R})$ which acts on the space $M(n, \mathbf{R})$ of $n \times n$ matrices by conjugation. The subgroup H is a maximal $\mathbf{Q}$-torus in G. For explicit formulas for calculating c_p in some cases (see Eskin, Mozes & Shah 1996 and references therein). Let us also note that for the case when V is affine symmetric, the asymptotic formula (21) had been earlier proved in Duke, Rudnick & Sarnak (1993) using harmonic analysis; subsequently a simpler proof using the mixing property of the geodesic flow appeared in Eskin & McMullen (1993) — a similar 'mixing property' approach had been also used in the author's thesis (Margulis 1970) to obtain asymptotic formulas for the number of closed geodesics in M^n and for the number of points from $\pi^{-1}(x)$ in balls of large radius in $\tilde{M}^n$, where M^n is a compact manifold of negative curvature, $x \in M^n$, $\tilde{M}^n$ is the universal covering space of M^N and $\pi : \tilde{M}^n \to M^n$ is the natural projection.

The proof of Theorem 20 is mostly based on the study of limit points of the set $\{g\mu_H \mid g \in G\}$ of the translates of the measure μ_H. It turns out that if a sequence $\{g_i \mu_H\}$ converges to a probability measure μ on G/Γ, then μ is a homogeneous measure (i.e. μ is Haar measure on a closed orbit of a subgroup). Here a key observation is that the limit measure μ is either a translate of μ_H or is invariant under some non-trivial unipotent elements, and the main tool is Ratner's measure classification theorem (Theorem 23 below). Another ingredient in the proof of Theorem 20 is to obtain conditions under which the sequence $\{g_i \mu_H\}$ of probability measures does not escape to infinity. This is done in Eskin, Mozes & Shah (1997) using, as in the proof of Theorem 2 above, a generalization of the argument from Dani (1986) and Margulis (1975).

For $T > 0$, define a function F_T on G by

$$F_T(g) = \sum_{\gamma \in \Gamma/(\Gamma \cap H)} \chi_T(g\gamma v_0),$$

where χ_T is the characteristic function of B_T. The function F_T is right Γ-invariant, and hence it can be treated as a function on G/Γ. The connection between counting and translates of measures is via the following two identities (see Duke, Rudnick & Sarnak 1993 and Eskin & McMullen 1993):

$$F_T(e) = \sum_{\gamma \in \Gamma/(\Gamma \cap H)} \chi_T(\gamma v_0) = N(T, V, \mathcal{O}) \tag{23}$$

and

$$\langle F_T, \psi \rangle = \int_{R_T} \left(\int_{G/\Gamma} \overline{\psi} d(g\mu_H) \right) d\lambda_{G/H}(gH), \tag{24}$$

where ψ is any function in $C_0(G/\Gamma)$. If the non-focusing assumption is satisfied then, as is shown in Eskin, Margulis & Mozes (1998), for 'most' values of g, the inner integral in (24) will approach $\int_{G/\Gamma} \overline{\psi} d\mu = \langle 1, \psi \rangle$. Now considering a sequence $\{\psi_m\}$ which 'converges' to the δ-function at e, it is not difficult to deduce (21) from (23) and (24).

5 Translates of submanifolds and unipotent flows on homogeneous spaces

Let G be Lie group, Γ a discrete subgroup of G and Y a (smooth) submanifold of G/Γ. In previous sections in connection with problems in Diophantine approximation and number theory, we essentially tried to answer in some cases the following general question:

(Q) What is the distribution of gY in G/Γ when g tends to infinity in G?

This question can be divided into two subquestions:

(Q1) What is the behavior of gY 'near infinity' of G/Γ?
(Q2) What is the distribution of gY in the 'bounded part' of G/Γ?

In Section 1 we considered two cases:

(a) $G = SL(n+1, \mathbf{R})$, $\Gamma = SL(n+1, \mathbf{Z})$, $g = g_\mathbf{t}$ and $Y = \{U_{\mathbf{f}(\mathbf{x})}\mathbf{Z}^{n+1} \mid \mathbf{x} \in V\}$ (Proposition 1);
(b) $G = SL(n+2, \mathbf{R})$, Γ is the stabilizer of the subgroup Λ of $\mathbf{Z}^{n+2}$, $g = D$ and $Y = \{U_x \mid x \in B\}$ (the end of Section 1), and in that section we were basically interested only in the question (Q1).

In Section 3 and implicitly in the part of Section 2 related to quadratic forms, G, Γ, g and Y were, respectively, $SL(n, \mathbf{R})$, $SL(n, \mathbf{Z})$, a_t and $Kg\Gamma$, where $g \in G$ and K is a maximal compact subgroup of $SO(p, q)$. Theorems 5 and 7 are related only to the question (Q2) but other statements about quadratic forms (Theorems 8, 9 and 11–15) are related to both questions (Q1) and (Q2). In Section 4 we were interested again in both questions (Q1) and (Q2) for $Y = H/(\Gamma \cap H)$.

Let $y \in Y$ and let W be a 'small' neighborhood of y in Y. Then for any $w \in W$ there exists an element $h \in G$ that is 'close' to e, such that $w = hy$. It is clear that $gw = (ghg^{-1})gy$ and that $\operatorname{Ad}(ghg^{-1})$ has the same eigenvalues as $\operatorname{Ad} h$ where Ad denotes the adjoint representation of G. Hence gW consists of translates of gy by 'almost' Ad-unipotent elements ($x \in G$ is called *Ad-unipotent* if $\operatorname{Ad} x$ is unipotent; that is, if all eigenvalues of $\operatorname{Ad} x$ are equal to 1). This is exactly the reason why results and methods from the theory of unipotent flows on homogeneous spaces play such an important role in topics which we discussed in Sections 1–4. We will state now some of these results.

Theorem 22 (Dani & Margulis 1993, Theorem 6.1) *Let G be a connected Lie group, Γ a lattice (i.e. a discrete subgroup with finite covolume) in G, F a compact subset of G/Γ and $\varepsilon > 0$. Then there exists a compact subset K of G/Γ such that for any* Ad*-unipotent one-parameter subgroup $\{u(t)\}$ of G, any $x \in F$, and $T \geq 0$,*

$$|\{t \in [0, T] \mid u(t)x \in K\}| \geq (1 - \varepsilon)T.$$

This theorem is essentially due to Dani. He proved it in Dani (1984) for semisimple groups G of $\mathbf{R}$-rank 1 and in Dani (1986) for arithmetic lattices. The general case can be easily reduced to these two cases using the arithmeticity theorem. In the case of arithmetic lattices, Theorem 22 can be considered as the quantitative version of the assertion that orbits of unipotent subgroups do not tend to infinity in $SL(n, \mathbf{R})/SL(n, \mathbf{Z})$. This assertion was proved in Margulis (1995) in connection with the proof of arithmeticity of nonuniform lattices in higher rank semisimple Lie groups. The proof given in Dani (1986) is similar to the proof from Margulis (1995) and, as was noticed in Sections 1 and 4, a generalization of the argument from these papers is basic for the proof of Theorem 2 and is also used in the proof of Theorem 20.

The following three fundamental Theorems 23–25 are due to Ratner.

Theorem 23 (Measure classification theorem, Ratner 1990a,b, 1991a) *Let G be a connected Lie group and Γ a discrete subgroup of G (not necessarily a lattice). Let H be a Lie subgroup of G that is generated by the* Ad*-unipotent* *subgroups contained in it. Then any finite H-ergodic H-invariant measure μ on G/Γ is homogeneous in the sense that there exists a closed subgroup F of G such that μ is F-invariant and* $\operatorname{supp}\mu = Fx$ *for some $x \in G/\Gamma$.*

Theorem 24 (Uniform distribution theorem, Ratner 1991b) *If G is a connected Lie group, Γ is a lattice in G, $\{u(t)\}$ is a one-parameter* Ad-*unipotent subgroup of G and $x \in G/\Gamma$, then the orbit $\{u(t)x\}$ is uniformly distributed with respect to a homogeneous probability measure μ_x on G/Γ in the sense that for any bounded continuous function f on G/Γ*

$$\frac{1}{T}\int_0^T f(u(t)x)dt \to \int_{G/\Gamma} f d\mu_x \quad as \quad T \to \infty.$$

Theorem 25 (Orbit closure theorem, Ratner 1991b) *Let G and H be the same as in Theorem 23. Let Γ be a lattice in G. Then for any $x \in G/\Gamma$, there exists a closed connected subgroup $L = L(x)$ containing H such that $\overline{Hx} = Lx$ and there is an L-invariant probability measure supported on Lx.*

Remarks (a) The proof of Theorem 23 is based on the polynomial divergence of unipotent flows in combination with multidimensional versions of Birkhoff's individual ergodic theorem. The proof of Theorem 24 uses Theorem 23 together with Theorem 22 and a simple result about the countability of a certain (depending on Γ) set of subgroups of G. Theorem 24 and the same countability result rather easily imply Theorem 25.

(b) Theorems 23 and 24 prove two conjectures of S.G. Dani, and Theorem 25 proves the Raghunathan conjecture. Before Ratner's work these theorems had been proved in some cases. In particular, Theorem 25 had been proved in Dani & Margulis (1990) in the case when $G = SL(3, \mathbf{R})$, $\Gamma = SL(3, \mathbf{Z})$ and $H = \{u(t)\}$ is a one-parameter unipotent subgroup of G such that $u(t) - 1$ has rank 2 for all $t \neq 0$; using this we proved a refinement of Theorems 7 and 8(a).

Concluding Remarks (I) Though the homogenous space approach allows to prove many new theorems, it has a serious defect. Namely it is not effective in the following sense. For the Oppenheim conjecture (Theorems 5 and 7) it does not give an estimate of the norm of the shortest vectors $v \in \mathbf{Z}^n$, $v \neq 0$ and $w \in \mathbf{Z}^n$ with $|Q(v)| < \varepsilon$ and $a < Q(w) < b$. In the asymptotic formulas for $N_{Q,\Omega}(a, b, T)$, it does not give estimates for error terms. These estimates should be expressed in terms of Diophantine properties of the quadratic from Q. My proof of the Oppenheim conjecture is not effective because it uses such notions as a minimal set of an action, and these notions have no effective analogs. Ratner's uniform distribution theorem, which is used in the proof of the asymptotic lower bound (10), is even 'less effective' because its proof uses ergodic theorems and such notions as the limit set of a set of measures. It is not clear either how to obtain error terms in the asymptotic formulas (21) and (22) because the proof of these formulas also uses Ratner's uniform distribution theorem. Note that a more detailed discussion of the problem of effective proofs for the homogeneous space approach is given in Margulis (2000).

(II) In connection with the previous remark let us mention that the upper estimates from Section 3 as well as all estimates from Section 1 are effective.

(III) Most of the topics considered in this paper are also discussed in surveys Borel (1995), Dani (1996), Margulis (1997, 2000), Ratner (1994a), Starkov (1997) and ICM addresses Dani (1994), Eskin (1998), Margulis (1990) and Ratner (1994b).

References

Baker, A. (1966), On a theorem of Sprindžuk, *Proc. Roy. Soc. London* **A292**, 92–104.

Baker, A. (1975), *Transcendental Number Theory*, Cambridge University Press.

Bentkus, V. & F. Götze, Lattice point problems and distribution of values of quadratic forms, *Ann. Math.* **150**, 977–1027.

Beresnevich, V. (1999), On approximation of real numbers by real algebraic numbers, *Acta Arith.* **90** (1999), 97–112.

Beresnevich, V. (2000), A Groshev type theorem for convergence on manifolds, *Acta Math. Hungar.*, (to appear).

Bernik, V. (1984), A proof of Baker's conjecture in the metric theory of transcendental numbers, *Doklady Akad. Nauk SSSR* **277**, 1036–1039 (in Russian).

Bernik, V., H. Dickinson & M. Dodson (1998), A Khintchine-type version of Schmidt's theorem for planar curves, *Proc. Roy. Soc. London* **A454**, 179–185.

Bernik, V., D. Kleinbock & G. Margulis (1999), Khintchine-type theorems on manifolds: convergence case for standard and multiplicative versions, the University of Bielefeld, Preprint, (1999).

Borel, A. (1995), Values of indefinite quadratic forms at integral points and flows on the space of lattices, *Bull. Amer. Math. Soc.* **32**, 184–204.

Borel, A. & G. Prasad (1992), Values of quadratic forms at S-integral points, *Compositio Mathematica* **83**, 347–372.

Brennan, T. (1994), *Distribution of values of diagonal quadratic forms at integer points*, Princeton University undergraduate thesis.

Cassels, J.W.S. & H.P.F. Swinnerton-Dyer (1995), On the product of three homogeneous forms and indefinite ternary quadratic forms, *Philos. Trans. Roy. Soc. London* **A248**, 73–96.

Dani, S.G. (1984), On orbits of unipotent flows on homogeneous spaces, *Ergod. Theor. Dynam. Syst.* **4**, 25–34.

Dani, S.G. (1985), Divergent trajectories of flows on homogeneous spaces and Diophantine approximation, *J. Reine Angew. Math.* **359**, 55–89.

Dani, S.G. (1986), On orbits of unipotent flows on homogeneous spaces II, *Ergod. Theor. Dynam. Syst.* **6**, 167–182.

Dani, S.G. (1994), Flows on homogeneous spaces and Diophantine approximation. In *Proc. ICM 1994*, 780–789.

Dani, S.G. (1996), Flows on homogeneous spaces: a review. In *Proc. of the Warwick Symposium on Ergodic Theory of* $\mathbf{Z}^d$*-action*, London Math. Soc. Lect. Notes Series **228**, Cambridge University Press, 63–112.

Dani, S.G. & G. Margulis (1989), Values of quadratic forms at primitive integral points, *Invent. Math.* **98**, 405–424.

Dani, S.G. & G. Margulis (1990), Orbit closures of generic unipotent flows on homogeneous spaces of $SL(3, \mathbf{R})$, *Math. Ann.* **286**, 143–174.

Dani, S.G. & G. Margulis (1993), Limit distribution of orbits of unipotent flows and values of quadratic forms, *Adv. Soviet Math.* **16**, 91–137.

Duke, W., Z. Rudnick & P. Sarnak (1993), Density of integer points on affine homogeneous varieties, *Duke Math. J.* **71**, 143–179.

Eskin, A. (1998), Counting problems and semisimple groups. In *Proc. ICM 1998*, **2**, 539–552.

Eskin, A. & C. McMullen (1993), Mixing, counting and equidistribution in Lie groups, *Duke Math. J.* **71**, 181–209.

Eskin, A., G. Margulis & S. Mozes (1998), Upper bounds and asymptotics in a quantitative version of the Oppenheim conjecture, *Ann. Math.* **147**, 93–141.

Eskin, A., G. Margulis & S. Mozes (2001), Quadratic forms of signature (2,2) and eigenvalue spacings on flat 2-tori. Preprint.

Eskin, A., S. Mozes & N. Shah, Unipotent flows and counting lattice points on homogeneous varieties, *Ann. Math.* **143**, 253–299.

Eskin, A., S. Mozes & N. Shah (1997), Nondivergence of translates of certain algebraic measures, *Geom. Functional Anal.* **7**, 93–141.

Kleinbock, D. (1996), *Nondense orbits of nonquasiunipotent flows and applications to Diophantine approximation*, PhD Thesis, Yale University.

Kleinbock, D. (1998), Flows on homogeneous spaces and Diophantine properties of matrices, *Duke Math. J.* **95**, 107–124.

Kleinbock, D. & G. Margulis (1998), Flows on homogeneous spaces and Diophantine approximation on manifolds, *Ann. Math.* **148**, 339–360.

Lindenstrauss, E. & B. Weiss (2001), On sets invariant under the action of the diagonal group, preprint.

Mahler, K. (1932), Über das Mass der Menge aller S-Zahlen, *Math. Ann.* **106**, 131–139.

Margulis, G. (1970), *On some problems in the theory of U-systems*, Thesis, Moscow University, (in Russian).

Margulis, G. (1975), On the action of unipotent groups in the space of lattices. In *Proc. of the Summer School on Group Representations* (Bolyai Janos Math. Soc., Budapest, 1971), 365–370; Akadémiai Kiado, Budapest.

Margulis, G. (1990), Dynamical and ergodic properties of subgroup actions on homogeneous spaces with applications to number theory. In *Proc. ICM 1990*, 193–215.

Margulis, G. (1997), Oppenheim conjecture. In *Fields Medalists' Lectures*, World Scientific, 272–327.

Margulis, G. (2000), Problems and conjectures in rigidity theory. In *International Mathematical Union. Mathematics: Frontiers and Perspectives*, Amer. Math. Soc., 161–174.

Ratner, M. (1990a), Strict measure rigidity for unipotent subgroups of solvable groups, *Invent. Math.* **101**, 449–482.

Ratner, M. (1990b), On measure rigidity of unipotent subgroups of semisimple groups, *Acta math.* **165**, 229–309.

Ratner, M. (1991a), On Raghunathan's measure conjecture, *Ann. Math.* **134**, 545–607.

Ratner, M. (1991b), Raghunathan's topological conjecture and distribution of unipotent flows, *Duke Math. J.* **63**, 235–280.

Ratner, M. (1994a), Invariant measures and orbit closures for unipotent actions on homogeneous spaces, *Geom. Functional Anal.* **4**, 236–256.

Ratner, M. (1994b), Interactions between ergodic theory, Lie groups and number theory. In *Proc. ICM 1994*, 157–182.

Sarnak, P. (1997), Values at integers of binary quadratic forms, in *Harmonic Analysis and Number Theory* (Montreal, PQ, 1996), CMS Conf. Proc., **21**, Amer. Math. Soc., 181–203.

Schmidt, W. (1960), A metrical theorem in Diophantine approximation, *Canadian J. Math.* **12**, 619–631.

Schmidt, W. (1964), Metrische Sätze über simultane Approximation abhänginger Grössen, *Monatsch. Math.* **68**, 154–166.

Sprindžuk, V. (1964), More on Mahler's conjecture, *Doklady Akad. Nauk. SSSR* **155**, 54–56 (Russian); English transl. in *Soviet math. Dokl.* **5** (1964), 361–363.

Sprindžuk, V. (1969), *Mahler's problem in metric number theory*, Translations of Mathematical Monographs, **25**, Amer. Math. Soc.

Sprindžuk, V. (1979), *Metric Theory of Diophantine Approximations*, Wiley.

Sprindžuk, V. (1980), Achievements and problems in Diophantine approximation theory, *Russian Math. Surveys* **35**, 1–80.

Starkov, A. (1997), New progress in the theory of homogeneous flows, *Russian Math. Surveys* **52**, 721–818.

19

On Linear Ternary Equations with Prime Variables – Baker's Constant and Vinogradov's Bound

Ming-Chit Liu & Tianze Wang

1 Baker's Constant

This part may be read as a continuation of the first author's survey (Liu & Tsang 1993) which is mainly on qualitative developments of Baker's Problem. In the present paper we shall discuss the latest progress on the Problem particularly with regard to numerical results. In order to make the paper largely self-contained, we shall first go through briefly the background to Baker's Problem, and then the relations between the Problem and Linnik's theorem on the smallest prime in an arithmetical progression. These relations indicate clearly the depth of Baker's Problem. In the last section of Part 1 we provide an outline of the proof of our recent numerical results. We hope that this may be useful for further developments on the quantitative part of Baker's Problem.

Introduction and Baker's Problem

Motivated mainly by the work of H. Davenport and H. Heilbronn (1946) on the solvability of some inequalities involving real quadratic diagonal forms in integer variables, A. Baker in his now well-known work (1967) considered the solvability of the Diophantine inequality

$$|\lambda_1 p_1 + \lambda_2 p_2 + \lambda_3 p_3| < \left(\log \max_{1\le j\le 3} p_j\right)^{-m}$$

in prime variables p_1, p_2, p_3 where m is any positive integer and $\lambda_1, \lambda_2, \lambda_3$ are nonzero real numbers, not all of the same sign and with at least one of the ratios λ_i/λ_j irrational. In the course of the investigation, Baker was led to consider a companion linear equation in three odd prime variables p_1, p_2, p_3

$$a_1 p_1 + a_2 p_2 + a_3 p_3 = b \tag{1}$$

where a_1, a_2, a_3 and b are given integers satisfying

$$a_1a_2a_3 \neq 0, \tag{2}$$

$$(a_1, a_2, a_3) := \gcd(a_1, a_2, a_3) = 1\ , \tag{3}$$

$$\text{not all } a_1, a_2, a_3 \text{ are of the same sign}\ , \tag{4}$$

$$a_1 + a_2 + a_3 \equiv b \pmod 2, \tag{5}$$

$$(a_i, a_j, b) = 1 \quad \text{for} \quad 1 \leq i < j \leq 3\ . \tag{6}$$

Solvability of some linear Diophantine equations in prime variables including (1) had been considered earlier in Richert (1953) but Baker's result (Baker 1967, p. 172) was the first that gave an upper bound to the small prime solutions of the equation (1). Baker's work (1967) stimulated research on the problem of obtaining the best possible bound in terms of a_j and b for small prime solutions p_j of the equation (1). We now call this problem *Baker's Problem.* As the culmination of the earlier discoveries in Liu (1985, 1987) in this context Theorem 1 was obtained by Liu & Tsang (1989).

Theorem 1 (Liu & Tsang 1989, Theorem 2). *Let a_1, a_2, a_3 and b satisfy* (2)–(6). *Then there is an effective absolute constant $B > 0$ such that equation* (1) *has a prime solution p_1, p_2, p_3 satisfying*

$$\max_{1\leq j\leq 3} p_j \leq 3|b| + \max\{3, |a_1|, |a_2|, |a_3|\}^B\ . \tag{7}$$

Note that the $\max\{3, |a_1|, |a_2|, |a_3|\}^B$ in (7) can be written as

$$C_0 \max\{|a_1|, |a_2|, |a_3|\}^B \tag{8}$$

for some absolute constant $C_0 > 0$.

Remark 1 *The constant B in* (7) *must satisfy*

$$B > 1\ .$$

Therefore, if we are not concerned about the numerical value of B then the form of the bound in (7) is best possible.

Proof Consider the simple example $a_1 = -a_2 = 1$, $a_3 < -3$, and

$$b = \begin{cases} 1 & \text{if } a_3 \text{ is odd}\ , \\ 0 & \text{if } a_3 \text{ is even}\ . \end{cases}$$

So conditions (2)–(6) are satisfied. Now any solution p_1, p_2, p_3 of (1) satisfies

$$p_1 = p_2 + |a_3|p_3 + b > 3b + 3|a_3| > 3b + \max\{3, |a_1|, |a_2|, |a_3|\}\ .$$

This violates (7) if $B \leq 1$. □

Remark 2 *Conditions* (2)–(6) *are either necessary or natural to the study of* (1). *This together with Remark 1 shows that Theorem 1 qualitatively settles Baker's Problem.*

Proof Inequality (2) is trivially necessary since we do not want to consider linear equations with less than three variables; (3) is natural since the solvability of (1) implies that b is divisible by (a_1, a_2, a_3) and then we may divide both sides of (1) by (a_1, a_2, a_3); (4) is clearly necessary for small solutions of (1); (5) is necessary for odd (prime) solutions of (1); (6) is necessary for the three prime variables problem for if $1 < d = (a_1, a_2, b)$ then by (3), $(d, a_3) = 1$. Hence (1) implies $d | p_3$ or $d = p_3$, and then the number of independent variables in (1) becomes at most 2. □

Because of Remark 2, it becomes of interest to determine numerical value of the constant B in (7).

Definition 1 The infimum $\mathcal{B}$ of all possible values of the constant B in (8) is called the *Baker Constant*.

Some Extensions of Baker's Problem

There are generalizations of Theorem 1 and some parallel results on Baker's Problem (see Liu & Tsang 1993, Section 4, and Liu & Wang 1999). In particular, the well-known Vinogradov theorem on the Three Primes Goldbach Conjecture has been generalized as follows.

Theorem 2 (Liu & Wang 1999, Theorem 1.) *Let k be any positive integer, and let a_1, a_2, a_3; ℓ_1, ℓ_2, ℓ_3 and b be integers satisfying* (2), (3), (5), (6) *and*

$$b \equiv a_1\ell_1 + a_2\ell_2 + a_3\ell_3 \pmod{k}\,; \quad (\ell_j, k) = 1 \quad \textit{for} \quad j = 1, 2, 3\,.$$

Put

$$K := \max\{|a_1|, |a_2|, |a_3|, k\}\,.$$

(i) *If all a_1, a_2, a_3 are positive then there are effective positive absolute constants υ and C_1 such that* (1) *is solvable in primes $p_j \equiv \ell_j \pmod{k}$, $1 \leq j \leq 3$ whenever*

$$b \geq C_1 K^{\upsilon}\,. \tag{9}$$

(ii) *If* a_1, a_2, a_3 *are not all of the same sign (i.e.* (4)*) then there are effective positive absolute constants* B *and* C_2 *such that* (1) *has a prime solution* $p_j \equiv \ell_j \pmod{k}$, $1 \le j \le 3$ *satisfying*

$$\max\{p_1, p_2, p_3\} \le C_2 \max\{|b|, K^B\}\,. \tag{10}$$

Remark 3 When $k = 1$ Theorem 2(ii) is Theorem 1. So in view of Remark 1 the form of the upper bound in (10) is best possible if we are not concerned about the exact value of B.

Remark 4 When $k = 1$, it was shown in (Liu & Tsang 1993, Remark 1.2) that the v in (9) must satisfy $v \ge 2$. So the form of the lower bound in (9) is best possible if we are not concerned about the exact value of v.

When $a_1 = a_2 = a_3 = 1$, Theorem 2(i) is a generalization of the well-known Goldbach–Vinogradov Theorem (see the second paragraph in Part 2). We reformulate the generalization as the following corollary.

Corollary 1 *Let* $k \ge 1$ *be any integer and* ℓ_j *be integers satisfying* $(\ell_j, k) = 1$ *for* $j = 1, 2, 3$. *Then there is an effective absolute constant* $\theta > 0$ *such that the equation* $N = p_1 + p_2 + p_3$ *with* $p_j \equiv \ell_j \pmod{k}$, $1 \le j \le 3$ *is solvable for sufficiently large odd* N *satisfying* $N \equiv \ell_1 + \ell_2 + \ell_3 \pmod{k}$ *and* $k \le N^\theta$.

Relations with Linnik's Theorem

Let ℓ, q be integers satisfying $1 \le \ell \le q$ and $(\ell, q) = 1$. Dirichlet's Theorem on primes in arithmetical progressions states that the sequence $\ell + kq$, $k = 1, 2, 3, \ldots$ contains infinitely many primes. Denote by $P(\ell, q)$ the smallest prime in $\ell + kq$. Linnik (1944) proved:

Theorem 3 (Linnik's Theorem) *There are absolute constants* $c > 0$ *and* $L > 0$ *such that*

$$P(\ell, q) < cq^L\,.$$

Definition 2 The infimum $\mathcal{L}$ of all possible values of the constant L is called the *Linnik Constant.*

Remark 5 Since the discovery of Linnik's Theorem much effort has been devoted to the determination of the value of $\mathcal{L}$. The first upper bound $\mathcal{L} \le 10,000$ was given by Pan (1957). This was subsequently sharpened by various authors as shown in Table 1.

Table 1.

$\mathcal{L} \leq$	Date	Author
5,448	1958	C.D. Pan
777	1965	J.-R. Chen
630	1971	M. Jutila
550	1970	M. Jutila
168	1977	J.-R. Chen
80	1977	M. Jutila
36	1977	S.W. Graham
20	1981	S.W. Graham
17	1979	J.-R. Chen
16	1986	W. Wang
13.5	1989	J.-R. Chen & J.-M. Liu
5.5	1992	D.R. Heath-Brown

Remark 6 *Theorem 1 contains Linnik's Theorem and hence* $\mathcal{B} \geq \mathcal{L}$.

Proof For given integers ℓ, q with $1 \leq \ell \leq q$ and $(\ell, q) = 1$, set $a_1 = 1$, $a_2 = -q$, $b = \ell$ and

$$a_3 = \begin{cases} -q & \text{if } \ell \text{ is odd ,} \\ -2q & \text{if } \ell \text{ is even .} \end{cases}$$

So conditions (2)–(6) are satisfied. By Theorem 1, (1) is solvable and

$$p_1 = (p_2 + p_3)q + \ell \quad \text{or} \quad p_1 = (p_2 + 2p_3)q + \ell \, .$$

That is, p_1 is in the arithmetical progression $\ell + kq$. By (7) where $B > 1$,

$$P(\ell, q) \leq p_1 \leq 3\ell + \max\{3,\ 2q\}^B \ll q^B .$$

□

This is Linnik's Theorem and so $\mathcal{B} \geq \mathcal{L}$.

Remark 7 *Corollary 1 implies Linnik's Theorem and hence* $\mathcal{L} \leq 1/\theta$.

Proof For any given integers q and ℓ with $1 \leq \ell \leq q$ and $(\ell, q) = 1$, we take the k in Corollary 1 to be q and specify the large odd N to be fixed which satisfies $N \equiv 3\ell \pmod q$ and $q^{1/\theta} \leq N \ll q^{1/\theta}$. Then Corollary 1 with $\ell_j = \ell$ asserts that there exist primes p_1, p_2 and p_3 in $\{kq + \ell\}_{k=0,1,\ldots}$ such that

$$p_1 + p_2 + p_3 = N \ll q^{1/\theta} .$$

So $P(\ell, q) \ll q^{1/\theta}$, which is Linnik's Theorem, and $\mathcal{L} \leq 1/\theta$. □

Some Recent Numerical Results on Baker's Constant

In view of Remarks 2 and 6, the problem of the determination of the numerical value of $\mathcal{B}$ becomes interesting. Furthermore, the results and techniques developed in the work for $\mathcal{L}$ by many authors as mentioned in Remark 5, reveal a feasible way for the investigation of the quantitative part of Baker's Problem. The first numerical bound for $\mathcal{B}$ was obtained by Choi (1990) who obtained

Theorem 4

$$\mathcal{B} \le 4,190\,.$$

Very recently, the authors improved upon this bound and obtained

Theorem 5 (c.f. Liu & Wang 1998, Theorem 1)

$$\mathcal{B} \le 44\,.$$

Outline of the Proof of Theorem 5

In this section we wish to bring out some key points in our proof of Theorem 5 which, we hope, will be helpful for further investigation on the quantitative part of Baker's Problem.

The work in Liu & Wang (1998) is an application of the Hardy–Littlewood Circle Method. Besides a very delicate refinement of Liu & Tsang (1989), the bulk of Liu & Wang (1998) consists of a careful estimate of the numerical bound for the constants appearing in zero-free regions and zero density of the Dirichlet L-functions $L(s, \chi)$.

Similar to previous work on Baker's Problem, in order to obtain a bound of type (7) we have to consider large major arcs $\mathcal{M}(h, q)$ (e.g., in comparison with the arcs described as in (i) below Theorem 6) as follows.

For large $N > 0$ let

$$Q := N^{\delta} \quad \text{and} \quad \tau := N^{-1}Q^{1+\varepsilon}\,.$$

Define the major arc $\mathcal{M}(h, q)$ to be the closed interval

$$\mathcal{M}(h, q) := [(h-\tau)/q\ ,\ (h+\tau)/q]$$

where the integers h and q satisfy $1 \le h \le q \le Q$ and $(h, q) = 1$. Let $\mathcal{M}$ be the union of all $\mathcal{M}(h, q)$. Dissect the interval $I := [\tau, 1+\tau]$ into two sets $\mathcal{M}$ and $C(\mathcal{M}) := I \setminus \mathcal{M}$. With this dissection, we shall describe now that in the proof in Liu & Wang (1998), Baker's Constant satisfies

$$\mathcal{B} \le \frac{3}{\delta} - 1 . \tag{11}$$

For a better explanation of the factor 3 in (11), let us consider a more general form of (1), namely

$$a_1 p_1 + \cdots + a_s p_s = b \tag{12}$$

where $s \ge 3$. Set

$$S_j(x) := \sum_{N_j' < n \le N_j} \Lambda(n) e(a_j n x) , \quad j = 1, \ldots, s, \tag{13}$$

where $\Lambda(n)$ is the von Mangoldt function, $e(\alpha) := \exp(i2\pi\alpha)$ for any real α and

$$N_j := N/|a_j| , \quad N_j' := N_j/(s+1) .$$

If $x \in C(\mathcal{M})$, Vinogradov's bound for trigonometric sum over primes (see, for example Davenport 1980, p. 143) gives

$$S_j(x) \ll N Q^{-1/2} |a_j|^{-1/2} \log^4 N . \tag{14}$$

Write

$$\int_I e(-xb) \prod_{j=1}^{s} S_j(x) dx =: \int_{\mathcal{M}} + \int_{C(\mathcal{M})} . \tag{15}$$

Applying (14) and

$$\int_0^1 |S_j(x)|^2 dx \ll N |a_j|^{-1} \log^2 N$$

we get

$$\begin{aligned} \int_{C(\mathcal{M})} &\ll \prod_{j=1}^{s} (N Q^{-1/2} |a_j|^{-1/2} \log^4 N)^{(s-2)/s} \Big(\int_0^1 |S_j(x)|^2 dx \Big)^{1/s} \\ &\ll N^{s-1} Q^{-(s-2)(1/2-\varepsilon)} |a_1 \cdots a_s|^{-1/2} . \end{aligned}$$

On the other hand, with additional log-factors the integral $\int_I$ in (15) represents the number of prime solutions $p_1, \ldots, p_s$ of the equation (12) with $p_j \le N_j$ and so the lower bound for $\int_{\mathcal{M}}$ is roughly of the form

$$N^{s-1} |a_1 \cdots a_s|^{-1} .$$

In order to obtain a positive value of the integral $\int_I$ in (15) we need basically

$$N^{s-1} |a_1 \cdots a_s|^{-1} \gg N^{s-1} Q^{-(s-2)(1/2-\varepsilon)} |a_1 \cdots a_s|^{-1/2} ,$$

that is,

$$N^{\delta} = Q \gg |a_1 \cdots a_s|^{(1+\varepsilon)/(s-2)} .$$

Now, since N_j is the upper bound for those integers n under the summation in (13), we have

$$\max_{1 \le j \le s} |a_j| p_j \ll |a_1 \cdots a_s|^{((s-2)\delta)^{-1}+\varepsilon} .$$

Then, besides an upper bound in terms of b as in (7), the corresponding constant B in (8) for equation (12) satisfies

$$B \le \frac{s}{(s-2)\delta} - 1 + \varepsilon .$$

This gives (11) when $s = 3$.

We now come to explain our choice of the value of δ in (11). In Liu & Wang (1998) we set

$$\delta = \frac{1}{15 - \varepsilon} . \tag{16}$$

Then the number 44 in Theorem 5 comes essentially from (11) and (16). The permissible value of δ is determined by some numerical requirements involving the upper bounds for certain triple sums of the form

$$\sum_{q \le Q} \ \sideset{}{^*}\sum_{\chi \,(\mathrm{mod}\, q)} \ \sideset{}{'}\sum_{|\gamma| \le Q^3} \tag{17}$$

where $*$ indicates that all Dirchlet characters χ (mod q) in the sum are primitive and $\sum'$ means that besides $|\gamma| \le Q^3$ the sum is over all nontrivial zeros $\rho = \beta + i\gamma$ of $L(s, \chi)$ with $1/2 \le \beta < 1$ and $\rho \ne \tilde{\beta}$, the Siegel zero (defined as in Lemma 1 below).

Brief explanations about the choice of δ are given as follows. For $x \in \mathcal{M}(h, q)$ write $x = h/q + \eta$. By the orthogonal relations of Dirichlet characters χ (mod q) we have

$$S_j(x) = (T_j + \tilde{T}_j - G_j)(x) + E_j$$

where T_j is the main term, the term $\tilde{T}_j$ exists if the $\tilde{\beta}$ exists, E_j is the error term and

$$G_j(x) := \varphi(q)^{-1} \sum_{\chi \,(\mathrm{mod}\, q)} \sum_{\ell=1}^{q} \overline{\chi}(\ell) e(a_j h\ell/q) \int_{N_j'}^{N_j} e(a_j \eta y) \sideset{}{'}\sum_{|\gamma| \le Q^3} y^{\rho-1} dy . \tag{18}$$

Here $\varphi(q)$ is the Euler function. Now to handle the integral $\int_{\mathcal{M}}$ on the right-hand side of (15), we have to deal with the product there with $s = 3$. If we ignore E_j, there are 19 terms (if $\tilde{\beta}$ exists) in $\int_{\mathcal{M}}$ having at least one $G_j(x)$ as factor. Since

$$\int_{\mathcal{M}} = \sum_{q \le Q} \sum_{\substack{h=1 \\ (h,q)=1}}^{q} \int_{\mathcal{M}(h,q)}$$

by (18) there is at least one triple sum described as in (17) in each of the corresponding 19 integrals.

In previous work (e.g. Liu & Tsang 1989) on the qualitative part of Baker's Problem, Gallagher's theorem (Gallagher 1970, Theorem 6) was successfully applied to handle triple sums similar to (17) though with constants in the upper bound estimate unspecified. Now, in the problem of quantitative part Liu & Wang (1998), the numerical requirements in the treatment of these triple sums force us to go much further. To accomplish the task we need to use numerical bounds for constants appearing in the zero-free regions and zero density of $L(s, \chi)$. In this connection we obtain the following Lemmas 1, 2 and 3 by the methods and results due to Chen (1979) and Heath-Brown (1992).

In what follows, $K(C)$ denotes a large positive number depending on C only.

Lemma 1 *For any constant $C > 0$, if $Q \ge K(C)$, then the function $\prod(\sigma + it) := \prod_{q \le Q} \prod^*_{\chi (\mathrm{mod}\, q)} L(\sigma + it, \chi)$ has at most one zero in the region $\sigma \ge 1 - 0.364/\log Q$, $|t| \le C$ where $*$ indicates that all χ are primitive. Such a zero $\tilde{\beta}$, if it exists, is called the Siegel zero.*

Lemma 2 *For any constant $C > 0$, if $Q \ge K(C)$, then the function $\prod(\sigma + it)$ has at most two zeros in the region $\sigma \ge 1 - 0.504/\log Q$, $|t| \le C$.*

Lemma 3 *For any constant $C > 0$, let $Q \ge K(C)$. Let $\alpha = 1 - \lambda/\log Q$ and $N(\chi, \alpha, C)$ denote the number of zeros of $L(\sigma + it, \chi)$ lying in the region: $\alpha \le \sigma < 1 - 0.364/\log Q$ and $|t| < C$. Then we have*

$$\sum_{q \le Q} \sum_{\chi (\mathrm{mod}\, q)} N(\chi, \alpha, C)$$
$$\le 42.54 \left(1 + \frac{35.385}{\lambda}\right)$$
$$\left(\exp(2.87538\lambda) - \frac{\exp(2.07176\lambda) - \exp(1.92136\lambda)}{0.1504\lambda}\right)$$

if $6 < \lambda \le \log\log\log Q$. *Similar results hold for different ranges of* λ *in the interval* $(0.504, 6]$.

Now, for any fixed constant $C > 0$, we split the innermost sum in (17) into two sums by the conditions $|\gamma| \le C$ and $C < |\gamma| \le Q^3$. Applying Lemmas 1, 2 and Liu & Wang (1998), Lemma 3.1, (containing the above Lemma 3) together with $\delta = 1/(15 - \varepsilon)$ as defined in (16), we obtain for sufficiently large Q,

$$\sum_{q \le Q} \sum_{\chi \, (\mathrm{mod}\, q)}^{*} \sum_{|\gamma| \le C}' N_j'^{(\beta-1)} \le \begin{cases} 0.096 & \text{if } \tilde{\beta} \text{ does not exist} \\ 0.5633((1-\tilde{\beta})\log Q)^3 & \text{if } \tilde{\beta} \text{ exists.} \end{cases}$$

With these numerical results we derive an upper bound for the sum of the 19 integrals involving $G_j(x)$ and hence obtain a satisfactory lower bound for

$$\int_{\mathcal{M}} e(-bx) \prod_{j=1}^{3} S_j(x) dx$$

which dominates the estimate for the integral $\int_{C(\mathcal{M})}$ in (15) with $s = 3$. The proof of Theorem 5 is then complete.

Clearly the crux of further improvement upon Theorem 5 lies in (11).

2 Vinogradov's Bound

In this part, we shall consider the equation (1) with all $a_j = 1$, namely, the equation (19) below.

The Three Primes Goldbach Conjecture (3GC) states that every odd integer ≥ 9 is a sum of three odd primes. Assuming the Generalized Riemann Hypothesis (GRH), Hardy & Littlewood (1923) proved the 3GC for all sufficiently large odd integers. Vinogradov (1937) successfully removed the GRH, namely, he proved that there is a positive integer V such that for any odd integer $b \ge V$ (so the above 'sufficiently large' condition is still assumed) we have

$$b = p_1 + p_2 + p_3 \tag{19}$$

where p_j are odd primes. The result is usually called the Goldbach–Vinogradov theorem or simply Vinogradov's theorem. It qualitatively settles the 3GC and it remains to consider the quantitative part, that is to remove the condition 'sufficiently large' also from the above Hardy–Littlewood result or equivalently to show that the V in the Vinogradov result can be 9. Although the 3GC is still not completely settled, Vinogradov's theorem is no doubt one of the most remarkable results in the 20th century. Because of its significance we call the value of V the *Vinogradov Bound*. A crude value for V, following from Vinogradov's work, is $V = 3^{3^{15}}$ $(= 10^{6,846,168.5\cdots})$. Thus, to accomplish the quantitative part of the 3GC we need to check all odd integers lying between 9 and V. Plainly, the numerical value for V is far from satisfactory and we should like to lower its value considerably until it falls in the range of the capacity of the latest powerful computer. In this direction Borozdkin (1956) showed that the V can be $e^{e^{16.038}}$ $(= 10^{4,008,659.9\cdots})$. The latest known result for V was obtained by Chen & Wang (1989). They showed that a permissible value is

$$V = e^{e^{11.503}} \quad (= 10^{43,000.5\cdots})\ .$$

Another way to investigate the quantitative part of the 3GC is, of course, to check as many odd integers $< V$ as possible. The latest result in this direction was obtained by Saouter (1998) who has showed that each odd integer $\leq 10^{20}$ can be expressed as in (19).

Zinoviev (1997) showed that under the GRH, V can be 10^{20}. So under the GRH, this together with the above result of Saouter settles the quantitative part of the 3GC. Independently, the same result was obtained by J Deshouillers, Effinger, Te Riele & Zinoviev (1997) without referring to Saouter's result. That is, under the GRH, the 3GC is now completely settled. These recent numerical developments stimulate a strong desire to lower the known Vinogradov Bound $10^{43,000}$ unconditionally. Very recently the authors proved (Liu & Wang 2002, Theorem 1):

Theorem 6 *Every odd integer* $\geq V = e^{3,100} (= 10^{1,346.3\cdots})$ *is a sum of three odd primes as in* (19).

The framework of our proof of Theorem 6 is again based on the Hardy–Littlewood Circle Method. One of the features of the latter is that it leads to asymptotic results and so it works well if some parameters are large enough. Therefore the 'sufficiently large' condition is essential and crucial in many steps of the Circle Method. Our goal in Theorem 6 is to lower the Vinogradov Bound V or to replace the 'sufficiently large' condition by explicit values of the large parameters. So during the proof there is absolutely no shelter to prevent the 'sufficiently large' condition from being numerically checked. Comparing with the previous works on the Vinogradov bound, besides some tricks and the help of the computer to obtain better numerical constants in many inequalities, we have mainly the following three novelties in Liu & Wang (2002) for the proof of Theorem 6.

(i) In contrast to the proof of Theorem 5, where the larger the major arcs $\mathcal{M}(h,q)$, the better bound for Baker's Constant, it is not difficult to see that in practice, in order to obtain small values for the essential parameters, the major arcs cannot be too large. We have to choose suitably the length of these arcs. Furthermore, we dissect the interval I into four disjoint subsets (Liu & Wang 2002, Section 6), instead of the usual two subsets $\mathcal{M}$ and $C(\mathcal{M})$ as in those above (11). We treat three of these four as minor arcs and the remaining one as major arcs.

(ii) We obtain a new numerical version of the explicit formula for $\sum_{n\le N} \Lambda(n)\chi(n)$ (Liu & Wang 2002, Lemma 4.1). We believe that this result is of general interest for problems involving prime number theorem when all constants concerned are required to be explicit.

(iii) We obtain a new numerical version of the Vinogradov estimate for trigonometric sums over primes (Liu & Wang 2002, Proposition 9.1) which should prove useful in the numerical treatment of minor arcs whenever the Hardy–Littlewood Circle Method is applied.

Acknowledgement

The work is partially supported by Hong Kong Government RGC (HKU518/96P & HKU7221/99P) research grants.

References

Baker, A. (1967), On some diophantine inequalities involving primes, *J. Reine Angew. Math.* **228**, 166–181.

Borozdkin, K.G. (1956), On I.M.Vinogradov's constant, *Proc. 3rd All-Union Math. Conf., Izdat. Akad. Nauk SSSR, Moscow* **1** (1956), 3.

Chen, J.R. (1979), On the least prime in an arithmetical progression and theorems concerning the zeros of Dirichlet's L-functions (II), *Sci. Sinica* **22**, 859–889.

Chen, J.R. & T.Z. Wang (1989), On odd Goldbach problem, *Acta Math. Sinica* **32**, 702–718.

Choi, S.K.K. (1990), *Some Explicit Estimates on Linear Diophantine Equations in Three Prime Variables*, Thesis, The University of Hong Kong.

Davenport, H. (1980), *Multiplicative Number Theory*, 2nd ed., Springer-Verlag.

Davenport, H. & H. Heilbronn (1946), On indefinite quadratic forms in five variables, *J. London Math. Soc.* **21**, 185–193.

Deshouillers, J.M., G. Effinger, H. Te Riele & D. Zinoviev (1997), A complete Vinogradov 3-primes theorem under the Riemann hypothesis, *ERA Amer. Math. Soc.* **3**, 99–104.

Gallagher, P.X. (1970), A large sieve density estimate near $\sigma = 1$, *Invent. Math.* **11**, 329–339.

Hardy, G.H. & J.E. Littlewood (1923), Some problems of *Partitio Numerorum*: III On the expression of a number as a sum of primes, *Acta Math.* **44**, 1–70.

Heath-Brown, D.R. (1992), Zero-free regions for Dirichlet L-functions, and the least prime in an arithmetic progression, *Proc. London Math. Soc.* **64**, 265–338.

Linnik, Yu.V. (1944), On the least prime in an arithmetic progression I, and II, *Rec. Math. (Mat. Sb.) N.S.* **15** (57), 139–178; 347–368.

Liu, M.-C. (1985), A bound for prime solutions of some ternary equations, *Math. Z.* **188**, 313–323.

Liu, M.-C. (1987), An improved bound for prime solutions of some ternary equations, *Math. Z.* **194**, 573–583.

Liu, M.-C. & K.-M. Tsang (1989), Small prime solutions of linear equations. In *Théorie des Nombres*, J.-M. de Koninck & C. Levesque (eds.), de Gruyter, 595–624.

Liu, M.-C. & K.-M. Tsang (1993), Recent progress on a problem of A. Baker. In *Séminaire de Théorie des Nombres, Paris 1991–1992*, Progress Math., **116**, Birkhäuser, 121–133.

Liu, M.-C. & T. Wang (1998), A numerical bound for small prime solutions of some ternary linear equations, *Acta Arith.* **86**, 343–383.

Liu, M.-C. & T. Wang (1999), On the equation $a_1p_1 + a_2p_2 + a_3p_3 = b$ with prime variables in arithmetic progressions. In *CRM Proceedings and Lecture Notes*, **19**, Amer. Math. Soc., 243–264.

Liu, M.-C. & T. Wang (2002), On the Vinogradov bound in the three-prime Goldbach conjecture, *Acta Arith.*, to appear. 31 pages.

Pan, C.-D. (1957), On the least prime in an arithmetical progression, *Sci. Record (N.S.)* **1**, 311–313; *Acta Sci. Natur. Univ. Pekinensis* **4** (1958), 1–34.

Richert, H.-E. (1953), Aus der additiven Primzahtheorie, *J. Reine Angew. Math.* **191**, 179–198.

Saouter, Y. (1998), Checking the odd Goldbach conjecture up to 10^{20}, *Math. Comp.* **67**, 863–866.

Vinogradov, I.M. (1937), Representation of an odd number as a sum of three primes, *C.R. (Dokl.) l'Acad. Sci. l'USSR* **15**, 291–294.

Zinoviev, D. (1997), On Vinogradov's constant in Goldbach's ternary problem, *J. Number Theory* **65**, 334–358.

20

Powers in Arithmetic Progression

T.N. Shorey

1 Cubes and higher powers

For an integer $\nu > 1$, we write $P(\nu)$ for the greatest prime factor of ν and we put $P(1) = 1$. Let $d > 0$, $n > 0$, $k \geq 2$, $t \geq 2$ and $r \in \{0, 1\}$ be integers such that $\gcd(n, d) = 1$ and $t = k - r$. Thus $k \geq 2$ if $r = 0$ and $k \geq 3$ if $r = 1$. Let $d_1 < d_2 < \cdots < d_t$ be integers in $[0, k)$. We put

$$\Delta = \Delta(n, d, k, d_1, \ldots, d_t) = (n + d_1 d) \cdots (n + d_t d).$$

If $r = 0$, then $d_i = i$ for $0 \leq i < k$ and $\Delta = n(n + d) \cdots (n + (k - 1)d)$. If $r = 1$, we see that $\Delta = n(n + d) \cdots (n + (k - 1)d)/(n + id)$ for some i with $0 \leq i < k$. We write

$$\Delta = \begin{cases} \Delta_0 & \text{if } r = 0 \\ \Delta_1 & \text{if } r = 1. \end{cases}$$

Let b and ℓ be positive integers such that $P(b) \leq k$ and ℓ is prime. We consider

$$\Delta = (n + d_1 d) \cdots (n + d_t d) = by^{\ell}. \tag{1}$$

Equation (1) with $r = 0$ is an old equation; it has been considered by Fermat, Euler, Goldbach and others. It has its origin in the result of Fermat that there are no four squares in arithmetic progression. Euler proved a more general result that a product of four terms in arithmetic progression is never a square. Goldbach showed that a product of three consecutive positive integers is not a square. As pointed out in Shorey (1988), section 8, equation (1) with $r = 1$ leads to (1) with $r = 0$ and b replaced by b times the power of a prime exceeding k. We refer to Shorey & Tijdeman (1997) and Shorey (1999a,b) for an account of results on (1).

Since $P(b) \leq k$, it is natural to suppose that the left-hand side of (1) is divisible by a prime exceeding k. The first result in this direction dates back to

Sylvester (1892) who proved that

$$P(\Delta_0) > k \text{ if } n \geq d + k.$$

Langevin (1977) improved this to

$$P(\Delta_0) > k \text{ if } n > k.$$

Shorey & Tijdeman (1990a) showed that

$$P(\Delta_0) > k \text{ if } d > 1 \text{ and } (n, d, k) \neq (2, 7, 3).$$

The assumptions in the preceding result are necessary since $P(1 \times 2 \times \cdots \times k) \leq k$ and $P(2 \times 9 \times 16) = 3$. Further, Saradha & Shorey (2001a) proved that

$$P(\Delta_1) > k \text{ if } d > 1 \text{ and } k \geq 4$$

unless

$$\begin{aligned}(n, d, k, d_1, \ldots, d_t) \in \{&(1, 5, 4, 0, 1, 3), (2, 7, 4, 0, 1, 2), (3, 5, 4, 0, 1, 3),\\ &(1, 2, 5, 0, 1, 2, 4), (2, 7, 5, 0, 1, 2, 4), (4, 7, 5, 0, 2, 3, 4),\\ &(4, 23, 5, 0, 1, 2, 4)\}. \end{aligned} \tag{2}$$

We observe that $P(\Delta_1) \leq k$ for each of the above 7 tuples. Therefore, it is necessary to exclude them in the above result. Further, we shall exclude them and $(n, d, k) = (2, 7, 3)$ for considering (1) with $r = 1$ and $r = 0$, respectively. If $d = 1$, we record the analogous inequality

$$P(\Delta) > k \text{ if } d = 1 \tag{3}$$

for later references. The above result on $P(\Delta_1)$ is equivalent to

Theorem 1 *Let $d > 1, k \geq 4$ and (n, d, k) be not given by* (2). *Then Δ_0 is divisible by at least two distinct primes which are greater than k.*

An account given above on Sylvester's theorem is enough for considering (1) but we conclude it by stating the following two refinements of Theorem 1. For $d > 1$ and $k \geq 6$, Saradha, Shorey & Tijdeman (2002) proved that

$$\omega(\Delta_0) \geq \pi(k) + \left[\frac{1}{5}\pi(k)\right] + 2$$

unless

$$\begin{aligned}(n, d, k) \in \{&(1, 2, 6), (1, 3, 6), (1, 2, 7), (1, 3, 7), (1, 4, 7), (2, 3, 7), (2, 5, 7),\\ &(3, 2, 7), (1, 2, 8), (1, 2, 11), (1, 3, 11), (1, 2, 13), (3, 2, 13), (1, 2, 14)\}\end{aligned}$$

in which cases the preceding inequality is not satisfied. Further, Saradha & Shorey (2001c) showed that for $d = 1$ and $n > k \geq 3$,

$$\omega(\Delta_0) \geq \pi(k) + \left[\frac{1}{3}\pi(k)\right] + 2$$

except when $n = 4, 6, 7, 8, 16$ if $k = 3$; $n = 6$ if $k = 4$; $n = 6, 7, 8, 9, 12, 14, 15, 16, 23, 24$ if $k = 5$; $n = 7, 8, 15$ if $k = 6$; $n = 8, 9, 10, 12, 14, 15, 24$ if $k = 7$; and $n = 9, 14$ if $k = 8$. Finally, we observe that it is necessary to exclude these cases.

Now I give a sketch of the proof of Theorem 1. First, we observe that $\gcd(\Delta_0, d) = 1$ since $\gcd(n, d) = 1$. Suppose that the assertion of Theorem 1 is not valid. Then

$$\omega(\Delta_0) = \sum_{p|\Delta_0} 1 \leq \pi_d(k) + 1$$

where

$$\pi_d(k) = \sum_{\substack{p \leq k \\ \gcd(p,d)=1}} 1.$$

Now we apply an argument of Erdős. For every prime $p \mid \Delta_0$, we take an $f(p)$ with $0 \leq f(p) < k$ such that

$$\mathrm{ord}_p(n + f(p)d) = \max_{0 \leq i < k} \mathrm{ord}_p(n + id).$$

We write S for the set obtained from $\{n, n + d, \ldots, n + (k-1)d\}$ by deleting all $n + f(p)d$ with $p \mid \Delta_0$. Let $p \mid \Delta_0$. For $n + id \in S$, we observe that

$$\begin{aligned} \mathrm{ord}_p(n + id) &= \min(\mathrm{ord}_p(n + id), \mathrm{ord}_p(n + f(p)d)) \\ &\leq \mathrm{ord}_p(n + id - (n + f(p)d)) = \mathrm{ord}_p(i - f(p)), \end{aligned}$$

since $\gcd(p, d) = 1$. Thus, for distinct $n + i_1 d, \ldots, n + i_{k'} d \in S$, we have

$$\begin{aligned} \mathrm{ord}_p((n + i_1 d) \ldots (n + i_{k'} d)) &\leq \mathrm{ord}_p((i_1 - f(p)) \cdots (i_{k'} - f(p))) \\ &\leq \mathrm{ord}_p\left(\prod_{\substack{j=0 \\ j \neq f(p)}}^{k-1} (j - f(p))\right) \\ &= \mathrm{ord}_p(f(p)!(k - 1 - f(p))!) \\ &\leq \mathrm{ord}_p((k-1)!). \end{aligned}$$

Hence $|S| \geq k - \pi_d(k) - 1$ and

$$\begin{aligned}\prod_{s \in S} s &\leq \prod_{\substack{p<k \\ \gcd(p,d)=1}} p^{[\frac{k-1}{p}]+[\frac{k-1}{p^2}]+\cdots} \\ &= (k-1)!\prod_{p|d} p^{-\mathrm{ord}_p((k-1)!)}.\end{aligned}$$

On the other hand

$$\prod_{s \in S} s \geq n(n+d)\cdots(n+(k-\pi_d(k)-2)d) \geq (k-\pi_d(k)-2)!\, d^{k-\pi_d(k)-2}.$$

Finally, we show that the above estimates for $\prod_{s \in S}$ are not consistent. We remark that the proof of Theorem 1 does not depend on results on primes in arithmetic progressions as was the case in the proof of Shorey & Tijdeman (1990a) and this is necessary for the refinement of Saradha, Shorey & Tijdeman (2001) stated above.

We always suppose that $\ell > 2$ in Section 1. Baker's sharpenings on linear forms in logarithms led Tijdeman to show that the exponential equation of Catalan has only finitely many solutions. Around the same time, Erdős & Selfridge developed an elementary method of Erdős (1955) to establish a striking theorem on an exponential diophantine equation that a product of two or more consecutive positive integers is never a cube or a higher power. This is a consequence of the following result.

Theorem 2 (Erdős & Selfridge 1975) *Equation* (1) *with* $r = 0$, $d = 1$, $P(b) < k$ *and* (3) *never holds.*

If $n(n+1)\cdots(n+k-1) = y^\ell$, we see from Theorem 2 and the theorem of Sylvester stated above that $n \leq k$ such that $n \leq (n+k)/2 \leq q \leq n+k-1$ for some prime q and $\mathrm{ord}_q(n(n+1)\cdots(n+k-1)) = 1$. This is a contradiction implying that a product of two or more consecutive positive integers is never a cube or a higher power. The first general result on Theorem 2 dates back to when Erdős (1939b) and Rigge (1939), independently, proved that a product of k consecutive positive integers is an ℓth power only if k is bounded by a number depending only on ℓ. The proof depends on the fundamental theorem of Thue on the approximations of algebraic numbers by rationals. We write d as

$$d = D_1 D_2$$

where D_1 is the maximal divisor of d such that all the prime divisors of D_1 are

$\equiv 1 \pmod{\ell}$. Thus $D_1 > 1$ implies that $P(d) \geq 2\ell + 1 \geq 7$ since $\ell \geq 3$. Now we give the following extension of Theorem 2.

Theorem 3 (Saradha & Shorey 2001a) *Equation* (1) *with* $r = 0$, $d > 1$ *and* $P(b) < k$ *implies that* $D_1 > 1$.

Thus (1) with $r = 0$ and $P(b) < k$ never holds if $d > 1$ is composed only of 2, 3 and 5. Thus it has been possible to solve completely (1) with $r = 0$ and $P(b) < k$ for infinitely many values of d. In view of the binomial equation

$$\binom{n+k-1}{n-1} = y^\ell$$

i.e.

$$n(n+1)\cdots(n+k-1) = k!\, y^\ell$$

solved completely by Erdős (1951) for $k \geq 4$, Győry (1997) for $k = 3$ and Darmon & Merel (1997) for $k = 2$, it is of some interest to relax the assumption $P(b) < k$ to $P(b) \leq k$ in Theorem 2. This was done by Saradha (1997) for $k \geq 4$ and Győry (1998) for $k = 2, 3$. Infact, as an immediate consequence of Theorem 6, it can be relaxed to $P(b) \leq P_k$ for $k \geq 6$ and P_k denotes the least prime exceeding k. The proof of Saradha depends on the method of Erdős & Selfridge whereas Győry derived it from the results of Ribet (1997) and Darmon & Merel (1997) that a generalised Fermat equation $x^\ell + 2^\alpha y^\ell + z^\ell = 0$ has no non-trivial solution. This also implies, as pointed out by Győry (1999), that (1) with $r = 0$, $d > 1$, $k = 3$ and $P(b) < k$ does not hold and the assertion of Theorem 3 with $k = 3$ follows. Let $k = 2$. We observe that

$$d = y_1^\ell - y_2^\ell = (y_1 - y_2)\left(\frac{y_1^\ell - y_2^\ell}{y_1 - y_2}\right)$$

and every prime factor of the second term on the right hand side is congruent to 1 (mod ℓ) except, possibly, ℓ which appears in its factorisation to the first power. This implies that $D_1 > 1$. The proof of Theorem 3 with $k \geq 4$ depends on the method of Erdős & Selfridge and Shorey (1988). Now we give an analogous statement for Theorem 3 where the assumption $P(b) < k$ has been relaxed to $P(b) \leq k$ and the case $r = 1$ is also covered.

Theorem 4 (Saradha & Shorey 2001a) *Assume* (1) *with* $r \in \{0, 1\}$, $d > 1$ *and* $k \geq 4$ *if* $r = 0$; $k \geq 9$ *if* $r = 1$. *Then* $D_1 > 1$.

Tijdeman (1988) observed that $64 \times 375 \times 686$ is 6 times a cube. In this example, $d = 311$ and $D_1 = 1$. Thus the assertion of Theorem 4 is not valid if $r = 0$ and $k = \ell = 3$. This is also the case when $r = 0$ and $k = 2$ in view of the examples $1 \times 54 = 2 \times 3^3$ and $1 \times 486 = 2 \times 3^5$. Further, non-trivial lower bounds for D_1 and $d \geq \theta^{-1} D_1$ have been given under the assumptions of Theorem 4; it is shown in Saradha & Shorey (2001a) that

$$\begin{array}{ll} D_1 > 1.59\theta k^{\frac{\ell}{2}-3-\frac{5}{2\ell}} & \text{for } \ell \geq 17, \\ D_1 > 1.1\theta k^{43/13} & \text{for } \ell = 13, \\ D_1 > .93\theta k^{25/11} & \text{for } \ell = 11, \\ D_1 > .73\theta k^{9/7} & \text{for } \ell = 7, \\ D_1 > .6\theta k^{7/5} & \text{for } \ell = 5, 5 \mid d, \\ D_1 > .65\theta k^{1/5} & \text{for } \ell = 5, 5 \nmid d \\ D_1 > .41\theta k^{1/3} & \text{for } \ell = 3 \end{array}$$

where

$$\theta = \begin{cases} 1 \text{ if } \ell \nmid d \\ 1/\ell \text{ if } \ell \mid d. \end{cases}$$

The first result on Theorems 3 and 4 appeared in Shorey (1988) where it is shown that (1) with $r = 0$ and $d > 1$ implies that

$$D_1 > 1 \text{ if } k \geq C_1$$

where C_1 and the subsequent letters $C_2, \ldots, C_5$ are effectively computable absolute constants. Further, Shorey & Tijdeman gave lower bounds for D_1 and d which are non-trivial only for large values of k. For example, it is shown in Shorey & Tijdeman (1990, 1992) that

$$d \geq k^{C_2 \log\log k} \tag{4}$$

whenever (1) with $r = 0$ and $d > 1$ holds. A conjecture on (1) states

Conjecture 1 (Erdős) *Let* $d > 1$. *Equation* (1) *with* $r = 0$ *implies that* $k \leq C_3$.

Shorey (1999a) applied (4) to show that the *abc* conjecture implies the above conjecture of Erdős for $\ell > 3$. We give some details of the proof. We may assume that $k \geq C_3$ where C_3 is sufficiently large. By (1), we write

$$n + id = A_i X_i^{\ell} \text{ for } 0 \leq i < k$$

where $P(A_i) \leq k$ and $\gcd\left(\prod_{p \leq k} p, X_i\right) = 1$. By a fundamental argument of Erdős mentioned in the proof of Theorem 1, we find positive integers $f < g < h$ such that

$$\max\,(A_f, A_g, A_h) \leq k^2.$$

We have

$$(g-f)(n+hd)+(h-g)(n+fd)=(h-f)(n+gd)$$

i.e.

$$(g-f)A_hX_h^\ell+(h-g)A_fX_f^\ell=(h-f)A_gX_g^\ell.$$

Now we observe that max $(X_f, X_h) < kX_g$ and we conclude from the *abc* conjecture that

$$n+gd=A_gX_g^\ell \le k^{C_4}X_g^4 \le k^{C_4}(A_gX_g^\ell)^{4/\ell}$$

which, since $\ell \ge 5$, implies that $n+gd \le k^{C_5}$ contradicting (4) if C_3 is sufficiently large.

Theorem 2 is on (1) with $r=0$, $d=1$ and Theorems 3 and 4 are on $r \in \{0,1\}$, $d>1$. Thus it remains to consider (1) with $r=d=1$ where we prove the following result.

Theorem 5 (Saradha & Shorey 2001a) *Equation* (1) *with* $r=d=b=1$ *is possible only if*

$$2\times 4=2^3, \quad 1\times 2\times 4=2^3.$$

This answers a question of Erdős & Selfridge (1975) p. 300. An analogue of Theorem 5 for $b>1$ is as follows.

Theorem 6 (Hanrot, Saradha & Shorey 2001) *Equation* (1) *with* $r=d=1$, $k\ge 6$ *and* (3) *does not hold. This is also the case for* $k=3,5$ *if* $P(b)<k$.

The cases $k=3,4,5$ if $P(b)\le k$ and $k=4$ if $P(b)<k$ remain open. It is possible to settle these cases if we solve a more general equation than the one dealt by Ribet, Darmon and Merel, for example, an equation of the form $Ax^\ell+By^\ell+Cz^\ell=0$ with $P(ABC)\le 3$. The proofs of Theorems 5 and 6 depend on the elementary method of Erdős & Selfridge and the contributions of Wiles, Ribet and others on Fermat equation. Further, Baker's method has been applied to find all the integral solutions of several Thue equations in the proof of Theorem 6. For applying Baker's method, we must keep a check on the degree and the coefficients of Thue equations. A check on the degree is given by the method of Erdős & Selfridge and a check on the coefficients is possible by the contributions on Fermat equation. The contributions on Fermat equation referred above have been applied via Saradha & Shorey (2001a), Lemma 13, that the above equation has no solution in non-zero integers X, Y, Z with

$\gcd(AX^\ell, BY^\ell, CZ^\ell) = 1$ whenever one of the terms in the equation is divisible by 16 and either $P(ABC) \leq 3$ or A, B, C are composed only of 2 and 5. This formulation has come from the work of Sander (1999).

If k is sufficiently large, Shorey (1986, 1987) showed that the assumption $r \in \{0, 1\}$ in Theorem 6 can be relaxed to $r \leq 9k/56$. Thus (1) with $r \leq 9k/56$, $d = 1$ and (3) implies that k is bounded by an absolute constant. Nesterenko & Shorey (1996) replaced $9k/56$ by $.51k$ if $\ell \geq 7$. For a more precise formulation of these results, we refer to the papers. The proofs depend on the theory of linear forms in logarithms, irrationality measures of Baker proved by hypergeometric method and the method of Roth & Halberstam on difference between consecutive ν-free integers. Here linear forms in logarithms with α_is close to 1 appear and the best possible estimates for these linear forms are crucial for the proof. The study of these special linear forms in logarithms with α_is close to 1 was initiated by the author in Shorey(1974). By integrating the auxiliary function on a circle of large radius, the author showed that it is possible to obtain the best possible lower bounds for linear forms in logarithms with α_is very close to 1. These special linear forms in logarithms find several applications and we refer to Shorey (1999a, b) for an account. Further, it was shown for the first time by the author in Shorey (1986) that lower bounds for linear forms in logarithms with α_is close to 1 combine well with the estimates given by hypergeometric method. This approach has led to several results and we refer again to Shorey (1999a, b) for a survey.

Finally, I would like to give an idea of the proof of Theorem 4. Suppose that the assumptions of Theorem 4 are satisfied and let $D_1 = 1$. By Theorem 1 and $P(b) \leq k$, we derive from (1) that $p^\ell \mid (n+id)$ for some i with $0 \leq i < k$ and for some prime $p > k$. Thus

$$n + (k-1)d \geq n + id \geq p^\ell > k^\ell$$

and we put

$$\delta = \frac{n + (k-1)d}{k^{\ell+1}}.$$

Then

$$\delta > \frac{1}{k}.$$

By (1), we have

$$n + d_i d = a_i x_i^\ell, \ \ P(a_i) \leq k, \quad a_i \text{ is } \ell\text{th power-free for } 1 \leq i \leq t.$$

Now we state the following result which is crucial to the proof of Theorem 4 and we refer to Saradha & Shorey (2001a) for its proof.

Lemma 1 *Let* $1 \leq \ell' \leq \ell - 1$, $\kappa > 0$ *and*

$$\kappa_0 = \min\left(\frac{\ell}{\ell'(\kappa+1)^{(\ell-\ell')/\ell}}, \frac{\kappa}{(\kappa+1)\ell'^{1/\ell}}\right).$$

Assume (1) *and*

$$D_1 \leq \kappa_0\, \theta\, \delta^{(\ell-\ell')/\ell} k^{\ell-\ell'-\frac{\ell'}{\ell}}. \tag{5}$$

Then for no distinct ℓ'*-tuples* $(i_1, \ldots, i_{\ell'})$ *and* $(j_1, \ldots, j_{\ell'})$ *with* $i_1 \leq i_2 \leq \cdots \leq i_{\ell'}$ *and* $j_1 \leq j_2 \leq \cdots \leq j_{\ell'}$*, the ratio of two products* $a_{i_1} \cdots a_{i_{\ell'}}$ *and* $a_{j_1} \cdots a_{j_{\ell'}}$ *is an* ℓth *power of a rational number.*

We recall that $D_1 = 1$. If $k > 11380$, we apply the Lemma with $\ell' = 2$ to conclude that $a_i a_j$ are distinct and we show that this is not possible. Thus $k \leq 11380$. Suppose that (5) is satisfied for a suitable value of ℓ' and κ. The improvement in the estimate $\delta > 1/k$ is necessary to secure that (5) is not very restrictive. Then the assertion of the Lemma is valid. This is not possible by a counting argument of Erdős & Selfridge. This is the case when $\ell \geq 7$. Thus we may assume that $\ell = 3, 5$ and (5) is not satisfied. It turns out that we need to consider only the cases $\ell = 5$, $k = 4$ and $\ell = 3$, $k \leq 70$. Since (5) is not satisfied, we obtain an upper bound for $\delta \leq \delta_0$ i.e. $n + (k-1)d \leq \delta_0 k^{\ell+1}$ in these cases and they are excluded by computations.

2 Squares

In this section, we consider (1) with $t \geq 3$, $P(b) \leq k$ and $\ell = 2$ i.e.

$$(n + d_1 d) \cdots (n + d_t d) = by^2. \tag{6}$$

Let $d = 1$ and $t = k$. If $P(b) < k$ and $k \geq 3$, Erdős & Selfridge (1975), developing on the work of Erdős (1939) and Rigge (1939), proved that (6) with (3) does not hold. Saradha (1997), (1998) relaxed the assumption $P(b) < k$ to $P(b) \leq k$ unless $(n, k) = (48, 3)$ in which case (6) is valid. Let $d > 1$ and $t = k$. Shorey & Tijdeman (1990b) showed that (6) is not possible whenever k exceeds an effectively computable number depending only on $\omega(d)$. They also showed that the assertion continues to be valid for (6) with $l > 2$ and, as pointed out in Shorey (1999a), the assumption $\gcd(n, d) = 1$ can be relaxed to $d \nmid n$. Further, Saradha & Shorey (2001b) gave infinitely many explicit values of d including all prime powers for which (6) can be solved completely. More precisely, they showed that (6) with $P(b) < k$ has no solution other than $(n, d, k, b, y) = (1, 24, 3, 1, 35)$ whenever d is given by 2^α or χp^α, $1 < \chi \leq 12$, $\chi \neq 11$, p prime, $\gcd(\chi, p^\alpha) = 1$. We suppose that $k \geq 4$ if $\chi = 7$, $p \neq 2$ in the preceding result and we refer to the paper for its formulation under more relaxed assumption $\gcd(n, \chi) = 1$ than $\gcd(n, d) = 1$. Furthermore, they proved that a product of four or more terms in an arithmetic progression with common difference a prime power is never a square. This is also not of the form by^2 with $P(b) < k$ under the necessary assumption that the common difference is not divisible by the initial term. The preceding assertion is proved in Shorey & Saradha (2001b) for $k > 9$ and in Mukhopadhyay & Shorey (2001) for $4 \leq k \leq 9$. These results imply that all the solutions of (6) with $1 < d \leq 104$ are given by

$$(n, d) \in \{(2, 7), (18, 7), (64, 17), (2, 23), (4, 23), (75, 23), (98, 23), (338, 23), (3675, 23), (1, 24), (800, 41), (2, 47), (27, 71), (50, 71), (96, 73), (864, 97)\}$$

if $k = 3$ and $(n, d) = (75, 23)$ if $k = 4$. This implies the results of Saradha (1998) and Filakovszky & Hajdu (2001) where all the solutions of (6) with $d \leq 22$ and $23 \leq d \leq 30$, respectively, have been determined. The result of Filakovszky & Hajdu utilises SIMATH package on solving elliptic equations in integers and this is useful for subsequent investigations.

Let $t = k - 1$. We may assume that $d_1 = 0$ and $d_t = k - 1$. First we suppose that $d = 1$. Then Saradha & Shorey (2001c) confirmed a conjecture of Erdős & Selfridge (1975), p. 300 analogous to Theorem 5 that there is no square other than $12^2 = \frac{6!}{5}$ and $720^2 = \frac{10!}{7}$ such that it can be written as product of $k - 1$ integers out of k consecutive positive integers. This follows from a more general result that (6) with (3) implies that $(n, k, y, b) = (24, 4, 90, 2)$. This is analogous to Theorem 6 for $l = 2$. Let $d > 1$. Then Saradha & Shorey (2001b) showed that (6) with infinitely many d listed above in this section implies that either $(n, d, k) = (1, 8, 4), (1, 40, 4), (25, 48, 4)$ or $d \in \{p^\alpha, 5p^\alpha, 7p^\alpha\}$ with $p > 2$ prime and $\alpha > 0$ such that $k \leq 29$ if $d = p^\alpha$ and $k \leq 5$ if $d = 5p^\alpha, 7p^\alpha$. This has been applied to solve (6) completely for $d \leq 67$.

References

Darmon, H. & L. Merel (1997), Winding quotient and some variants of Fermat's Last Theorem, *Jour. Reine Angew. Math.* **490**, 81–100.

Erdős, P. (1939a), Note on the product of consecutive integers (I), *Jour. London Math. Soc.* **14**, 194–198.

Erdős, P. (1939b), Note on the product of consecutive integers (II), *Jour. London Math. Soc.* **14**, 245–249.

Erdős, P. (1951), On a diophantine equation, *Jour. London Math. Soc.* **26**, 176–178.

Erdős, P. (1955), On the product of consecutive integers III, *Indag. Math.* **17**, 85–90.

Erdős, P. & J.L. Selfridge (1975), The product of consecutive integers is never a power, *Illinois Jour. Math.* **19**, 292–301.

Filakovszky, P. & L. Hajdu (2001), The resolution of the diophantine equation $x(x+d)\cdots(x+(k-1)d) = by^2$ for fixed d, *Acta Arith.* **98**, 151–154.

Győry, K. (1997), On the diophantine equation $\binom{n}{k} = x^\ell$, *Acta Arith.* **80**, 289–295.

Győry, K. (1998), On the diophantine equation $n(n+1)\cdots(n+k+1) = bx^\ell$, *Acta Arith.* **83**, 87–92.

Győry, K. (1999), Power values of products of consecutive integers and binomial coefficients. In *Number Theory and its Applications*, S. Kanemitsu & K. Győry (eds.), Kluwer, 145–156.

Hanrot, G., N. Saradha & T.N. Shorey (2001), Almost perfect powers in consecutive integers, *Acta Arith.*, **99**, 13–25.

Langevin, M. (1977), Plus grand facteur premier d'entiers en progression arithmétique, *Sém. Delange–Poitou,* **18** année, 1976/77, No. 3, 6pp.

Mukhopadhyay, Anirban & T.N. Shorey (2001), Almost squares in arithmetic progression (II), to appear.

Ribet, K.A. (1997), On the equation $a^p + 2^\alpha b^p + c^p = 0$, *Acta Arith.* **79**, 7–16.

Rigge, O. (1939), Über ein diophantisches Problem. In *9th Congress Math. Scand. Helsingfors, 1938*, Mercator, 155–160.

Sander, J.W. (1999), Rational points on a class of super elliptic curves, *J. London Math. Soc.* **59**, 422–434.

Saradha, N. (1997), On perfect powers in products with terms from arithmetic progressions, *Acta Arith.* **82**, 147–172.

Saradha, N. (1998), Squares in products with terms in an arithmetic progression, *Acta Arith.* **86**, 27–43.

Saradha, N. & T.N. Shorey (2001a), Almost perfect powers in arithmetic progression, *Acta Arith.*, **99**, 363–388.

Saradha, N. & T.N. Shorey (2001b), Almost squares in arithmetic progression, *Compositio Math.*, to appear.

Saradha, N. & T.N. Shorey (2001c), Almost squares and factorisations in consecutive integers, *Compositio Math.*, to appear.

Saradha, N., T.N. Shorey & R. Tijdeman (2002), Extensions and improvements of a theorem of Sylvester, *Acta Arith.*, to appear.

Shorey, T.N. (1974), Linear forms in the logarithms of algebraic numbers with small coefficients I, *Jour. Indian Math. Soc.* **38**, 271–284.

Shorey, T.N. (1986), Perfect powers in values of certain polynomials at integer points, *Math. Proc. Camb. Philos. Soc.* **99**, 195–207.

Shorey, T.N. (1987), Perfect powers in products of integers from a block of consecutive integers, *Acta Arith.* **49**, 71–79.

Shorey, T.N. (1988), Some exponential diophantine equations. In *New Advances in Transcendence Theory*, A. Baker (ed.), Cambridge University Press, 352–365.

Shorey, T.N. (1999a), Exponential diophantine equations involving products of consecutive integers and related equations. In *Number Theory*, R.P. Bambah, V.C. Dumir & R.J. Hans-Gill (eds.), Hindustan Book Agency, 463–495.

Shorey, T.N. (1999b), Mathematical Contributions, *Bull. Bombay Math. Colloq.* **15** (1999), 1–19.

Shorey, T.N. & Yu.V. Nesterenko (1996), Perfect powers in products of integers from a block of consecutive integers (II), *Acta Arith.* **76**, 191–198.

Shorey, T.N. & R. Tijdeman (1990a), On the greatest prime factor of an arithmetical progression. In *A Tribute to Paul Erdős*, A. Baker, B. Bollobás & A. Hajnal (eds.), Cambridge University Press, 385–389.

Shorey, T.N. & R. Tijdeman (1990b), Perfect powers in products of terms in an arithmetical progression, *Compositio Math.* **75**, 307–344.

Shorey, T.N. & R. Tijdeman (1992), Perfect powers in products of terms in an arithmetical progression (II), *Compositio Math.* **82**, 119–136.

Shorey, T.N. & R. Tijdeman (1997), Some method of Erdős applied to finite arithmetic progressions. In *The Mathematics of Paul Erdős I*, R.L. Graham & J. Nešetřil (eds.), Springer, 251–267.

Sylvester, J.J. (1892), On arithmetical series, *Messenger Math.* **21**, 1–19; 87–120.

Tijdeman, R. (1988), Diophantine equations and Diophantine approximations. In *Number Theory and Applications*, Richard A. Mollin (ed.), Kluwer, 215–233.

21

On the Greatest Common Divisor of Two Univariate Polynomials, I

A. Schinzel

P. Weinberger proposed at the West Coast Number Theory Meeting in 1976 the following problem. Does there exist a function $A(r, s)$ such that if polynomials f, g have exactly r and s non-zero coefficients, respectively, then the greatest common divisor (f, g) has at most $A(r, s)$ non-zero coefficients? We are going to study this problem in the case where $f, g \in K[x]$ and K is a field. Accordingly, we denote by $A(r, s, K)$ the supremum of the number of non-zero coefficients of (f, g), where f, g run over all univariate polynomials over K with r and s non-zero coefficients, respectively. Clearly, $A(r, s, K) = A(s, r, K)$, hence we may assume $r \le s$ and trivially $A(1, s, K) = 1$. We shall denote by K_0 the prime field of K, by p its characteristic, by $\overline{K}$ its algebraic closure and by ${}^p\zeta_q$ a generator of the group of qth roots of unity in $\overline{K}$. We set $K^q = \{a^q : a \in K\}$. Moreover, for a Laurent polynomial F over K,

$$F(x_1, \dots, x_k) = F_0(x_1, \dots, x_k) \prod_{i=1}^{k} x_i^{\alpha_i},$$

where $F_0 \in K[x_1, \dots, x_k]$ is prime to $\prod_{i=1}^{k} x_i$, we set

$$JF = F_0.$$

We shall prove the following two theorems.

Theorem 1 *If m, n, q are positive integers with $(m, n, q) = 1$ and $a, b, c \in K^*$, then $(x^n + ax^m + b,\ x^q - c)$ is of degree at most 1, if $a^{-n/(m,n)} b^{(n-m)/(m,n)} \notin K_0({}^p\zeta_q)$ and of degree 0, if, additionally, $c \notin K^q$. Moreover, if $p = 0$ or $p > 6^{\varphi(q)}$, then $(x^n + ax^m + b,\ x^q - c)$ is of degree at most 2 and of degree 0, if, as well, $c^2 \notin K^q$.*

Theorem 2 *If* $1 < r \le s$ *and* $\langle r, s, p\rangle \ne \langle 3, 3, 0\rangle$ *then*

$$A(r, s, K) = \begin{cases} 2, & \text{if } r = s = 2, \\ 3, & \text{if } r = 2,\ s = 3,\ p = 0, \\ \infty, & \text{otherwise.} \end{cases}$$

The case $\langle r, s, p\rangle = \langle 3, 3, 0\rangle$ has been studied in Schinzel (2001).

Lemma 1 *Let* z_i, $(1 \le i \le 4)$ *be roots of unity in* $\overline{K}$ *such that* $z_i^q = 1$, *and*

$$\begin{vmatrix} 1 & 1 & 1 \\ z_1 & z_2 & 1 \\ z_3 & z_4 & 1 \end{vmatrix} = 0. \tag{1}$$

If either $p = 0$ *or* $p > 6^{\varphi(q)}$, *then either two rows or two columns of the determinant are equal.*

Proof In the case $p = 0$ this is Lemma 9 of Győry & Schinzel (1994). The proof outlined there was by a tedious consideration of cases. J. Browkin has supplied the following proof for $p = 0$ (it is enough to take $K = \mathbb{C}$), which is no longer tedious and works for arbitrary unimodular z_i (cf. Schlickewei & Wirsing 1997, Corollary 3.3). The equation (1) gives

$$(z_1 - 1)(z_4 - 1) = (z_2 - 1)(z_3 - 1). \tag{2}$$

If $z_1 = 1$, then $z_2 = 1$ and the rows $1, 2$ are equal, or $z_3 = 1$ and the columns $1, 3$ are equal. Similarly, if $z_i = 1$ for $i \le 4$. If $z_i \ne 1$ for all i we take the complex conjugates of both sides of (2) and obtain

$$z_1^{-1} z_4^{-1}(z_1 - 1)(z_4 - 1) = z_2^{-1} z_3^{-1}(z_2 - 1)(z_3 - 1), \tag{3}$$

hence, dividing side by side (2) and (3), we get

$$z_1 z_4 = z_2 z_3. \tag{4}$$

The formulae (2) and (4) give

$$z_1 + z_4 = z_2 + z_3, \tag{5}$$

while (4) and (5) give either $z_1 = z_3$, $z_2 = z_4$ (the rows 2 and 3 are equal), or $z_1 = z_2$, $z_3 = z_4$ (the columns 1 and 2 are equal).

The case $p > 6^{\varphi(q)}$ is reduced to the case $p = 0$ as follows. Let $\mathfrak{p}$ be a prime ideal factor of p in $\mathbb{Q}({}^0\zeta_q)$. The residues $\bmod \mathfrak{p}$ form a subfield of $\overline{K}$ containing q distinct zeros of $x^q - 1$, since $p \nmid q$, represented by residues of ${}^0\zeta_q^r$ $(0 \le r < q)$. Hence

$$z_i \equiv {}^0\zeta_q^{r_i} \bmod \mathfrak{p}\ (1 \le i \le 4) \tag{6}$$

and equation (1) gives

$$D := \begin{vmatrix} 1 & 1 & 1 \\ {}^0\zeta_q^{r_1} & {}^0\zeta_q^{r_2} & 1 \\ {}^0\zeta_q^{r_3} & {}^0\zeta_q^{r_4} & 1 \end{vmatrix} \equiv 0 \,(\mathrm{mod}\,\mathfrak{p}); \quad N_{\mathbb{Q}({}^0\zeta_q)/\mathbb{Q}} D \equiv 0 \,(\mathrm{mod}\,\mathfrak{p}). \tag{7}$$

However D is the sum of six complex roots of unity. Hence each conjugate of D over $\mathbb{Q}$ does not exceed 6 in absolute value and

$$\left| N_{\mathbb{Q}({}^0\zeta_q)/\mathbb{Q}} D \right| \le 6^{\varphi(q)} < p.$$

Since D is an algebraic integer, $N_{\mathbb{Q}({}^0\zeta_q)/\mathbb{Q}} D$ is an integer and the above inequality together with the second congruence of (7) gives

$$N_{\mathbb{Q}({}^0\zeta_q)/\mathbb{Q}} D = 0; \;\; D = 0.$$

By the already settled case $p = 0$ the determinant defining D has two rows or two columns equal and by (6) the same applies to the determinant

$$\begin{vmatrix} 1 & 1 & 1 \\ z_1 & z_2 & 1 \\ z_3 & z_4 & 1 \end{vmatrix}.$$

□

Proof of Theorem 1. Let $(n, m) = d$, $n = dn'$, $m = dm'$, $(x^n + ax^m + b,\ x^q - c)$ be of degree δ and assume first that

$$a^{-n'} b^{n'-m'} \notin K_0\left({}^p\zeta_q\right) \tag{8}$$

and

$$\delta \ge 2. \tag{9}$$

If $(x^n + ax^m + b,\ x^q - c)$ has a multiple zero in $\overline{K}$, then $p > 0$, $p \mid q$ and since $(n, m, q) = 1$, $p \nmid d$. Moreover,

$$\Delta := \operatorname{disc}\left(x^n + ax^m + b\right) = 0. \tag{10}$$

However (see Lefton 1979)

$$\Delta = (-1)^{\frac{1}{2}n(n-1)} b^{m-1} \left(n^{n'} b^{n'-m'} + (-1)^{n'-1} (n-m)^{n'-m'} m^{m'} a^{n'}\right)^d. \tag{11}$$

It follows from $p \nmid d$

$$a^{-n'} b^{n'-m'} = (-1)^{n'} (n-m)^{n'-m'} m^{m'} n^{-n'} \in K_0,$$

contrary to (8). Thus, by (9), $(x^n + ax^m + b,\ x^q - c)$ has two distinct zeros in $\overline{K}$. Denoting them by ξ_i $(i = 1, 2)$ we have for $i = 1, 2$, $\xi_i^q = c$ and

$$\xi_i^n + a\xi_i^m + b = 0. \tag{12}$$

If $\xi_1^m = \xi_2^m$, then also $\xi_1^n = \xi_2^n$, and, since $\xi_1^q = \xi_2^q$, it follows from $(m, n, q) = 1$ that $\xi_1 = \xi_2$, a contradiction. Thus $\xi_1^m \neq \xi_2^m$ and solving the system (12) for a, b we find

$$a = \frac{\xi_2^n - \xi_1^n}{\xi_2^m - \xi_1^m}, \quad b = \frac{\xi_1^n \xi_2^m - \xi_1^m \xi_2^n}{\xi_2^m - \xi_1^m}.$$

Since $\xi_2 = {}^p\zeta_q^r \xi_1$ for a certain r, it follows that ${}^p\zeta_q^{rm} \neq 1$ and

$$a = \xi_1^{n-m} \frac{{}^p\zeta_q^{rn} - 1}{1 - {}^p\zeta_q^{rm}}, \quad b = \xi_1^n \frac{{}^p\zeta_q^{rm} - {}^p\zeta_q^{rn}}{1 - {}^p\zeta_q^{rm}}; \tag{13}$$

$$\begin{aligned} &a^{-n'} b^{n'-m'} \\ &= \left(1 - {}^p\zeta_q^{rm}\right)^{m'} \left({}^p\zeta_q^{rm} - {}^p\zeta_q^{rn}\right)^{n'-m'} \left({}^p\zeta_q^{rn} - 1\right)^{-n'} \in K_0\left({}^p\zeta_q\right), \end{aligned} \tag{14}$$

contrary to (8). Thus (8) implies that $\delta \leq 1$. If $\delta \neq 0$, then

$$\left(x^n + ax^m + b,\ x^q - c\right) = x - \xi,\ \xi \in K,$$

hence $c = \xi^q \in K^q$.

It remains to consider the case where $p = 0$ or $p > 6^{\varphi(q)}$. Then $x^q - c$ has no multiple zeros and $\delta \geq 3$ implies the existence of three distinct zeros ξ_i of $x^q - c$ such that (12) holds for $i = 1, 2, 3$. Putting $z_1 = \left(\frac{\xi_2}{\xi_1}\right)^n$, $z_2 = \left(\frac{\xi_2}{\xi_1}\right)^m$, $z_3 = \left(\frac{\xi_3}{\xi_1}\right)^n$, $z_4 = \left(\frac{\xi_3}{\xi_1}\right)^m$ we can rewrite the system (12) in the form

$$\begin{aligned} \xi_1^n + a\xi_1^m + b &= 0, \\ z_1\xi_1^n + z_2 a\xi_1^m + b &= 0, \\ z_3\xi_1^n + z_4 a\xi_1^m + b &= 0, \end{aligned} \tag{15}$$

hence

$$\begin{vmatrix} 1 & 1 & 1 \\ z_1 & z_2 & 1 \\ z_3 & z_4 & 1 \end{vmatrix} = 0,$$

and by Lemma 1, either two rows or two columns of the determinant are equal. If two rows are equal we infer from $(m, n, q) = 1$ that $\xi_2 = \xi_1$, or $\xi_3 = \xi_1$, or $\xi_3 = \xi_2$, a contradiction. If two columns are equal, then equations (15) imply, since $ab \neq 0$ that $z_1 = z_2 = z_3 = z_4 = 1$, hence $\xi_3 = \xi_2 = \xi_1$, again a contradiction.

Hence $\delta \leq 2$. If $\delta = 1$, $(x^n + ax^m + b,\ x^q - c) = x - \xi$, where $\xi \in K$ and $c = \xi^q \in K^q$. If $\delta = 2$, $(x^n + ax^m + b,\ x^q - c) = (x - \xi_1)(x - \xi_2)$, hence $[K(\xi_1) : K] \leq 2$ and $\xi_1^q = c$ implies $\left(N_{K(\xi_1)/K}\xi_1\right)^q = N_{K(\xi_1)/K}c = c$ or c^2; $c^2 \in K^q$.

Lemma 2 *Let $0 = a_0 < a_1 < \cdots < a_r$ and $0 = b_0 < b_1 < \cdots < b_s$ be integers and set*

$$R(t) = \sum_{t=a_i+b_j} 1.$$

If there exist at most two positive integers t such that $R(t) = 1$, then there exist $l \leq 2$ integers u_j, $(1 \leq j \leq l)$ such that

$$a_i = \sum_{j=1}^{l} \alpha_{ij} u_j \quad (0 \leq i \leq r), \qquad b_i = \sum_{j=1}^{l} \beta_{ij} u_j \quad (0 \leq i \leq s),$$

where α_{ij}, β_{ij} are integers and

$$\prod_{j=1}^{l} \max\left\{\max_{0\leq i\leq r}|\alpha_{ij}|,\ \max_{0\leq i\leq s}|\beta_{ij}|\right\} \leq 2^{r+s-l}.$$

Proof Clearly, we have

$$R(a_r + b_s) = 1,$$

thus, by the assumption, there exists at most one pair $\langle r_1, s_1\rangle \neq \langle 0, 0\rangle,\ \langle r, s\rangle$ such that

$$R\left(a_{r_1} + b_{s_1}\right) = 1.$$

If $0 \leq i \leq r$, $0 \leq j \leq s$ and $\langle i, j\rangle \neq \langle 0, 0\rangle,\ \langle r, s\rangle, \langle r_1, s_1\rangle$, there exists a pair $\langle g_{ij}, h_{ij}\rangle \neq \langle i, j\rangle$ such that

$$a_i + b_j = a_{g_{ij}} + b_{h_{ij}}. \tag{16}$$

Let us consider the system of equations for $r+s+2$ unknowns x_i $(0 \leq i \leq r)$, y_j $(0 \leq j \leq s)$:

$$\begin{aligned} x_0 &= 0, \\ y_0 &= 0, \\ x_r + y_s &= 0, \\ x_{r_1} + y_{s_1} &= 0, \\ x_i + y_j - x_{g_{ij}} - y_{h_{ij}} &= 0 \quad (\langle i, j\rangle \neq \langle 0, 0\rangle,\ \langle r, s\rangle,\ \langle r_1, s_1\rangle). \end{aligned} \tag{17}$$

We assert that the system has only the zero solution. Indeed, suppose that $\langle c_0, \dots, c_r, d_0, \dots, d_s\rangle$ is a solution of this system and let

$$\begin{aligned} i_1 \text{ be the least } i \text{ such that } c_i &= \min c_k, \\ i_2 \text{ be the least } i \text{ such that } c_i &= \max c_k, \\ j_1 \text{ be the least } j \text{ such that } d_j &= \min d_k, \\ j_2 \text{ be the least } j \text{ such that } d_j &= \max d_k. \end{aligned}$$

If for $\nu = 1$ or 2 we have $\langle i_\nu, j_\nu\rangle \neq \langle 0, 0\rangle,\ \langle r, s\rangle,\ \langle r_1, s_1\rangle$ let

$$g_\nu = g_{i_\nu j_\nu}, \quad h_\nu = h_{i_\nu j_\nu}.$$

The equations (17) give

$$c_{i_\nu} + d_{j_\nu} = c_{g_\nu} + d_{h_\nu},$$

hence $c_{g_\nu} = c_{i_\nu}, d_{h_\nu} = d_{j_\nu}$; $g_\nu \geq i_\nu, h_\nu \geq j_\nu$ and since $\langle g_\nu, h_\nu\rangle \neq \langle i_\nu, j_\nu\rangle$ it follows that $a_{g_\nu} + b_{h_\nu} > a_{i_\nu} + b_{j_\nu}$, contrary to (16). Therefore, $\langle i_\nu, j_\nu\rangle \in \{\langle 0, 0\rangle,\ \langle r, s\rangle,\ \langle r_1, s_1\rangle\}$ for $\nu \leq 2$ and thus

$$c_{i_\nu} + d_{j_\nu} = 0 \quad (\nu = 1, 2).$$

However $c_{i_2} \geq c_{i_1}, d_{j_2} \geq d_{j_1}$, thus $c_{i_2} = c_{i_1}, d_{j_2} = d_{j_1}$ and, by the definition of c_{i_ν} and d_{j_ν}, all c_i are equal $(0 \leq i \leq r)$ and all d_j are equal $(0 \leq j \leq s)$. Since $c_0 = d_0 = 0$ we infer that $c_i = 0$ $(0 \leq i \leq r)$ and $d_j = 0$ $(0 \leq j \leq s)$. It follows from the proved assertion that the rank of the matrix of the system (17) is $r + s + 2$ and thus the rank of the matrix of the reduced system

$$\begin{aligned} x_0 &= 0, \\ y_0 &= 0, \\ x_i + y_j - x_{g_{ij}} - y_{h_{ij}} &= 0 \quad (\langle i, j\rangle \neq \langle 0, 0\rangle,\ \langle r, s\rangle,\ \langle r_1, s_1\rangle) \end{aligned} \tag{18}$$

is $r + s + 2 - l$, where $l \leq 2$. By (16) we have $l > 0$.

Let Δ be a submatrix of the matrix of the system (18) consisting of $r + s + 2 - l$ linearly independent rows. By Steinitz's lemma we may assume that the submatrix contains the first two rows. By the Bombieri–Vaaler theorem (Bombieri & Vaaler 1983, Theorem 2) there exists a system of l linearly independent integer solutions $\mathbf{v}_j$ $(j = l)$ of the equation

$$\mathbf{x}\Delta = 0 \tag{19}$$

satisfying the inequality

$$\prod_{j=1}^{l} h(\mathbf{v}_j) \leq \sqrt{\det \Delta\Delta^T},$$

where $h(\mathbf{v}_j)$ is the maximum of the absolute values of the coordinates of $\mathbf{v}_j$. However, by an inequality of Fischer generalizing Hadamard's inequality (see Bombieri & Vaaler 1983, formula (2.6)) $\sqrt{\det \Delta\Delta^T}$ does not exceed the product of the Euclidean lengths of the rows of Δ, i.e. 2^{r+s-l}.

Now, from the system $\mathbf{v}_j$ $(j \le l)$ of $l \le 2$ linearly independent integer solutions of the equation (19) one can obtain a basis $\mathbf{w}_j$ $(j \le l)$ of all integer solutions satisfying

$$h(\mathbf{w}_j) \le h(\mathbf{v}_j) \quad (j \le l)$$

(see Cassels 1959, Chapter V, Lemma 8). It suffices now to take

$$\mathbf{w}_j = \left[\alpha_{0j}, \dots, \alpha_{rj}, \beta_{0j}, \dots, \beta_{sj}\right] \quad (j \le l).$$

□

Remark In the same way one can prove the following generalization of Lemma 2. If, with the same notation, $R(t) = 1$ for at most k positive integers t, then there exist $l \le k$ integers u_j, $(1 \le j \le l)$ such that

$$a_i = \sum_{j=1}^{l} \alpha_{ij} u_j \ \ (0 \le i \le r), \qquad b_i = \sum_{j=1}^{l} \beta_{ij} u_j \ \ (0 \le i \le s),$$

where α_{ij}, β_{ij} are integers and

$$\prod_{j=1}^{l} \max\left\{\max_{0\le i\le r} |\alpha_{ij}|,\ \max_{0\le i\le s} |\beta_{ij}|\right\} \le 2^{r+s-l}\frac{(l+m+1)!}{4^{l-m}(2m+1)!},$$

where $m = \left[\frac{1+\sqrt{16l+7}}{4}\right]$.

Instead of a result quoted from Cassels (1959) one has to use an argument from Schinzel (1987), pp. 701–702, due essentially to H. Weyl (1942).

It is also possible to generalize Lemma 2 to the case of more than two increasing sequences of integers.

Lemma 3 *Let $a, b \in K^*$, $n > m > 0$. If $(n, m) \not\equiv 0 \bmod p$ and*

$$x^n + ax^m + b = g(x)h(x),$$

where $g, h \in K[x] \setminus K$ and g, h have exactly $r+1$ and $s+1$ non-zero coefficients, respectively, then

$$2^{r+s+3} + 1 \ge \frac{n}{(n, m)}. \tag{20}$$

Proof Let us put

$$g(x)=\sum_{i=0}^{r} g_i x^{a_i}, \quad h(x)=\sum_{j=0}^{s} h_j x^{b_j}, \tag{21}$$

where $0 < a_0 < a_1 < \cdots < a_r$, $0 < b_0 < b_1 < \cdots < b_s$ and $g_i \neq 0$ $(0 \le i \le r)$, $h_j \neq 0$ $(0 \le j \le s)$. In the notation of Lemma 2 for each positive integer $t \neq m, n$ we have $R(t) \neq 1$. Hence, by Lemma 2, there exist $l \le 2$ integers u_j $(1 \le j \le l)$ such that

$$a_i=\sum_{j=1}^{l} \alpha_{ij} u_j \ \ (0 \le j \le r), \qquad b_i=\sum_{j=1}^{l} \beta_{ij} u_j \ \ (0 \le j \le s)\,.$$

where α_{ij}, β_{ij} are integers and

$$\prod_{j=1}^{l} \max\left\{\max_{0\le i\le r} |\alpha_{ij}|,\ \max_{0\le i\le s} |\beta_{ij}|\right\} \le 2^{r+s-l}. \tag{22}$$

Clearly,

$$\begin{aligned} n &= \sum_{j=1}^{l} u_j\left(\alpha_{rj}+\beta_{sj}\right), \\ m &= \sum_{j=1}^{l} u_j\left(\alpha_{r'j}+\beta_{s'j}\right), \end{aligned} \tag{23}$$

where $0 \le r' \le r$, $0 \le s' \le s$.

If $l=1$, then $u_1 \mid (m,n)$ and by (22) and (23)

$$n \le (n,m)2^{r+s}$$

which is stronger than (20).

If $l=2$, let us put for $j=1,2$

$$\begin{aligned} \nu_j &= \alpha_{rj}+\beta_{sj}, \\ \mu_j &= \alpha_{r'j}+\beta_{s'j}, \end{aligned} \tag{24}$$

$$\begin{aligned} F(x_1,x_2) &= J\left(x_1^{\nu_1}x_2^{\nu_2}+ax_1^{\mu_1}x_2^{\mu_2}+b\right), \\ G(x_1,x_2) &= J\left(\sum_{i=0}^{r} g_i x_1^{\alpha_{i1}} x_2^{\alpha_{i2}}\right), \\ H(x_1,x_2) &= J\left(\sum_{i=0}^{s} h_i x_1^{\beta_{i1}} x_2^{\beta_{i2}}\right), \end{aligned}$$

the notation being explained in the introduction.

By (21) and (23), (24)

$$x^n + ax^m + b = JF\left(x^{u_1}, x^{u_2}\right), \tag{25}$$

$$g(x) = JG\left(x^{u_1}, x^{u_2}\right), \ h(x) = JH\left(x^{u_1}, x^{u_2}\right), \tag{26}$$

while, by (22)

$$\prod_{j=1}^{2} \max\left\{\left|\mu_j\right|, \left|\nu_j\right|\right\} \le 2^{r+s}. \tag{27}$$

It follows that

$$\begin{aligned} \deg_{x_j} F &\le \left|\mu_j\right| + \left|\nu_j\right| \le 2\max\left\{\left|\mu_j\right|, \left|\nu_j\right|\right\} \\ &\le 4\max\left\{\max_i \left|\alpha_{ij}\right|, \max_i \left|\beta_{ij}\right|\right\}, \\ \deg_{x_j} G &= \max_i \alpha_{ij} - \min_i \alpha_{ij} \le 2\max_i \left|\alpha_{ij}\right|, \\ \deg_{x_j} H &= \max_i \beta_{ij} - \min_i \beta_{ij} \le 2\max_i \left|\beta_{ij}\right|. \end{aligned} \tag{28}$$

If $\nu_1\mu_2 - \nu_2\mu_1 = 0$, then by (23) and (24)

$$\frac{u_1\nu_1 + u_2\nu_2}{(\nu_1, \nu_2)} \,\Big|\, (n, m),$$

hence, by (27), $n \le (n, m)(\nu_1, \nu_2) \le (n, m)2^{r+s/2}$, which is stronger than (20).

If $\nu_1\mu_2 - \nu_2\mu_1 \ne 0$, $F(x_1, x_2)$ is irreducible over K, by Theorem 23 of Schinzel (2000). Indeed, the only assumption of this theorem that needs to be verified is that $F(x_1, x_2)$ is not of the form cF_0^p, where $c \in K$, $F_0 \in K[x_1, x_2]$. If it were the case, we should have $\nu_j \equiv 0$, $\mu_j \equiv 0 \bmod p$, $(j = 1, 2)$, hence by (23) and (24) $(n, m) \equiv 0 \bmod p$, contrary to the assumption of the lemma.

If now $(F, G) \ne 1$, it follows by the irreducibility of F that $F \mid G$, hence, by (25) and (26), $x^n + ax^m + b \mid g(x)$ and by, (19), $h(x) \in K$, contrary to the assumption of the lemma. Therefore $(F, G) = 1$ and by Lemma 5 of Schinzel (1969) the number of solutions in $\overline{K}^2$ of the system of equations $F(x_1, x_2) = G(x_1, x_2) = 0$ does not exceed the degree of the resultant R of F and G with respect to x_1.

From the form of the resultant as the determinant of the Sylvester matrix we infer by (28) and (22)

$$\begin{aligned} \deg R &\le \deg_{x_1} F \cdot \deg_{x_2} G + \deg_{x_2} F \cdot \deg_{x_1} G \\ &\le 16\prod_{j=1}^{2} \max\left\{\max_i \left|\alpha_{ij}\right|, \max_i \left|\beta_{ij}\right|\right\} \\ &\le 2^{r+s+2}. \end{aligned}$$

Thus the number of solutions in $\overline{K}^2$ of the system of equations $F(x_1, x_2) = G(x_1, x_2) = 0$ does not exceed 2^{r+s+2} and the same applies to the system $F(x_1, x_2) = H(x_1, x_2) = 0$. Since ξ^{u_1}, ξ^{u_2} determine the value of $\xi^{(u_1,u_2)}$, they give (u_1, u_2) possibilities for ξ. Hence the systems of equations $F(\xi^{u_1}, \xi^{u_2}) = G(\xi^{u_1}, \xi^{u_2})$ and $F(\xi^{u_1}, \xi^{u_2}) = H(\xi^{u_1}, \xi^{u_2})$ have each at most $(u_1, u_2)2^{r+s+2}$ distinct solutions in $\overline{K}^2$. In view of (19), (25) and (26) it follows that $x^n + ax^m + b$ has at most $2^{r+s+3}(n, m)$ distinct zeros in $\overline{K}^2$. Since each zero of $x^n + ax^m + b$ is at most double, and the number of double zeros is at most (m, n), we get

$$n - (m, n) \leq 2^{r+s+3}(m, n),$$

which gives the lemma. □

Lemma 4 *For every prime field $K_0 \neq \mathbb{F}_2$ and every integer $k > 1$ there exists a polynomial $f_k \in K_0[x]$ of degree at most k with exactly k non-zero coefficients, such that $f_k(0) = 1$, $f_k(1) = 0$ and $f_k'(1) \neq 0$. For $K_0 = \mathbb{F}_2$ such a polynomial exists, if k is even.*

Proof We set

$$f_k(x) = \sum_{i=0}^{k-1}(-1)^i x^i \quad \text{if } k \text{ is even, } k \not\equiv 0 \bmod 2p;$$

$$f_k(x) = \sum_{i=0}^{k-2}(-1)^i x^i - x^k \quad \text{if } k \text{ is even, } k \equiv 0 \bmod 2p;$$

$$f_k(x) = \sum_{i=0}^{k-3}(-1)^i x^i - 2x^{k-2} + x^{k-1} \quad \text{if } k \text{ is odd, } k \not\equiv 3 \bmod 2p;$$

$$f_k(x) = \sum_{i=0}^{k-3}(-1)^i x^i - 2x^{k-2} + x^k \quad \text{if } k \text{ is odd, } k \equiv 3 \bmod 2p.$$

□

Definition For convenience we set $f_1(x) = 0$.

Lemma 5 *For every $K \neq \mathbb{F}_2$, every $f \in K[x]$ and every positive integer k there exists a polynomial $h = h(x; k, f) \in K[x]$ with exactly k non-zero coefficients such that $(h(x^l), xf(x)) = 1$ for every positive integer l. For $K = \mathbb{F}_2$ such a polynomial exists if k is odd, and, moreover, with the weaker property $(h(x), xf(x)) = 1$ also if $f(1) \neq 0$.*

Proof If K contains $\mathbb{Q}$ or $\mathbb{F}_p(t)$ with t transcendental over $\mathbb{F}_p$, then the multiplicative group of K contains a free abelian group of infinite rank. Hence, denoting the zeros of f by $\xi_1, \ldots, \xi_n$ we can choose $a \in K^*$ such that for all $\nu \leq n$ and all l the element $a\xi_\nu^{-l}$ is not a root of unity, and then

$$h(x) = \frac{x^k - a^k}{x - a}$$

has the desired property.

If K contains neither $\mathbb{Q}$ nor $\mathbb{F}_p(t)$, then $K \subset \overline{\mathbb{F}}_p$, hence there exists an exponent $e > 0$ such that $\xi_\nu^e = 1$ for every $\xi_\nu \neq 0$ $(1 \leq \nu \leq n)$. Then we write $k = p^\kappa k_1$, where $(k_1, p) = 1$ and set

$$h(x) = \frac{x^{k_1 e} - 1}{x^e - 1} \text{ if } \kappa = 0, \tag{29}$$

$$h(x) = \left(\frac{x^{k_1 e} - 1}{x^e - 1}\right)^{p^\kappa} \left(x^e + a\right)^{p^\kappa - 1}, \quad \text{if } \kappa > 0,\ K \neq \mathbb{F}_2,\ a \in K \setminus \{0, -1\}, \tag{30}$$

$$h(x) = \left(\frac{x^{k_1 e} - 1}{x^e - 1}\right)^{2^\kappa} (x + 1)^{2^\kappa - 1}, \text{ if } \kappa > 0,\ K = \mathbb{F}_2,\ f(1) \neq 0. \tag{31}$$

It is easy to see that $h(x)$ has exactly k non-zero coefficients and in cases (29), (30) $h(\xi_\nu^l) \neq 0$, in case (31) $h(\xi_\nu) \neq 0$ for all $\nu \leq n$. □

Lemma 6 *If $n \equiv 1 \bmod 6$, over $\mathbb{F}_2$, then the trinomial*

$$T_n(x) = x^{2^{n+1}+1} + x^{2^n - 1} + 1$$

is the product of two non-constant factors, one of which divides $x^{2^{2n}-1} + 1$. and the other $x^{2^{3n}-1} + 1$; both are prime to $x^{2^n-1} + 1$.

Proof This is a special case of the result of Mills & Zierler (1969), the case admitting a shorter proof. Let $r = 2^n$. By the identity of Mills & Zierler

$$T_n\left(x^r\right) + x^{r^2 - r} T_n(x) = \left(x^{r^2-1} + 1\right)\left(x^{r^2+r+1} + 1\right),$$

hence every irreducible factor of $T_n(x)$ divides one of the relevant binomials. Since $T_n(x)$ has no multiple zeros, $T_n(1) \neq 0$ and 1 is the only common zero of the two binomials, we have

$$T_n(x) = \left(T_n(x),\ x^{r^2-1} + 1\right)\left(T_n(x),\ x^{r^2+r+1} + 1\right).$$

In order to show that the factors are non-constant let us observe that for $n \equiv 1 \pmod 6$

$$2r+1 \equiv 5 \pmod{21}, \ \ r-1 \equiv 1 \pmod{21},$$

$$x^5+x+1 = \left(x^2+x+1\right)\left(x^3+x^2+1\right)$$

and

$$x^2+x+1 \mid x^3+1 \mid x^{r^2-1}+1,$$

$$x^3+x^2+1 \mid x^7+1 \mid x^{r^2+r+1}+1 \mid x^{r^3-1}+1,$$

hence

$$x^2+x+1 \mid \left(T_n(x),\ x^{r^2-1}+1\right),$$

$$x^3+x^2+1 \mid \left(T_n(x),\ x^{r^2+r+1}+1\right).$$

Finally,

$$\left(T_n(x),\ x^{r-1}+1\right) \mid T_n(x)+x^{r-1}+1 = x^{2r+1},$$

hence

$$\left(T_n(x),\ x^{r-1}+1\right) = 1$$

and we can also write

$$T_n(x) = \left(T_n(x),\ x^{r^2-1}+1\right)\left(T_n(x),\ x^{r^3-1}+1\right).$$

□

Lemma 7 *If* $n \equiv 1 \bmod 6$ *and* $T_n(x) = x^{2^{n+1}+1} + x^{2^n-1} + 1 \in \mathbb{F}_2[x]$, *then there exists* $c = c(n) \in \{2, 3\}$ *such that* $\left(T_n(x),\ x^{2^{cn}-1}+1\right)$ *has at least* $n/2$ *non-zero coefficients.*

Remark If 2 and 3 both have the required property, we put $c(n) = 2$.

Proof For $n \equiv 1 \bmod 6$ we have $(2^{n+1}+1, 2^n-1) = 1$. Hence, denoting by $r(i,n)$ $(i = 2, 3)$ the number of non-zero coefficients of $(T_n(x), x^{2^{in}-1}+1)$, we have by Lemmas 3 and 6

$$2^{r(2,n)+r(3,n)+1} + 1 \geq 2^{n+1}+1,$$

hence $\max\{r(2,n), r(3,n)\} \geq n/2$. □

Proof of Theorem 2 Consider first the case $r = s = 2$. It is nearly obvious that if $a_1, a_2 \in K^*$ and n_1, n_2 are positive integers, then

$$\left(x^{n_1} - a_1,\ x^{n_2} - a_2\right) = \begin{cases} 1, & \text{if } a_1^{n_2/(n_1,n_2)} \neq a_2^{n_1/(n_1,n_2)} \\ x^{(n_1,n_2)} - c & \text{if } a_1^{n_2/(n_1,n_2)} = a_2^{n_1/(n_1,n_2)} \text{ and } a_i = c^{n_i/(n_1,n_2)}. \end{cases}$$

This proves that $A(2, 2, K) = 2$.

Consider next the case $r = 2$, $s = 3$, $p = 0$. By Theorem 1 we have $A(2, 3, K) \le 3$ and since $(x^3 - 1, x^2 + x + 1) = x^2 + x + 1$, $A(2, 3, K) = 3$. Therefore, we assume $\langle r, s\rangle \neq \langle 2, 2\rangle$, $\langle r, s, p\rangle \neq \langle 2, 3, 0\rangle$, $\langle 3, 3, 0\rangle$ and we have to prove $A(r, s, K) = \infty$.

Consider first the case $p \neq 2$.

If $r = 2$, $s = 3$, $p > 0$, we take $f(x) = x^{p^{(n-2)!}-1} - 1$, $g(x) = x^n - nx + n - 1$, where $n \not\equiv 0, 1 \bmod p$. The trinomial $g(x)$ has exactly one multiple zero in $\overline{K}$, namely $x = 1$ and this is a double zero. All other zeros are of degree at most $n - 2$, hence they are zeros of $f(x)$. Since 1 is not a multiple zero of this binomial, we obtain

$$(f, g) = \frac{x^n - nx + n - 1}{x - 1} = x^{n-1} + \cdots + 1 - n, \tag{32}$$

where on the right hand side we have n non-zero coefficients. Thus $A(2, 3, K) = \infty$.

If $r = 2$, $s \ge 4$, we take

$$f = x^{ab} - 1,\ g = \left(x^a - 1\right)\left(x^b - 1\right) + f_{s-3}\left(x^{ab}\right),$$

where $1 < a < b$, $(a, b) = 1$, $ab \not\equiv 0 \bmod p$ and f_{s-3} has the meaning of Lemma 4. Since $f \mid f_{s-3}(x^{ab})$ we have $(f, g) = (f, (x^a - 1)(x^b - 1))$. However f has no multiple zeros, and $(x^a - 1)(x^b - 1)$ has just one such zero, namely 1, which is a double zero. Hence

$$(f, g) = \frac{(x^a - 1)\left(x^b - 1\right)}{x - 1} = x^{b+a-1} + \cdots + x^b - x^{a-1} - \cdots - 1 \tag{33}$$

has $2a$ non-zero coefficients and we obtain $A(2, 3, K) = \infty$.

If $r = 3$, $s \ge 3$, $p > 0$, we take

$$f = x^n - nx + n - 1,\ g = f_s\left(x^{p^{(n-2)!}-1}\right).$$

Since $f \left| x^{p^{(n-2)!}-1} - 1 \right| f_s\left(x^{p^{(n-2)!}-1}\right)$ and $f_s'(1) \neq 0$ we have again (32), hence $A(r, s, K) = \infty$.

If $r = 3$, $s > 3$, $p = 0$, we take

$$f = x^{2ab} - 3x^{ab} + 2,\ g(x) = \left(x^a - 1\right)\left(x^b - 1\right) + f_{s-3}\left(x^{ab}\right),$$

where again $1 < a < b$, $(a, b) = 1$. We have $f = (x^{ab} - 1)(x^{ab} - 2)$. It follows from the irreducibility of $x^{ab} - 2$ over $\mathbb{Q}$ that

$$\left(x^{ab} - 2, \left(x^a - 1\right)\left(x^b - 1\right) + f_{s-3}(2)\right) = 1,$$

hence

$$\left(x^{ab} - 2, \left(x^a - 1\right)\left(x^b - 1\right) + f_{s-3}\left(x^{ab}\right)\right) = 1,$$

and we obtain again (33), thus $A(3, s, K) = \infty$.

If $r \geq 4$, $s \geq r$, we take

$$f = (x^a - 1)\left(x^b - 1\right) + f_{r-3}\left(x^{ab}\right),\ h = h\left(x; \left[\tfrac{s-r}{2}\right] + 1, f\right),$$

$$g = f(0)\left(x^{ab} - 1\right) h\left(x^{rab}\right) + dh(0) f(x),$$

where

$$d = \begin{cases} 2 & \text{if } s \equiv r + 1 \bmod 2, \\ 1 & \text{if } s \equiv r \bmod 2 \end{cases}$$

and obtain (33), hence $A(r, s, K) = \infty$.

Consider now $p = 2$.

If $r = 2$, $s \geq 3$, $s \equiv 0 \bmod 2$, we take

$$f = x^{ab} + 1,\ g = \left(x^a + 1\right)\left(x^b + 1\right) + f_{s-2}\left(x^{ab}\right),$$

where $1 < a < b$, $(a, b) = 1$, $ab \not\equiv 0 \bmod 2$, and obtain (33), hence $A(2, 3, K) = \infty$.

If $r = 2$, $s \geq 3$, $s \equiv 1 \bmod 2$, we take

$$f = x^{2^{cn}-1} + 1,\ g = T_n(x) + f_{s-1}\left(x^{2^{cn}-1}\right),$$

where $n \equiv 1 \bmod 6$ and $c = c(n)$ is the number defined in Lemma 7. By that lemma

$$(f, g) = \left(x^{2^{cn}-1} + 1,\ T_n(x)\right) \tag{34}$$

has at least $n/2$ non-zero coefficients, hence $A(2, s, K) = \infty$.

If $r = 3$, $s \geq 3$, $s \equiv 0 \bmod 2$, we write $s = 2^\sigma s_1$, s_1 odd, and take $n \equiv 1 \bmod 6$,

$$f = T_n(x),\ g = g_s := \left(x^{2^{cn}-1} + 1\right)^{2^{sn}(2^\sigma - 1)} \frac{x^{(2^{(5-c)n}-1)s_1} + 1}{x^{2^{(5-c)n}-1} + 1}.$$

Since $x^{2^{cn}-1}+1 \mid g_s$ we have

$$\left(f, x^{2^{cn}-1}+1, g_s\right) = \left(f, x^{2^{cn}-1}+1\right).$$

On the other hand, since s_1 is odd

$$\left(x^{2^{(5-c)n}-1}+1,\ \frac{x^{\left(2^{(5-c)n}-1\right)s_1}+1}{x^{2^{(5-c)n}-1}+1}\right) = 1;$$

$$\left(x^{2^{(5-c)n}-1}+1,\ g_s\right) = x^{2^n-1}+1$$

hence, by Lemma 6,

$$\left(f, x^{2^{(5-c)n}-1}+1,\ g_s\right) = 1$$

and, again by Lemma 6,

$$(f, g_s) = \left(f, x^{2^{cn}-1}+1\right).$$

Hence, by Lemma 7, (f, g_s) has at least $n/2$ non-zero coefficients and $A(3, s, K) = \infty$.

If $r = 3$, $s \geq 3$, $s \equiv 1 \bmod 2$, we take $n \equiv 1 \bmod 6$,

$$f = T_n(x),\ \ g = f + g_{s-1}$$

and obtain that $(f, g) = (f, g_{s-1})$ has at least $n/2$ non-zero coefficients, hence $A(3, s, K) = \infty$.

If $r \geq 4$, $s \geq r$, $r \equiv 0 \bmod 2$, $s \equiv r \bmod 4$, we take $1 < a < b$, $(a, b) = 1$, $ab \equiv 1 \bmod 2$,

$$f = (x^a+1)\left(x^b+1\right) + f_{r-2}\left(x^{ab}\right),\ \ h = h\left(x; \tfrac{s-r}{2}+1, f\right),$$

$$g = \left(x^{ab}+1\right) h\left(x^{rab}\right) x^a + h(0) f(x)$$

to obtain

$$(f, g) = \frac{(x^a+1)\left(x^b+1\right)x^a}{x+1} = x^{b+2a-1}+\cdots+x^{b+a-1}+x^{2a-1}+\cdots+x^{a-1},$$

hence $A(r, s, K) = \infty$.

If $r \geq 4$, $s \geq r$, $r \equiv 0 \bmod 2$, $s \equiv r+2 \bmod 4$, we take $1 < a < b$, $(a, b) = 1$, $ab \equiv 1 \bmod 2$,

$$\begin{aligned} f &= (x^a+1)\left(x^b+1\right) + f_{r-2}\left(x^{ab}\right), \\ g &= \left(x^{ab}+1\right) h\left(x^{rab}; \frac{s-r}{2}, f\right) + f \end{aligned}$$

and obtain (33), hence $A(r, s, K) = \infty$.

If $r \geq 4$, $s \geq r$, $r \equiv 1 \bmod 2$, $s \equiv 0 \bmod 2$, we take $n \equiv 1 \bmod 6$,

$$f = T_n(x) + f_{r-1}\left(x^{2^{cn}-1}\right), \quad g = \left(x^{2^{cn}-1} + 1\right) h\left(x^{2^{cn}}; \frac{s}{2}, f\right)$$

and again obtain (33), hence $A(r, s, K) = \infty$.

Finally, if $r \geq 4$, $s \geq r$, $r \equiv s \equiv 1 \bmod 2$, we take $n \equiv 1 \bmod 6$,

$$f = T_n(x) + f_{r-1}\left(x^{2^{cn}-1}\right), \quad h = h\left(x; \tfrac{s-r}{2} + 1, f\right),$$

$$g = \left(x^{2^{cn}-1} + 1\right) h\left(x^{2^{cn+r}}\right) x^{2^n-1} + h(0) f(x)$$

and infer that

$$(f, g) = x^{2^n-1}\left(x^{2^{cn}-1}, T_n(x)\right)$$

has at least $n/2$ non-zero coefficients, hence $A(r, s, K) = \infty$.

References

Bombieri, E. & J. Vaaler (1983), On Siegel's lemma, *Invent. Math.* **73**, 11–32; Addendum, *ibid.* **75** (1984), 377.

Cassels, J.W.S. (1959), *An Introduction to the Geometry of Numbers*, Springer-Verlag.

Győry, K. & A. Schinzel (1994), On a conjecture of Posner and Rumsey, *J. Number Theory* **47**, 63–78.

Lefton, P. (1979), On the Galois group of cubics and trinomials, *Acta Arith.* **35**, 239–246.

Mills, W.H. & N. Zierler (1969), On a conjecture of Golomb, *Pacific J. Math.* **28**, 635–640.

Schinzel, A. (1969), Reducibility of lacunary polynomials, I, *Acta Arith.* **16**, 123–159.

Schinzel, A. (1987), A decomposition of integer vectors, III, *Bull. Polish Acad. Sci. Math.* **35**, 693–703.

Schinzel, A. (2000), *Polynomials with Special Regard to Reducibility*, Cambridge University Press.

Schinzel, A. (2001), On the greatest common divisor of two univariate polynomials, II, *Acta Arith.* **98**, 95–106.

Schlickewei, H.P. & E. Wirsing (1997), Lower bounds for the heights of solutions of linear equations, *Invent. math.* **129**, 1–10.

Weyl, H. (1942), On geometry of numbers, *Proc. London Math. Soc.* **47** (8), 268–289.

22

Heilbronn's Exponential Sum and Transcendence Theory

D.R. Heath-Brown

Let p be a prime, and set $e(x) = \exp(2\pi i x)$. Heilbronn's exponential sum is defined to be

$$S(a, p) = \sum_{n=1}^{p} e\left(\frac{an^p}{p^2}\right),$$

for any integer a coprime to p. Although the sum appears to be defined modulo p^2, one may observe that if $n \equiv n' \bmod p$, then $n^p \equiv n'^p \bmod p^2$. Thus the summand in $S(a, p)$ in fact has period p with respect to n. Heilbronn's sum is therefore a 'complete sum' to modulus p.

Heilbronn asked whether $S(a, p) = o(p)$ as $p \to \infty$. Methods based on algebraic geometry, in the spirit of Weil or Deligne, appear to be ineffectual for this problem, and elementary techniques have also failed to provide an answer. Nonetheless we can now answer Heilbronn's question with the following theorem.

Theorem 1 *If p is a prime and $p \nmid a$ then $S(a, p) \ll p^{7/8}$, uniformly in a.*

This result is due to Heath-Brown and Konyagin (2000), there being an earlier estimate, due to Heath-Brown (1996), with an exponent $11/12$.

To prove the theorem one begins with some elementary manipulations using the sum

$$S_0(a) = \sum_{n=1}^{p-1} e\left(\frac{an^p}{p^2}\right).$$

Since $S_0(a) = S_0(am^p)$ when $p \nmid m$ it follows that

$$(p-1)\sum_{r=1}^{p} |S_0(a+rp)|^4 = \sum_{r=1}^{p}\sum_{m=1}^{p-1} |S_0((a+rp)m^p)|^4 \le \sum_{n=1}^{p^2} |S_0(n)|^4,$$

because each value of n arises at most once. We deduce that

$$\begin{aligned}(p-1)&\sum_{r=1}^{p}|S_0(a+rp)|^4\\ &\le \sum_{m_1,\ldots,m_4=1}^{p-1}\sum_{n=1}^{p^2} e_{p^2}\big((m_1^p+m_2^p-m_3^p-m_4^p)n\big)\\ &= p^2\#\{1\le m_1,\ldots,m_4\le p-1:\ m_1^p+m_2^p\equiv m_3^p+m_4^p \bmod p^2\}.\end{aligned}$$

The final congruence implies that $m_1+m_2\equiv m_3+m_4 \bmod p$, and hence $m_1-m_3\equiv m_4-m_2\equiv b \bmod p$, say. The case $p|b$ makes a negligible contribution. When $p\nmid b$ we write $m_1\equiv v_1 b \bmod p$, whence $m_3\equiv (v_1-1)b \bmod p$. Thus

$$m_1^p-m_3^p\equiv (v_1^p-(v_1-1)^p)b^p \bmod p^2.$$

Similarly we find that

$$m_4^p-m_2^p\equiv (v_2^p-(v_2-1)^p)b^p \bmod p^2,$$

where $m_4\equiv v_2 b \bmod p$.

The congruence $m_1^p+m_2^p\equiv m_3^p+m_4^p \bmod p^2$ now produces

$$(v_1^p-(v_1-1)^p)b^p\equiv (v_2^p-(v_2-1)^p)b^p \bmod p^2.$$

Since

$$v^p-(v-1)^p=\sum_{l=1}^{p}(-1)^{l-1}v^{p-l}\binom{p}{l}\equiv 1-pf(v) \bmod p^2,$$

it then follows, on allowing for the various possibilities for b, that

$$\begin{aligned}&\#\{1\le m_1,\ldots,m_4\le p-1:\ m_1^p+m_2^p\equiv m_3^p+m_4^p \bmod p^2\}\\ &\quad\le (p-1)^2+(p-1)\#\{2\le v_1,v_2\le p-1:\ f(v_1)\equiv f(v_2) \bmod p\},\end{aligned}$$

where

$$f(X)=X+\frac{X^2}{2}+\frac{X^3}{3}+\cdots+\frac{X^{p-1}}{p-1}\in\mathbb{Z}_p[X].$$

Thus

$$(p-1)\sum_{r=1}^{p}|S_0(a+rp)|^4\le p^2\left\{(p-1)^2+(p-1)\sum_{r=1}^{p}N_r^2\right\},$$

where N_r is the number of solutions $k\in\mathbb{Z}_p-\{0,1\}$ of $f(k)=r$. This suffices for the following result.

Lemma 1 *We have*

$$S(a, p) \ll p^{1/2} \left\{ \sum_{r=1}^{p} N_r^2 \right\}^{1/4}.$$

Trivially one has $\sum_{r=1}^{p} N_r = p - 2$ and hence $\sum_{r=1}^{p} N_r^2 \ll p^2$. Since this leads to the estimate $S(a, p) \ll p$, we see that nothing has been lost up to this point. On the other hand, it is not so clear how any non-trivial estimate for N_r may be obtained.

It turns out that ideas from the work of Stepanov (1969) are the key to handling N_r. Stepanov established Weil's theorem on the number of points on a curve over a finite field. However, his ideas can be applied to bound the number of zeros of a polynomial in one variable. A simple bound for N_r was obtained in this way by Mit'kin (1992). One begins by constructing an auxiliary polynomial

$$\Phi(X, Y, Z) \in \mathbb{Z}_p[X, Y, Z]$$

such that

$$\Psi(X) = \Phi(X, f(X), X^p)$$

vanishes to high order at roots of $f(X) = r$. This is a principle familiar from transcendence theory. Indeed the link goes much further, for the similarity between $f(X)$ and the function

$$-\log(1 - X) = X + \frac{X^2}{2} + \frac{X^3}{3} + \cdots \in \mathbb{Q}[[X]]$$

is of crucial importance in the details of the argument. Thus the construction of the auxiliary polynomial $\Phi(X, Y, Z)$ depends on the fact that $f(X)$ satisfies some simple differential equations. These are given by the following result.

Lemma 2 *For any positive integer r there exist polynomials $q_r(X)$ and $h_r(X)$ in $\mathbb{Z}_p[X]$, of degrees at most r and $r - 1$ respectively, such that*

$$\{X(1 - X)\}^r \left(\frac{d}{dX}\right)^r f(X) = q_r(X) + (X^p - X)h_r(X).$$

Thus, although $f(X)$ has large degree, its derivatives may, in effect, be replaced by $q_r(X)$, which has small degree.

It is essential for the proof that $\Psi(X)$ should not vanish identically. Again this is a familiar aspect of transcendence arguments. For our situation we are motivated by the fact that $-\log(1 - X)$ is a transcendental function, and hence cannot satisfy a polynomial relation. Since the function $f(X)$ is almost equal to $-\log(1-X)$, we expect that $f(X)$ similarly should not satisfy a polynomial relation of small degree. In fact one has the following result.

Lemma 3 *Let $F(X, Y) \in \mathbb{Z}_p[X, Y]$ have degree less than A with respect to X, and degree less than B with respect to Y. Then if F does not vanish identically we will have $X^p \nmid F(X, f(X))$, providing only that $AB < p$.*

As soon as $AB \geq p$, the polynomial F will have enough coefficients to ensure that $X^p | F(X, f(X))$ is possible. Thus the above result is surprisingly sharp.

In Heath-Brown (1996), Stepanov's method was applied in a simple-minded way, to show that $N_r = O(p^{2/3})$. This result had in fact been obtained earlier by Mit'kin (1992). Using Lemma 1, the above bound for N_r immediately produces the estimate $S(a, p) \ll p^{11/12}$. However later, in Heath-Brown & Konyagin (2000), the auxiliary polynomial was constructed so as to vanish for the roots of several different equations $f(X) = r$, thereby producing a bound for a sum

$$\sum_{r \in \mathcal{R}} N_r.$$

This leads to the superior exponent $7/8$ quoted in our theorem.

Two questions naturally arise. First: the sum $\sum_r N_r^2$ counts points on the curve $f(X) = f(Y)$. Is there a way of attacking this directly, rather than handling individual values of N_r? Second: the function $-\log(1 - X)$ satisfies a first-order differential equation. Can one handle problems in which the function corresponding to $f(X)$ is related to a solution of a second- (or higher-) order equation?

References

Heath-Brown, D.R. (1996), An estimate for Heilbronn's exponential sum. In *Analytic Number Theory: Proceedings of a Conference in Honor of Heini Halberstam*, B.C Berndt, H.G Diamond & A.J. Hildebrand (eds.), Birkhäuser, 451–463.

Heath-Brown, D.R. & S. Konyagin (2000), New bounds for Gauss sums derived from kth powers, and for Heilbronn's exponential sum, *Quart. J. Math. Oxford Ser. (2)*, **51**, 221–235.

Mit'kin, D.A. (1992), An estimate for the number of roots of some comparisons by the Stepanov method, *Mat. Zametki*, **51**, 52–58, 157. (Translated as *Math. Notes*, **51** (1992), 565–570.)

Stepanov, S.A. (1969), The number of points of a hyperelliptic curve over a prime field, *Izv. Akad. Nauk SSSR Ser. Mat.*, **33**, 1171–1181.

Lightning Source UK Ltd.
Milton Keynes UK
UKHW02n2352171018
330742UK00013B/290/P

9 780521 807999